Banach
Algebra Techniques
in Operator Theory

Banach
Algebra Techniques
in Operator Theory

RONALD G. DOUGLAS

Department of Mathematics
State University of New York at Stony Brook
Stony Brook, New York

1972

ACADEMIC PRESS   New York and London

ACADEMIC PRESS, INC.
111 Fifth Avenue, New York, New York 10003

United Kingdom Edition published by
ACADEMIC PRESS, INC. (LONDON) LTD.
24/28 Oval Road, London NW1

LIBRARY OF CONGRESS CATALOG CARD NUMBER: 78-187253

AMS (MOS) 1970 Subject Classifications: 47B05, 47B15, 47B30,
47B35, 46L05, 46J15, 46-02, 47-02

PRINTED IN THE UNITED STATES OF AMERICA

To my family

Nan, Mike, Kevin, Kristin

Contents

3. Geometry of Hilbert Space

4. Operators on Hilbert Space and C^*-Algebras

5. Compact Operators, Fredholm Operators, and Index Theory

6. The Hardy Spaces

7. Toeplitz Operators

Preface

Operator theory is a diverse area of mathematics which derives its impetus and motivation from several sources. It began as did practically all of modern analysis with the study of integral equations at the end of the last century. It now includes the study of operators and collections of operators arising in various branches of physics and mechanics as well as other parts of mathematics and indeed is sufficiently well developed to have a logic of its own. The appearance of several monographs on recent studies in operator theory testifies both to its vigor and breadth.

The intention of this book is to discuss certain advanced topics in operator theory and to provide the necessary background for them assuming only the standard senior–first year graduate courses in general topology, measure theory, and algebra. There is no attempt at completeness and many "elementary" topics are either omitted or mentioned only in the problems. The intention is rather to obtain the main results as quickly as possible.

The book begins with a chapter presenting the basic results in the theory of Banach spaces along with many relevant examples. The second chapter concerns the elementary theory of commutative Banach algebras since these techniques are essential for the approach to operator theory presented in the later chapters. Then after a short chapter on the geometry of Hilbert space, the study of operator theory begins in earnest. In the fourth chapter operators on Hilbert space are studied and a rather sophisticated version of the spectral theorem is obtained. The notion of a C^*-algebra is introduced and used throughout the last half of this chapter. The study of compact operators and Fredholm operators is taken up in the fifth chapter along with certain ancillary results concerning ideals in C^*-algebras. The approach here is a bit unorthodox but is suggested by modern developments.

The last two chapters are of a slightly different character and present a systematic development including recent research of the theory of Toeplitz operators. This latter class of operators has attracted the attention of several mathematicians recently and occurs in several rather diverse contexts.

In the sixth chapter certain topics from the theory of Hardy spaces are developed. The selection is dictated by needs of the last chapter and proofs are based on the techniques obtained earlier in the book. The study of Toeplitz operators is taken up in the seventh chapter. Most of what is known in the scalar case is presented including Widom's result on the connectedness of the spectrum.

At the end of each chapter there are source notes which suggest additional reading along with giving some comments on who proved what and when. Although a reasonable attempt has been made in the latter chapters at citing the appropriate source for important results, omissions have undoubtedly occurred. Moreover, the absence of a reference should not be construed to mean the result is due to the author.

In addition, following each chapter is a large number of problems of varying difficulty. The purposes of these are many: to allow the reader to test his understanding; to indicate certain extensions of the theory which are now accessible; to alert the reader to certain important and related results of which he should be aware along with a hint or a reference for the proof; and to point out certain questions for which the answer is not known. These latter questions are indicated by a double asterisk; a single asterisk indicates a difficult problem.

Acknowledgments

This book began as a set of lecture notes for a course given at the University of Michigan in Spring, 1968 and again at SUNY at Stony Brook in the academic year, 1969–1970.

I am indebted to many people in the writing of this book. Most of all I would like to thank Bruce Abrahamse who prepared the original notes and who has been a constant source of suggestions and constructive criticism ever since. Also I would like to thank Stuart Clary and S. Pattanayak for writing portions of later versions and for their many suggestions. In addition, I would like to thank many friends and colleagues and, in particular, Paul Halmos, Carl Pearcy, Pasquale Porcelli, Donald Sarason, and Allen Shields with whom I have learned many of the things presented in this book. Special thanks are due to Berrien Moore III who read and criticized the entire manuscript and to Joyce Lemen, Dorothy Lentz, and Carole Alberghine for typing the various versions of this manuscript. Lastly, I would like to thank the National Science Foundation and the Alfred E. Sloan Foundation for various support during the writing of this book.

Symbols and Notation

$M_{\mathfrak{B}}$	39	T_{φ}	177
M_{φ}	87	T_K	126
M_{∞}	155	$\mathfrak{T}(\ )$	179
$M(X)$	20		
		U	89
P	177	U_+	96
$\mathscr{P}$	51	$U_+^{(n)}$	136
$\mathscr{P}_+$	51		
PC	176	$w(\ )$	115
$P_{\mathscr{M}}$	86	$W(\ )$	115
		$\mathfrak{W}_T$	101
QC	176		
		$\mathbb{Z}$	7
$\mathbb{R}$	4		
$r(\ )$	41	$\|\ \|_{\infty}$	1
$\mathscr{R}(\ )$	57	$*$	6, 82, 119
ran	83	$(\ )_1$	10
$\rho(\ )$	41	$\|\ \|_p$	24
$\rho_{\mathrm{e}}(\ )$	191	$\oplus$	29
		$\perp$	30, 71
$\sigma(\ )$	41	$\otimes$	31, 62, 79, 118
Σ	4	$(\ ,\ )$	64
		$'$	108
$\mathbb{T}$	26		

1 Banach Spaces

1.1 We begin by introducing the most representative example of a Banach space. Let X be a compact Hausdorff space and let $C(X)$ denote the set of continuous complex-valued functions on X. For f_1 and f_2 in $C(X)$ and λ a complex number, we define:

(1) $(f_1 + f_2)(x) = f_1(x) + f_2(x)$;
(2) $(\lambda f_1)(x) = \lambda f_1(x)$; and
(3) $(f_1 f_2)(x) = f_1(x) f_2(x)$.

With these operations $C(X)$ is a commutative algebra with identity over the complex field $\mathbb{C}$.

Each function f in $C(X)$ is bounded, since it follows from the fact that f is continuous and X is compact that the range of f is a compact subset of $\mathbb{C}$. Thus the least upper bound of $|f|$ is finite; we call this number the norm of f and denote it by

$$\|f\|_\infty = \sup\{|f(x)| : x \in X\}.$$

The following properties of the norm are easily verified:

(1) $\|f\|_\infty = 0$ if and only if $f = 0$;
(2) $\|\lambda f\|_\infty = |\lambda| \, \|f\|_\infty$;
(3) $\|f + g\|_\infty \leqslant \|f\|_\infty + \|g\|_\infty$; and
(4) $\|fg\|_\infty \leqslant \|f\|_\infty \|g\|_\infty$.

We define a metric ρ on $C(X)$ by $\rho(f,g) = \|f-g\|_\infty$. The properties of a metric, namely,

(1) $\rho(f,g) = 0$ if and only if $f = g$,
(2) $\rho(f,g) = \rho(g,f)$, and
(3) $\rho(f,h) \leqslant \rho(f,g) + \rho(g,h)$,

follow immediately from properties (1)–(3) of the norm. It is easily seen that convergence with respect to the metric ρ is just uniform convergence. An important property of this metric is that $C(X)$ is complete with respect to it.

1.2   Proposition   If X is a compact Hausdorff space, then $C(X)$ is a complete metric space.

Proof   If $\{f_n\}_{n=1}^\infty$ is a Cauchy sequence, then

$$|f_n(x) - f_m(x)| \leqslant \|f_n - f_m\|_\infty = \rho(f_n, f_m)$$

for each x in X. Hence, $\{f_n(x)\}_{n=1}^\infty$ is a Cauchy sequence of complex numbers for each x in X, so we may define $f(x) = \lim_{n\to\infty} f_n(x)$. We need to show that f is in $C(X)$ and that $\lim_{n\to\infty} \|f - f_n\|_\infty = 0$. To that end, given $\varepsilon > 0$, choose N such that $n, m \geqslant N$ implies $\|f_n - f_m\|_\infty < \varepsilon$. For x_0 in X there exists a neighborhood U of x_0 such that $|f_N(x_0) - f_N(x)| < \varepsilon$ for x in U. Therefore,

$$|f(x_0) - f(x)| \leqslant \lim_{n\to\infty} |f_n(x_0) - f_N(x_0)| + |f_N(x_0) - f_N(x)|$$
$$+ \lim_{n\to\infty} |f_N(x) - f_n(x)|$$
$$\leqslant 3\varepsilon$$

which implies f is continuous. Further, for $n \geqslant N$ and x in X, we have

$$|f_n(x) - f(x)| = |f_n(x) - \lim_{m\to\infty} f_m(x)| = \lim_{m\to\infty} |f_n(x) - f_m(x)|$$
$$\leqslant \limsup_{m\to\infty} \|f_n - f_m\|_\infty \leqslant \varepsilon.$$

Thus, $\lim_{n\to\infty} \|f_n - f\|_\infty = 0$ and hence $C(X)$ is complete. ■

We next define the notion of Banach space which abstracts the salient properties of the preceding example. We shall see later in this chapter that every Banach space is isomorphic to a subspace of some $C(X)$.

1.3   Definition   A Banach space is a complex linear space $\mathscr{X}$ with a norm $\| \ \|$ satisfying

(1) $\|f\| = 0$ if and only if $f = 0$,

(2) $\|\lambda f\| = |\lambda| \, \|f\|$ for λ in $\mathbb{C}$ and f in $\mathscr{X}$, and

(3) $\|f+g\| \leqslant \|f\| + \|g\|$ for f and g in $\mathscr{X}$,

such that $\mathscr{X}$ is complete in the metric given by this norm.

1.4 Proposition Let $\mathscr{X}$ be a Banach space. The functions

$a: \mathscr{X} \times \mathscr{X} \to \mathscr{X}$ defined $a(f,g) = f+g$,

$s: \mathbb{C} \times \mathscr{X} \to \mathscr{X}$ defined $s(\lambda, f) = \lambda f$, and

$n: X \to \mathbb{R}^+$ defined $n(f) = \|f\|$

are continuous.

Proof Obvious. ∎

1.5 Directed Sets and Nets The topology of a metric space can be described in terms of the sequences in it that converge. For more general topological spaces a notion of generalized sequence is necessary. In what follows it will often be convenient to describe a topology in terms of its convergent generalized sequences. Thus we proceed to review for the reader the notion of net.

A directed set A is a partially ordered set having the property that for each pair α and β in A there exists γ in A such that $\gamma \geqslant \alpha$ and $\gamma \geqslant \beta$. A net is a function $\alpha \to \lambda_\alpha$ on a directed set. If the λ_α all lie in a topological space X, then the net is said to converge to λ in X if for each neighborhood U of λ there exists α_U in A such that λ_α is in U for $\alpha \geqslant \alpha_U$. Two topologies on a space X coincide if they have the same convergent nets. Lastly, a topology can be defined on X by prescribing the convergent nets. For further information concerning nets and subnets, the reader should consult [71].

We now consider the convergence of Cauchy nets in a Banach space.

1.6 Definition A net $\{f_\alpha\}_{\alpha \in A}$ in a Banach space X is said to be a Cauchy net if for every $\varepsilon > 0$, there exists α_0 in A such that $\alpha_1, \alpha_2 \geqslant \alpha_0$ implies $\|f_{\alpha_1} - f_{\alpha_2}\| < \varepsilon$.

1.7 Proposition In a Banach space each Cauchy net is convergent.

Proof Let $\{f_\alpha\}_{\alpha \in A}$ be a Cauchy net in the Banach space $\mathscr{X}$. Choose α_1 such that $\alpha \geqslant \alpha_1$ implies $\|f_\alpha - f_{\alpha_1}\| < 1$. Having chosen $\{\alpha_k\}_{k=1}^n$ in A, choose $\alpha_{n+1} \geqslant \alpha_n$ such that $\alpha \geqslant \alpha_{n+1}$ implies

$$\|f_\alpha - f_{\alpha_{n+1}}\| < \frac{1}{n+1}.$$

The sequence $\{f_{\alpha_n}\}_{n=1}^{\infty}$ is clearly Cauchy and, since $\mathscr{X}$ is complete, there exists f in $\mathscr{X}$ such that $\lim_{n \to \infty} f_{\alpha_n} = f$.

It remains to prove that $\lim_{\alpha \in A} f_\alpha = f$. Given $\varepsilon > 0$, choose n such that $1/n < \varepsilon/2$ and $\|f_{\alpha_n} - f\| < \varepsilon/2$. Then for $\alpha \geqslant \alpha_n$ we have

$$\|f_\alpha - f\| \leqslant \|f_\alpha - f_{\alpha_n}\| + \|f_{\alpha_n} - f\| < 1/n + \varepsilon/2 < \varepsilon. \quad \blacksquare$$

We next consider a general notion of summability in a Banach space which will be used in Chapter 3.

1.8   Definition   Let $\{f_\alpha\}_{\alpha \in A}$ be a set of vectors in the Banach space $\mathscr{X}$. Let $\mathscr{F} = \{F \subset A : F \text{ finite}\}$. If we define $F_1 \leqslant F_2$ for $F_1 \subset F_2$, then $\mathscr{F}$ is a directed set. For each F in $\mathscr{F}$, let $g_F = \sum_{\alpha \in F} f_\alpha$. If the net $\{g_F\}_{F \in \mathscr{F}}$ converges to some g in $\mathscr{X}$, then the sum $\sum_{\alpha \in A} f_\alpha$ is said to converge and we write $g = \sum_{\alpha \in A} f_\alpha$.

1.9   Proposition   If $\{f_\alpha\}_{\alpha \in A}$ is a set of vectors in the Banach space $\mathscr{X}$ such that $\sum_{\alpha \in A} \|f_\alpha\|$ converges in $\mathbb{R}$, then $\sum_{\alpha \in A} f_\alpha$ converges in $\mathscr{X}$.

Proof   It suffices to show, in the notation of Definition 1.8, that the net $\{g_F\}_{F \in \mathscr{F}}$ is Cauchy. Since $\sum_{\alpha \in A} \|f_\alpha\|$ converges, for $\varepsilon > 0$, there exists F_0 in $\mathscr{F}$ such that $F \geqslant F_0$ implies

$$\sum_{\alpha \in F} \|f_\alpha\| - \sum_{\alpha \in F_0} \|f_\alpha\| < \varepsilon.$$

Thus for $F_1, F_2 \geqslant F_0$ we have

$$\begin{aligned}
\|g_{F_1} - g_{F_2}\| &= \Big\| \sum_{\alpha \in F_1} f_\alpha - \sum_{\alpha \in F_2} f_\alpha \Big\| \\
&= \Big\| \sum_{\alpha \in F_1 \backslash F_2} f_\alpha - \sum_{\alpha \in F_2 \backslash F_1} f_\alpha \Big\| \\
&\leqslant \sum_{\alpha \in F_1 \backslash F_2} \|f_\alpha\| + \sum_{\alpha \in F_2 \backslash F_1} \|f_\alpha\| \\
&\leqslant \sum_{\alpha \in F_1 \cup F_2} \|f_\alpha\| - \sum_{\alpha \in F_0} \|f_\alpha\| < \varepsilon.
\end{aligned}$$

Therefore, $\{g_F\}_{F \in \mathscr{F}}$ is Cauchy and $\sum_{\alpha \in A} f_\alpha$ converges by definition. $\quad \blacksquare$

We now state an elementary criterion for a normed linear space (that is, a complex linear space with a norm satisfying (1)–(3) of Definition 1.3) to be complete and hence a Banach space. This will prove very useful in verifying that various examples are Banach spaces.

1.10 Corollary A normed linear space $\mathscr{X}$ is a Banach space if and only if for every sequence $\{f_n\}_{n=1}^{\infty}$ of vectors in $\mathscr{X}$ the condition $\sum_{n=1}^{\infty} \|f_n\| < \infty$ implies the convergence of $\sum_{n=1}^{\infty} f_n$.

Proof If $\mathscr{X}$ is a Banach space, then the conclusion follows from the preceding proposition. Therefore, assume that $\{g_n\}_{n=1}^{\infty}$ is a Cauchy sequence in a normed linear space $\mathscr{X}$ in which the series hypothesis is valid. Then we may choose a subsequence $\{g_{n_k}\}_{k=1}^{\infty}$ such that $\sum_{k=1}^{\infty} \|g_{n_{k+1}} - g_{n_k}\| < \infty$ as follows: Choose n_1 such that for $i,j \geq n_1$ we have $\|g_i - g_j\| < 1$; having chosen $\{n_k\}_{k=1}^{N}$ choose $n_{N+1} > n_N$ such that $i,j \geq n_{N+1}$ implies $\|g_i - g_j\| < 2^{-N}$. If we set $f_k = g_{n_k} - g_{n_{k-1}}$ for $k > 1$ and $f_1 = g_{n_1}$, then $\sum_{k=1}^{\infty} \|f_k\| < \infty$, and the hypothesis implies that the series $\sum_{k=1}^{\infty} f_k$ converges. It follows from the definition of convergence that the sequence $\{g_{n_k}\}_{k=1}^{\infty}$ converges in $\mathscr{X}$ and hence so also does $\{g_n\}_{n=1}^{\infty}$. Thus $\mathscr{X}$ is complete and hence a Banach space. ∎

In the study of linear spaces the notion of a linear functional is extremely important. The collection of linear functionals defined on a given linear space is itself a linear space and this duality is a powerful tool for studying either space. In the study of Banach spaces the corresponding notion is that of a continuous linear functional.

1.11 Definition Let $\mathscr{X}$ be a Banach space. A function φ from $\mathscr{X}$ to $\mathbb{C}$ is a bounded linear functional if:

(1) $\varphi(\lambda_1 f_1 + \lambda_2 f_2) = \lambda_1 \varphi(f_1) + \lambda_2 \varphi(f_2)$ for f_1, f_2 in $\mathscr{X}$ and λ_1, λ_2 in $\mathbb{C}$;
and

(2) There exists M such that $|\varphi(f)| \leq M \|f\|$ for every f in $\mathscr{X}$.

1.12 Proposition Let φ be a linear functional on the Banach space $\mathscr{X}$. The following statements are equivalent:

(1) φ is bounded;
(2) φ is continuous;
(3) φ is continuous at 0.

Proof (1) implies (2). If $\{f_\alpha\}_{\alpha \in A}$ is a net in $\mathscr{X}$ converging to f, then $\lim_{\alpha \in A} \|f_\alpha - f\| = 0$. Hence,

$$\lim_{\alpha \in A} |\varphi(f_\alpha) - \varphi(f)| = \lim_{\alpha \in A} |\varphi(f_\alpha - f)| \leq \lim_{\alpha \in A} M \|f_\alpha - f\| = 0,$$

which implies that the net $\{\varphi(f_\alpha)\}_{\alpha \in A}$ converges to $\varphi(f)$. Thus φ is continuous.

(2) implies (3). Obvious.

(3) implies (1). If φ is continuous at 0, then there exists $\delta > 0$ such that $\|f\| < \delta$ implies $|\varphi(f)| < 1$. Hence, for any nonzero g in X we have

$$|\varphi(g)| = \frac{2\|g\|}{\delta} \left| \varphi\left(\frac{\delta}{2\|g\|} g \right) \right| < \frac{2}{\delta} \|g\|,$$

and thus φ is bounded. ■

We next define a norm on the space of bounded linear functionals which makes it into a Banach space.

1.13   Definition   Let $\mathscr{X}^*$ be the set of bounded linear functionals on the Banach space $\mathscr{X}$. For φ in $\mathscr{X}^*$, let

$$\|\varphi\| = \sup \left\{ \frac{|\varphi(f)|}{\|f\|} : f \neq 0 \right\}.$$

Then $\mathscr{X}^*$ is said to be the conjugate or dual space of $\mathscr{X}$.

1.14   Proposition   The conjugate space $\mathscr{X}^*$ is a Banach space.

Proof   That $\mathscr{X}^*$ is a linear space is obvious, as are properties (1) and (2) for the norm. To prove (3) we compute

$$\|\varphi_1 + \varphi_2\| = \sup_{f \neq 0} \frac{|(\varphi_1 + \varphi_2)(f)|}{\|f\|} = \sup_{f \neq 0} \frac{|\varphi_1(f) + \varphi_2(f)|}{\|f\|}$$

$$\leqslant \sup_{f \neq 0} \frac{|\varphi_1(f)|}{\|f\|} + \sup_{f \neq 0} \frac{|\varphi_2(f)|}{\|f\|}$$

$$= \|\varphi_1\| + \|\varphi_2\|.$$

Finally, we must show that $\mathscr{X}^*$ is complete. Thus, suppose $\{\varphi_n\}_{n=1}^{\infty}$ is a Cauchy sequence in $\mathscr{X}^*$. For each f in $\mathscr{X}$ we have $|\varphi_n(f) - \varphi_m(f)| \leqslant \|\varphi_n - \varphi_m\| \|f\|$ so that the sequence of complex numbers $\{\varphi_n(f)\}_{n=1}^{\infty}$ is Cauchy for each f in $\mathscr{X}$. Hence, we can define $\varphi(f) = \lim_{n \to \infty} \varphi_n(f)$. The linearity of φ follows from the corresponding linearity of the functionals φ_n. Further, if N is chosen so that $n, m \geqslant N$ implies $\|\varphi_n - \varphi_m\| < 1$, then for f in $\mathscr{X}$ we have

$$|\varphi(f)| \leqslant |\varphi(f) - \varphi_N(f)| + |\varphi_N(f)|$$

$$\leqslant \lim_{n \to \infty} |\varphi_n(f) - \varphi_N(f)| + |\varphi_N(f)|$$

$$\leqslant \limsup_{n \to \infty} \|\varphi_n - \varphi_N\| \|f\| + \|\varphi_N\| \|f\|$$

$$\leqslant (1 + \|\varphi_N\|) \|f\|.$$

Thus φ is in $\mathcal{X}^*$ and it remains only to show that $\lim_{n\to\infty}\|\varphi-\varphi_n\|=0$. Given $\varepsilon>0$, choose N such that $n,m\geqslant N$ implies $\|\varphi_n-\varphi_m\|<\varepsilon$. Then for f in $\mathcal{X}$ and $m,n\geqslant N$, we have

$$|(\varphi-\varphi_n)(f)|\leqslant|(\varphi-\varphi_m)(f)|+|(\varphi_m-\varphi_n)(f)|\leqslant|(\varphi-\varphi_m)(f)|+\varepsilon\|f\|.$$

Since $\lim_{m\to\infty}|(\varphi-\varphi_m)(f)|=0$, we have $\|\varphi-\varphi_n\|<\varepsilon$. Thus the sequence $\{\varphi_n\}_{n=1}^\infty$ converges to φ and $\mathcal{X}^*$ is complete. ∎

The reader should compare the preceding proof to that of Proposition 1.2. We now want to consider some further examples of Banach spaces and to compute their respective conjugate spaces.

1.15 Examples Let $l^\infty(\mathbb{Z}^+)$ denote the collection of all bounded complex functions on the nonnegative integers $\mathbb{Z}^+$. Define addition and multiplication pointwise and set $\|f\|_\infty=\sup\{|f(n)|:n\in\mathbb{Z}^+\}$. It is not difficult to verify that $l^\infty(\mathbb{Z}^+)$ is a Banach space with respect to this norm, and this will be left as an exercise. Further, the collection of all functions f in $l^\infty(\mathbb{Z}^+)$ such that $\lim_{n\to\infty}f(n)=0$ is a closed subspace of $l^\infty(\mathbb{Z}^+)$ and hence a Banach space; we denote this space by $c_0(\mathbb{Z}^+)$.

In addition, let $l^1(\mathbb{Z}^+)$ denote the collection of all complex functions φ on $\mathbb{Z}^+$ such that $\sum_{n=0}^\infty|\varphi(n)|<\infty$. Define addition and scalar multiplication pointwise and set $\|\varphi\|_1=\sum_{n=0}^\infty|\varphi(n)|$. Again we leave as an exercise the task of showing that $l^1(\mathbb{Z}^+)$ is a Banach space for this norm.

We consider now the problem of identifying conjugate spaces and we begin with $c_0(\mathbb{Z}^+)$. For φ in $l^1(\mathbb{Z}^+)$ we define the functional $\hat\varphi$ on $c_0(\mathbb{Z}^+)$ such that $\hat\varphi(f)=\sum_{n=0}^\infty\varphi(n)f(n)$ for f in $c_0(\mathbb{Z}^+)$; the latter sum converges, since

$$|\hat\varphi(f)|=\left|\sum_{n=0}^\infty\varphi(n)f(n)\right|\leqslant\sum_{n=0}^\infty|\varphi(n)|\,|f(n)|$$

$$\leqslant\|f\|_\infty\sum_{n=0}^\infty|\varphi(n)|=\|f\|_\infty\|\varphi\|_1.$$

Moreover, since $\hat\varphi$ is obviously linear, this latter inequality shows that $\hat\varphi$ is in $c_0(\mathbb{Z}^+)^*$ and that $\|\varphi\|_1\geqslant\|\hat\varphi\|$, where the latter is the norm of $\hat\varphi$ as an element of $c_0(\mathbb{Z}^+)^*$. Thus the map $\alpha(\varphi)=\hat\varphi$ from $l^1(\mathbb{Z}^+)$ to $c_0(\mathbb{Z}^+)^*$ is well defined and is a contraction. We want to show that α is isometric and onto $c_0(\mathbb{Z}^+)^*$.

To that end let L be an element of $c_0(\mathbb{Z}^+)^*$ and define the function φ_L on $\mathbb{Z}^+$ so that $\varphi_L(n)=L(e_n)$ for n in $\mathbb{Z}^+$, where e_n is the element of $c_0(\mathbb{Z}^+)$

defined to be 1 at n and 0 otherwise. We want to show that $\hat{\varphi}_L = L$, and that $\|\varphi_L\|_1 \leqslant \|L\|$. For each N in $\mathbb{Z}^+$ consider the element

$$f_N = \sum_{n=0}^{N} \frac{\overline{L(e_n)}}{|L(e_n)|} e_n$$

of $c_0(\mathbb{Z}^+)$, where $0/0$ is taken to be 0. Then $\|f_N\|_\infty \leqslant 1$ and an easy computation yields

$$\|L\| \geqslant |L(f_N)| = \left| \sum_{n=0}^{N} \frac{\overline{L(e_n)}}{|L(e_n)|} L(e_n) \right| = \sum_{n=0}^{N} |L(e_n)| = \sum_{n=0}^{N} |\varphi_L(n)|;$$

hence φ_L is in $l^1(\mathbb{Z}^+)$ and $\|\varphi_L\|_1 \leqslant \|L\|$. Thus the map $\beta(L) = \varphi_L$ from $c_0(\mathbb{Z}^+)^*$ to $l^1(\mathbb{Z}^+)$ is also well defined and contractive. Moreover, let L be in $c_0(\mathbb{Z}^+)^*$ and g be in $c_0(\mathbb{Z}^+)$; then

$$\lim_{N \to \infty} \left\| g - \sum_{n=0}^{N} g(n) e_n \right\|_\infty = 0$$

and hence we have

$$L(g) = \lim_{N \to \infty} \left\{ \sum_{n=0}^{N} g(n) L(e_n) \right\} = \lim_{N \to \infty} \left\{ \sum_{n=0}^{N} g(n) \varphi_L(n) \right\}$$

$$= \sum_{n=0}^{\infty} g(n) \varphi_L(n) = \hat{\varphi}_L(g).$$

Therefore, the composition $\alpha \circ \beta$ is the identity on $c_0(\mathbb{Z}^+)^*$. Lastly, since $\hat{\varphi} = 0$ implies $\varphi = 0$, we have that α is one-to-one. Thus α is an isometrical isomorphism of $l^1(\mathbb{Z}^+)$ onto $c_0(\mathbb{Z}^+)^*$.

Consider now the problem of identifying the conjugate space of $l^1(\mathbb{Z}^+)$. For f in $l^\infty(\mathbb{Z}^+)$ we can define an element $\hat{f}$ of $l^1(\mathbb{Z}^+)^*$ as follows: $\hat{f}(\varphi) = \sum_{n=0}^{\infty} f(n) \varphi(n)$. We leave as an exercise the verification that this identifies $l^1(\mathbb{Z}^+)^*$ as $l^\infty(\mathbb{Z}^+)$.

1.16 We return now to considering abstract Banach spaces. If a sequence of bounded linear functionals $\{\varphi_n\}_{n=0}^{\infty}$ in $\mathscr{X}^*$ converges in norm to φ, then it must also converge pointwise, that is, $\lim_{n \to \infty} \varphi_n(f) = \varphi(f)$ for each f in $\mathscr{X}$. The following example shows that the converse is not true.

For k in $\mathbb{Z}^+$ and f in $l^1(\mathbb{Z}^+)$ define $L_k(f) = f(k)$. Then L_k is in $l^1(\mathbb{Z}^+)^*$ and $\|L_k\| = 1$ for each k. Moreover, $\lim_{k \to \infty} L_k(f) = 0$ for each f in $l^1(\mathbb{Z}^+)$. Thus, the sequence $\{L_k\}_{k=0}^{\infty}$ converges "pointwise" to the zero functional 0 but $\|L_k - 0\| = 1$ for each k in $\mathbb{Z}^+$.

Thus, pointwise convergence in $\mathcal{X}^*$ is, in general, weaker than norm convergence; that is, it is easier for a sequence to converge pointwise than it is for it to converge in the norm. Since the notion of pointwise convergence is a natural one, we might expect it to be useful in the study of Banach spaces. That is indeed correct and we shall define the topology of pointwise convergence after recalling a few facts about weak topologies.

1.17 Weak Topologies   Let X be a set, Y be a topological space, and $\mathcal{F}$ be a family of functions from X into Y. The weak topology on X induced by $\mathcal{F}$ is the weakest or smallest topology $\mathcal{T}$ on X for which each function in $\mathcal{F}$ is continuous. Thus $\mathcal{T}$ is the topology generated by the sets $\{f^{-1}(U) : f \in \mathcal{F}, U \text{ open in } Y\}$. Convergence of nets in this topology is completely characterized by $\lim_{\alpha \in A} x_\alpha = x$ if and only if $\lim_{\alpha \in A} f(x_\alpha) = f(x)$ for every f in $\mathcal{F}$. Thus $\mathcal{T}$ is the topology of pointwise convergence.

If Y is Hausdorff and $\mathcal{F}$ separates the points of X, then the weak topology is Hausdorff.

1.18 Definition   For each f in $\mathcal{X}$ let $\hat{f}$ denote the function on $\mathcal{X}^*$ defined $\hat{f}(\varphi) = \varphi(f)$. The w^*-topology on $\mathcal{X}^*$ is the weak topology on $\mathcal{X}^*$ induced by the family of functions $\{\hat{f} : f \in \mathcal{X}\}$.

1.19 Proposition   The w^*-topology on $\mathcal{X}^*$ is Hausdorff.

Proof   If $\varphi_1 \neq \varphi_2$, then there exists f in $\mathcal{X}$ such that $\varphi_1(f) \neq \varphi_2(f)$. Hence, $\hat{f}(\varphi_1) \neq \hat{f}(\varphi_2)$ so that the functions $\{\hat{f} : f \in \mathcal{X}\}$ separate the points of $\mathcal{X}^*$. The proposition now follows from the remark at the end of Section 1.17. ∎

We point out that the w^*-topology is not, in general, metrizable (see Problem 1.13). Next we record the following easy proposition for reference.

1.20 Proposition   A net $\{\varphi_\alpha\}_{\alpha \in A}$ in $\mathcal{X}^*$ converges to φ in $\mathcal{X}^*$ in the w^*-topology if and only if $\lim_{\alpha \in A} \varphi_\alpha(f) = \varphi(f)$ for every f in $\mathcal{X}$.

The following shows that the w^*-topology is determined on bounded subsets of $\mathcal{X}^*$ by a dense subset of $\mathcal{X}$ and this fact will be used in subsequent chapters.

1.21 Proposition   Suppose $\mathcal{M}$ is a dense subset of $\mathcal{X}$ and $\{\varphi_\alpha\}_{\alpha \in A}$ is a uniformly bounded net in $\mathcal{X}^*$ such that $\lim_{\alpha \in A} \varphi_\alpha(f) = \varphi(f)$ for f in $\mathcal{M}$.

Then the net $\{\varphi_\alpha\}_{\alpha \in A}$ converges to φ in the w^*-topology.

Proof   Given g in $\mathscr{X}$ and $\varepsilon > 0$, choose f in $\mathscr{M}$ such that $\|f - g\| < \varepsilon/3M$, where $M = \sup\{\|\varphi\|, \|\varphi_\alpha\| : \alpha \in A\}$. If α_0 is chosen such that $\alpha \geqslant \alpha_0$ implies $|\varphi_\alpha(f) - \varphi(f)| < \varepsilon/3$, then for $\alpha \geqslant \alpha_0$, we have

$$|\varphi_\alpha(g) - \varphi(g)| \leqslant |\varphi_\alpha(g) - \varphi_\alpha(f)| + |\varphi_\alpha(f) - \varphi(f)| + |\varphi(f) - \varphi(g)|$$

$$\leqslant \|\varphi_\alpha\| \|f - g\| + \varepsilon/3 + \|\varphi\| \|f - g\| < \varepsilon.$$

Thus $\{\varphi_\alpha\}_{\alpha \in A}$ converges to φ in the w^*-topology.   ■

1.22   Definition   The unit ball of a Banach space $\mathscr{X}$ is the set $\{f \in X : \|f\| \leqslant 1\}$ and is denoted $(\mathscr{X})_1$.

1.23   Theorem (Alaoglu)   The unit ball $(\mathscr{X}^*)_1$ of the dual of a Banach space is compact in the w^*-topology.

Proof   The proof is accomplished by identifying $(\mathscr{X}^*)_1$ with a closed subset of a large product space the compactness of which follows from Tychonoff's theorem (see [71]).

For each f in $(\mathscr{X})_1$ let $\mathbb{C}_1{}^f$ denote a copy of the closed unit disk in $\mathbb{C}$ and let $\mathbf{P}$ denote the product space $\underset{f \in (\mathscr{X})_1}{\mathsf{X}} \mathbb{C}_1{}^f$. By Tychonoff's theorem $\mathbf{P}$ is compact. Define Λ from $(\mathscr{X}^*)_1$ to $\mathbf{P}$ by $\Lambda(\varphi) = \varphi \,|\, (\mathscr{X})_1$. Since $\Lambda(\varphi_1) = \Lambda(\varphi_2)$ implies that the restrictions of φ_1 and φ_2 to the unit ball of $\mathscr{X}$ are identical, it follows that Λ is one-to-one. Further, a net $\{\varphi_\alpha\}_{\alpha \in A}$ in $\mathscr{X}^*$ converges in the w^*-topology to a φ in $\mathscr{X}^*$ if and only if $\lim_{\alpha \in A} \varphi_\alpha(f) = \varphi(f)$ for f in $\mathscr{X}$ if and only if $\lim_{\alpha \in A} \varphi_\alpha(f) = \varphi(f)$ for f in $(\mathscr{X})_1$ if and only if $\lim_{\alpha \in A} \Lambda(\varphi_\alpha)(f) = \Lambda(\varphi)(f)$ for f in $(\mathscr{X})_1$. This latter statement is equivalent to $\lim_{\alpha \in A} \Lambda(\varphi_\alpha) = \Lambda(\varphi)$ in the topology of $\mathbf{P}$. Thus, Λ is a homeomorphism between $(\mathscr{X}^*)_1$ and the subset $\Lambda[(\mathscr{X}^*)_1]$ of $\mathbf{P}$.

We complete the proof by showing that $\Lambda[(\mathscr{X}^*)_1]$ is closed in $\mathbf{P}$. Suppose $\{\Lambda(\varphi_\alpha)\}_{\alpha \in A}$ is a net in $\Lambda[(\mathscr{X}^*)_1]$ that converges in the product topology to ψ in $\mathbf{P}$. If f, g, and $f + g$ are in $(\mathscr{X})_1$, then

$$\psi(f + g) = \lim_{\alpha \in A} \Lambda(\varphi_\alpha)(f + g) = \lim_{\alpha \in A} \Lambda(\varphi_\alpha)(f) + \lim_{\alpha \in A} \Lambda(\varphi_\alpha)(g)$$

$$= \psi(f) + \psi(g).$$

Further, if f and λf are in $(\mathscr{X})_1$, then

$$\psi(\lambda f) = \lim_{\alpha \in A} \Lambda(\varphi_\alpha)(\lambda f) = \lim_{\alpha \in A} \varphi_\alpha(\lambda f) = \lambda \lim_{\alpha \in A} \varphi_\alpha(f) = \lambda \psi(f).$$

Hence ψ determines an element $\tilde{\psi}$ of $(\mathscr{X}^*)_1$ by the relationship

$$\tilde{\psi}(f) = \|f\|\psi(f/\|f\|)$$

for f in $\mathscr{X}$. Since $\tilde{\psi}(f) = \psi(f)$ for f in $(\mathscr{X})_1$, we see not only that $\tilde{\psi}$ is in $(\mathscr{X}^*)_1$ but, in addition, $\Lambda(\tilde{\psi}) = \psi$. Thus $\Lambda[(\mathscr{X}^*)_1]$ is a closed subset of **P**, and therefore $(\mathscr{X}^*)_1$ is compact in the w^*-topology. ∎

The importance of the preceding theorem lies in the fact that compact spaces possess many pleasant properties. We shall also use it to show that every Banach space is isomorphic to a subspace of some $C(X)$. Before doing this we need to know something about how many continuous linear functionals there are on a Banach space. This and more is contained in the Hahn–Banach theorem. Although we are only interested in Banach spaces in this chapter, it is more illuminating to state and prove the Hahn–Banach theorem in slightly greater generality. To do this we need the following definition.

1.24   Definition   Let $\mathscr{E}$ be a real linear space and p be a real-valued function defined on $\mathscr{E}$. Then p is said to be a sublinear functional on $\mathscr{E}$ if $p(f+g) \leqslant p(f)+p(g)$ for f and g in $\mathscr{E}$ and $p(\lambda f) = \lambda p(f)$ for f in $\mathscr{E}$ and positive λ.

1.25   Theorem (Hahn–Banach)   Let $\mathscr{E}$ be a real linear space and let p be a sublinear functional on $\mathscr{E}$. Let $\mathscr{F}$ be a subspace of $\mathscr{E}$ and φ be a real linear functional on $\mathscr{F}$ such that $\varphi(f) \leqslant p(f)$ for f in $\mathscr{F}$. Then there exists a real linear functional Φ on $\mathscr{E}$ such that $\Phi(f) = \varphi(f)$ for f in $\mathscr{F}$ and $\Phi(g) \leqslant p(g)$ for g in $\mathscr{E}$.

Proof   We may assume without loss of generality that $\mathscr{F} \neq \{0\}$. Take f not in $\mathscr{F}$ and let $\mathscr{G} = \{g + \lambda f : \lambda \in \mathbb{R}, g \in \mathscr{F}\}$. We first extend φ to $\mathscr{G}$ and to do this it suffices to define $\Phi(f)$ appropriately. We want $\Phi(g + \lambda f) \leqslant p(g + \lambda f)$ for all λ in $\mathbb{R}$ and g in $\mathscr{F}$. Dividing by $|\lambda|$ this can be written $\Phi(f - h) \leqslant p(f - h)$ and $\Phi(-f + h) \leqslant p(-f + h)$ for all h in $\mathscr{F}$ or equivalently,

$$-p(-f + h) + \varphi(h) \leqslant \Phi(f) \leqslant p(f - h) + \varphi(h)$$

for all h in $\mathscr{F}$. Thus a value can be chosen for $\Phi(f)$ such that the resultant Φ on $\mathscr{G}$ has the required properties if and only if

$$\sup_{h \in \mathscr{F}} \{-p(-f + h) + \varphi(h)\} \leqslant \inf_{k \in \mathscr{F}} \{p(f - k) + \varphi(k)\}.$$

However, for h and k in $\mathscr{F}$, we have

$$\varphi(h) - \varphi(k) = \varphi(h - k) \leqslant p(h - k) \leqslant p(f - k) + p(h - f),$$

so that

$$-p(h-f) + \varphi(h) \leqslant p(f-k) + \varphi(k).$$

Therefore, φ can be extended to Φ on $\mathscr{G}$ such that $\Phi(h) \leqslant p(h)$ for h in $\mathscr{G}$.

Our problem now is to somehow obtain a maximal extension of φ. To that end let $\mathscr{P}$ denote the class of extensions of φ to larger subspaces satisfying the required inequality. Hence an element of $\mathscr{P}$ consists of a subspace $\mathscr{G}$ of $\mathscr{E}$ which contains $\mathscr{F}$ and a linear functional $\Phi_{\mathscr{G}}$ on $\mathscr{G}$ which extends φ and satisfies $\Phi_{\mathscr{G}}(g) \leqslant p(g)$ for g in $\mathscr{G}$. There is a natural partial order defined on $\mathscr{P}$, where $(\mathscr{G}_1, \Phi_{\mathscr{G}_1}) \leqslant (\mathscr{G}_2, \Phi_{\mathscr{G}_2})$ if $\mathscr{G}_1 \subset \mathscr{G}_2$ and $\Phi_{\mathscr{G}_2}(f) = \Phi_{\mathscr{G}_1}(f)$ for f in $\mathscr{G}_1$. To apply Zorn's lemma to the class $\mathscr{P}$, we must show that for every chain $\{(\mathscr{G}_\alpha, \Phi_{\mathscr{G}_\alpha})\}_{\alpha \in A}$ in $\mathscr{P}$ there is a maximal element in $\mathscr{P}$. (Recall that a chain is just a linearly ordered set.) If $\{(\mathscr{G}_\alpha, \Phi_{\mathscr{G}_\alpha})\}_{\alpha \in A}$ is a chain in $\mathscr{P}$, let $\mathscr{G} = \bigcup_{\alpha \in A} \mathscr{G}_\alpha$ and define Φ on $\mathscr{G}$ by $\Phi(f) = \Phi_{\mathscr{G}_\alpha}(f)$, where f is in $\mathscr{G}_\alpha$. It is easily verified that $\mathscr{G}$ is a subspace of $\mathscr{E}$ which contains $\mathscr{F}$; that Φ is well defined, linear, and satisfies $\Phi(f) \leqslant p(f)$ for f in $\mathscr{G}$; and that $(\mathscr{G}_\alpha, \Phi_{\mathscr{G}_\alpha}) \leqslant (\mathscr{G}, \Phi)$ for each α in A. Thus the chain has a maximal element in $\mathscr{P}$ and Zorn's lemma implies that $\mathscr{P}$ itself has a maximal element $(\mathscr{G}_0, \Phi_0)$. If $\mathscr{G}_0$ were not $\mathscr{E}$, then the argument of the preceding paragraph would yield a strictly greater element in $\mathscr{P}$ which would contradict the maximality of $(\mathscr{G}_0, \Phi_{\mathscr{G}_0})$. Thus $\mathscr{G}_0 = \mathscr{E}$ and we have the desired extension of φ to $\mathscr{E}$. ■

The form of this result which we need in this chapter is the following.

1.26   Theorem (Hahn–Banach)   Let $\mathscr{M}$ be a subspace of the Banach space $\mathscr{X}$. If φ is a bounded linear functional on $\mathscr{M}$, then there exists Φ in $\mathscr{X}^*$ such that $\Phi(f) = \varphi(f)$ for f in $\mathscr{M}$ and $\|\Phi\| = \|\varphi\|$.

Proof   If we consider $\mathscr{X}$ as the real linear space $\tilde{\mathscr{X}}$, then the norm is a sublinear functional on $\tilde{\mathscr{X}}$ and $\psi = \operatorname{Re} \varphi$ is a real linear functional on the real subspace $\tilde{\mathscr{M}}$. It is evident that $\|\psi\| \leqslant \|\varphi\|$. Setting $p(f) = \|\varphi\| \|f\|$ we have $\psi(f) \leqslant p(f)$ and hence from the preceding theorem we obtain a real linear functional Ψ on $\tilde{\mathscr{X}}$ that extends ψ and satisfies $\Psi(f) \leqslant \|\varphi\| \|f\|$ for f in $\mathscr{X}$. If we now define Φ on $\mathscr{X}$ by $\Phi(f) = \Psi(f) - i\Psi(if)$, then we want to show that Φ is a bounded complex linear functional on $\mathscr{X}$ that extends φ and has norm $\|\varphi\|$.

For f and g in $\mathscr{X}$, we have

$$\begin{aligned}
\Phi(f+g) &= \Psi(f+g) - i\Psi(i(f+g)) \\
&= \Psi(f) + \Psi(g) - i\Psi(if) - i\Psi(ig) \\
&= \Phi(f) + \Phi(g).
\end{aligned}$$

Further, if λ_1 and λ_2 are real and f is in $\mathscr{X}$, then $\Phi(if) = \Psi(if) - i\Psi(-f) = i\Phi(f)$ and hence

$$\Phi((\lambda_1 + i\lambda_2)f) = \Phi(\lambda_1 f) + \Phi(i\lambda_2 f) = \lambda_1 \Phi(f) + i\lambda_2 \Phi(f)$$
$$= (\lambda_1 + i\lambda_2)\Phi(f).$$

Thus, Φ is a complex linear functional on $\mathscr{X}$. Moreover, for f in $\mathscr{M}$ we have

$$\Phi(f) = \Psi(f) - i\Psi(if) = \psi(f) - i\psi(if) = \operatorname{Re}\varphi(f) - i\operatorname{Re}\varphi(if)$$
$$= \operatorname{Re}\varphi(f) - i\operatorname{Re}(i\varphi(f)) = \operatorname{Re}\varphi(f) - i(-\operatorname{Im}\varphi(f)) = \varphi(f).$$

Lastly, to prove $\|\Phi\| = \|\Psi\|$ it suffices to show that $\|\Phi\| \leqslant \|\Psi\|$ in view of the fact that $\|\Psi\| = \|\psi\|$ and Φ is an extension of φ. For f in $\mathscr{X}$ write $\Phi(f) = re^{i\theta}$. Then

$$|\Phi(f)| = r = e^{-i\theta}\Phi(f) = \Phi(e^{-i\theta}f) = \Psi(e^{-i\theta}f)$$
$$\leqslant |\Psi(e^{-i\theta}f)| \leqslant \|\Psi\|\,\|f\|,$$

so that Φ has been shown to be an extension of φ to $\mathscr{X}$ having the same norm. ■

1.27   Corollary   If f is an element of a Banach space $\mathscr{X}$, then there exists φ in $\mathscr{X}^*$ of unit norm so that $\varphi(f) = \|f\|$.

Proof   We may assume $f \neq 0$. Let $\mathscr{M} = \{\lambda f : \lambda \in \mathbb{C}\}$ and define ψ on $\mathscr{M}$ by $\psi(\lambda f) = \lambda\|f\|$. Then $\|\psi\| = 1$ and an extension of ψ to $\mathscr{X}$ given by the Hahn–Banach theorem has the desired properties. ■

1.28   Corollary   If $\varphi(f) = 0$ for each φ in $\mathscr{X}^*$, then $f = 0$.

Proof   Obvious. ■

We give two applications of the Hahn–Banach theorem. First we prove a theorem of Banach showing that $C(X)$ is a universal Banach space and then we determine the conjugate space of the Banach space $C([0, 1])$.

1.29   Theorem (Banach)   Every Banach space $\mathscr{X}$ is isometrically isomorphic to a closed subspace of $C(X)$ for some compact Hausdorff space X.

Proof   Let X be $(\mathscr{X}^*)_1$ with the w^*-topology and define β from $\mathscr{X}$ to $C(X)$ by $(\beta f)(\varphi) = \varphi(f)$. For f_1 and f_2 in $\mathscr{X}$ and λ_1 and λ_2 in $\mathbb{C}$, we have

$$\beta(\lambda_1 f_1 + \lambda_2 f_2)(\varphi) = \varphi(\lambda_1 f_1 + \lambda_2 f_2) = \lambda_1 \varphi(f_1) + \lambda_2 \varphi(f_2)$$
$$= \lambda_1 \beta(f_1)(\varphi) + \lambda_2 \beta(f_2)(\varphi).$$

and thus β is a linear map. Further, for f in $\mathscr{X}$ we have

$$\|\beta(f)\|_\infty = \sup_{\varphi \in (\mathscr{X}^*)_1} |\beta(f)(\varphi)| = \sup_{\varphi \in (\mathscr{X}^*)_1} |\varphi(f)| \leqslant \sup_{\varphi \in (\mathscr{X}^*)_1} \|\varphi\| \|f\| \leqslant \|f\|,$$

and since by Corollary 1.27 there exists φ in $(\mathscr{X}^*)_1$ such that $\varphi(f) = \|f\|$, we have that $\|\beta(f)\|_\infty = \|f\|$. Thus β is an isometrical isomorphism. ∎

The preceding construction never yields an isomorphism of $\mathscr{X}$ onto $C((\mathscr{X}^*)_1)$ even if $\mathscr{X}$ is $C(Y)$ for some Y. If $\mathscr{X}$ is separable, then topological arguments can be used to show that X can be taken to be the closed unit interval.

Although this theorem can be viewed as a structure theorem for Banach spaces, the absence of a canonical X associated with each $\mathscr{X}$ vitiates its usefulness.

1.30 We now consider the problem of identifying the conjugate space of $C([0,1])$. By this we mean finding some concrete realization of the elements of $C([0,1])^*$ analogous with the identification obtained in Section 1.15 of the conjugate space of $l^1(\mathbb{Z}^+)$ as the space $l^\infty(\mathbb{Z}^+)$ of bounded complex functions on $\mathbb{Z}^+$. We shall identify $C([0,1])^*$ with a space of functions of bounded variation on $[0,1]$. We shall comment on $C(X)^*$ for general X later in the chapter. We begin by recalling a definition.

1.31 Definition If φ is a complex function on $[0,1]$, then φ is said to be of bounded variation if there exists $M \geqslant 0$ such that for every partition $0 = t_0 < t_1 < \cdots < t_n < t_{n+1} = 1$, it is true that

$$\sum_{i=0}^{n} |\varphi(t_{i+1}) - \varphi(t_i)| \leqslant M.$$

The greatest lower bound of the set of all such M will be denoted by $\|\varphi\|_v$.

An important property of a function of bounded variation is that it possesses limits from both the right and the left at all points of $[0,1]$.

1.32 Proposition A function of bounded variation possesses a limit from the left and right at each point.

Proof Let φ be a function on $[0,1]$ not having a limit from the left at some t in $(0,1]$; we shall show that φ is not of bounded variation on $[0,1]$.

If φ does not have a limit from the left at t, then for some $\varepsilon > 0$, it is true that for each $\delta > 0$ there exist s and s' in $[0,1]$ such that $t - \delta < s < s' < t$ and $|\varphi(s) - \varphi(s')| \geqslant \varepsilon$. Thus we can choose inductively sequences $\{s_n\}_{n=1}^{\infty}$ and

$\{s_n'\}_{n=1}^{\infty}$ such that $0 < s_1 < s_1' < \cdots < s_n < s_n' < t$ and $|\varphi(s_n) - \varphi(s_n')| \geq \varepsilon$. Now consider the partition $t_0 = 0$; $t_{2k+1} = s_k$ for $k = 0, 1, \ldots, N-1$; $t_{2k} = s_k'$ for $k = 1, 2, \ldots, N$; and $t_{2N+1} = 1$. Then

$$\sum_{k=0}^{2N} |\varphi(t_{k+1}) - \varphi(t_k)| \geq \sum_{n=1}^{N} |\varphi(s_n') - \varphi(s_n)| \geq N\varepsilon,$$

which implies that φ is not of bounded variation on $[0, 1]$. The proof that φ has a limit from the right proceeds analogously. ∎

Thus, if φ is a function of bounded variation on $[0, 1]$, the limit $\varphi(t^-)$ of φ from the left and the limit $\varphi(t^+)$ of φ from the right are well defined for t in $[0, 1]$. (We set $\varphi(0^-) = \varphi(0)$ and $\varphi(1^+) = \varphi(1)$.) Moreover, a function of bounded variation can have at most countably many discontinuities.

1.33 Corollary If φ is a function of bounded variation on $[0, 1]$, then φ has at most countably many discontinuities.

Proof Observe first that φ fails to be continuous at t in $[0, 1]$ if and only if $\varphi(t) \neq \varphi(t^+)$ or $\varphi(t) \neq \varphi(t^-)$. Moreover, if $t_0, t_1, \ldots, t_n$ are distinct points of $[0, 1]$, then

$$\sum_{i=0}^{N} |\varphi(t_i) - \varphi(t_i^+)| + \sum_{i=0}^{N} |\varphi(t_i) - \varphi(t_i^-)| \leq \|\varphi\|_v.$$

Thus for each $\varepsilon > 0$ there exists at most finitely many points t in $[0, 1]$ such that $|\varphi(t) - \varphi(t^+)| + |\varphi(t) - \varphi(t^-)| \geq \varepsilon$. Hence the set of discontinuities of φ is at most countable. ∎

We next recall the definition of the Riemann–Stieltjes integral. For f in $C([0, 1])$ and φ of bounded variation on $[0, 1]$, we denote by $\int_0^1 f\,d\varphi$, the integral of f with respect to φ; that is, $\int_0^1 f\,d\varphi$ is the limit of sums of the form $\sum_{i=0}^{n} f(t_i')[\varphi(t_{i+1}) - \varphi(t_i)]$, where $0 = t_0 < t_1 < \cdots < t_n < t_{n+1} = 1$ is a partition of $[0, 1]$ and t_i' is a point in the interval $[t_i, t_{i+1}]$. (The limit is taken over partitions for which $\max_{1 \leq i \leq n} |t_{i+1} - t_i|$ tends to zero.) In the following proposition we collect the facts about the Riemann–Stieltjes integral which we will need.

1.34 Proposition If f is in $C([0, 1])$ and φ is of bounded variation on $[0, 1]$, then $\int_0^1 f\,d\varphi$ exists. Moreover:

(1) $\int_0^1 (\lambda_1 f_1 + \lambda_2 f_2)\,d\varphi = \lambda_1 \int_0^1 f_1\,d\varphi + \lambda_2 \int_0^1 f_2\,d\varphi$ for f_1 and f_2 in $C([0, 1])$, λ_1 and λ_2 in $\mathbb{C}$, and φ of bounded variation on $[0, 1]$;

(2) $\int_0^1 f\,d(\lambda_1\varphi_1 + \lambda_2\varphi_2) = \lambda_1\int_0^1 f\,d\varphi_1 + \lambda_2\int_0^1 f\,d\varphi_2$ for f in $C([0,1])$, λ_1 and λ_2 in $\mathbb{C}$, and φ_1 and φ_2 of bounded variation on $[0,1]$; and

(3) $|\int_0^1 f\,d\varphi| \leqslant \|f\|_\infty \|\varphi\|_v$ for f in $C([0,1])$ and φ of bounded variation on $[0,1]$.

Proof Compare [65, p. 107]. ∎

Now for φ of bounded variation on $[0,1]$, let $\hat\varphi$ be the function defined by $\hat\varphi(f) = \int_0^1 f\,d\varphi$ for f in $C([0,1])$. That $\hat\varphi$ is an element of $C([0,1])^*$ follows from the preceding proposition. However, if φ is a function of bounded variation on $[0,1]$, t_0 is a point in $[0,1)$, and we define the function ψ on $[0,1]$ such that $\psi(t) = \varphi(t)$ for $t \neq t_0$ and $\psi(t_0) = \varphi(t_0^-)$, then an easy computation shows that $\int_0^1 f\,d\varphi = \int_0^1 f\,d\psi$ for f in $C([0,1])$. Thus if one is interested only in the linear functional which a function of bounded variation defines on $C([0,1])$, then φ and ψ are equivalent, or more precisely, $\hat\varphi = \hat\psi$. In order to avoid identifying the conjugate space of $C([0,1])$ with equivalence classes of functions of bounded variation, we choose a normalized representative from each class by requiring that the distinguished function be left continuous on $(0,1)$.

1.35 Proposition Let φ be of bounded variation on $[0,1]$ and ψ be the function defined $\psi(t) = \varphi(t^-)$ for t in $(0,1)$, $\psi(0) = \varphi(0)$, and $\psi(1) = \varphi(1)$. Then ψ is of bounded variation on $[0,1]$, $\|\psi\|_v \leqslant \|\varphi\|_v$, and

$$\int_0^1 f\,d\varphi = \int_0^1 f\,d\psi \qquad \text{for } f \text{ in } C([0,1]).$$

Proof From Corollary 1.33 it follows that we can list $\{s_i\}_{i\geqslant 1}$ the points of $[0,1]$ at which φ is discontinuous from the left. Moreover, from the definition of ψ we have $\psi(t) = \varphi(t)$ for $t \neq s_i$ for $i \geqslant 1$. Now let $0 = t_0 < t_1 < \cdots < t_n < t_{n+1} = 1$ be a partition of $[0,1]$ having the property that if t_i is in $S = \{s_i : i \geqslant 1\}$, then neither t_{i-1} nor t_{i+1} is. To show that ψ is of bounded variation and $\|\psi\|_v \leqslant \|\varphi\|_v$, it is sufficient to prove that

$$\sum_{i=0}^n |\psi(t_{i+1}) - \psi(t_i)| \leqslant \|\varphi\|_v.$$

Fix $\varepsilon > 0$. If t_i is not in S or $i = 0$ or $n+1$, then set $t_i' = t_i$. If t_i is in S and $i \neq 0, n+1$, choose t_i' in (t_{i-1}, t_i) such that $|\varphi(t_i^-) - \varphi(t_i')| < \varepsilon/2n+2$. Then $0 = t_0' < t_1' < \cdots < t_n' < t_{n+1}' = 1$ is a partition of $[0,1]$ and

$$\sum_{i=0}^{n} |\psi(t_{i+1}) - \psi(t_i)| = \sum_{i=0}^{n} |\varphi(t_{i+1}^-) - \varphi(t_i^-)|$$

$$\leqslant \sum_{i=0}^{n} |\varphi(t_{i+1}^-) - \varphi(t'_{i+1})| + \sum_{i=0}^{n} |\varphi(t'_{i+1}) - \varphi(t_i')|$$

$$+ \sum_{i=0}^{n} |\varphi(t_i') - \varphi(t_i^-)|$$

$$\leqslant \varepsilon/2 + \|\varphi\|_v + \varepsilon/2.$$

Since ε is arbitrary, we have that ψ is of bounded variation and that $\|\psi\|_v \leqslant \|\varphi\|_v$.

To complete the proof for N an integer, let η_N be the function defined $\eta_N(t) = 0$ for t not in $\{s_1, s_2, \ldots, s_N\}$ and $\eta_N(s_i) = \varphi(s_i) - \psi(s_i)$ for $1 \leqslant i \leqslant N$. Then it is easy to show that $\lim_{N\to\infty} \|\varphi - (\psi + \eta_N)\|_v = 0$ and $\int_0^1 f \, d\eta_N = 0$ for f in $C([0,1])$. Thus, we have from Proposition 1.34 that

$$\int_0^1 f \, d\varphi = \int_0^1 f \, d\psi + \lim_{N\to\infty} \int_0^1 f \, d\eta_N = \int_0^1 f \, d\psi. \quad\blacksquare$$

Let $BV[0,1]$ denote the space of all complex functions on $[0,1]$ which are of bounded variation on $[0,1]$, which vanish at 0, and which are continuous from the left on $(0,1)$. With respect to pointwise addition and scalar multiplication, $BV[0,1]$ is a linear space, and $\| \ \|_v$ defines a norm.

1.36   Theorem   The space $BV[0,1]$ is a Banach space.

Proof   We shall make use of Corollary 1.10 to show that $BV[0,1]$ is complete and hence a Banach space. Suppose $\{\varphi_n\}_{n=1}^{\infty}$ is a sequence of functions in $BV[0,1]$ such that $\sum_{n=1}^{\infty} \|\varphi_n\|_v < \infty$. Since

$$|\varphi_n(t)| \leqslant |\varphi_n(t) - \varphi_n(0)| + |\varphi_n(1) - \varphi_n(t)| \leqslant \|\varphi_n\|_v$$

for t in $[0,1]$, it follows that $\sum_{n=1}^{\infty} \varphi_n(t)$ converges absolutely and uniformly to a function φ defined on $[0,1]$. It is immediate that $\varphi(0) = 0$ and that φ is continuous from the left on $(0,1)$. It remains to show that φ is of bounded variation and that $\lim_{N\to\infty} \|\varphi - \sum_{n=1}^{N} \varphi_n\|_v = 0$.

If $0 = t_0 < t_1 < \cdots < t_k < t_{k+1} = 1$ is any partition of $[0,1]$, then

$$\sum_{i=0}^{k} |\varphi(t_{i+1}) - \varphi(t_i)| = \sum_{i=0}^{k} \left| \sum_{n=1}^{\infty} \varphi_n(t_{i+1}) - \sum_{n=1}^{\infty} \varphi_n(t_i) \right|$$

$$\leqslant \sum_{i=0}^{k} \sum_{n=1}^{\infty} |\varphi_n(t_{i+1}) - \varphi_n(t_i)|$$

$$\leqslant \sum_{n=1}^{\infty} \left[\sum_{i=0}^{k} |\varphi_n(t_{i+1}) - \varphi_n(t_i)| \right] \leqslant \sum_{n=1}^{\infty} \|\varphi_n\|_v.$$

Therefore, φ is of bounded variation and hence in $BV[0,1]$. Moreover, since the inequality

$$\sum_{i=0}^{k} \left| \left(\varphi - \sum_{n=1}^{N} \varphi_n \right)(t_{i+1}) - \left(\varphi - \sum_{n=1}^{N} \varphi_n \right)(t_i) \right|$$

$$= \sum_{i=0}^{k} \left| \sum_{n=N+1}^{\infty} \varphi_n(t_{i+1}) - \sum_{n=N+1}^{\infty} \varphi_n(t_i) \right|$$

$$\leqslant \sum_{i=0}^{k} \sum_{n=N+1}^{\infty} |\varphi_n(t_{i+1}) - \varphi_n(t_i)| \leqslant \sum_{n=N+1}^{\infty} \|\varphi_n\|_v$$

holds for every partition of $[0,1]$, we see that

$$\left\| \varphi - \sum_{n=1}^{N} \varphi_n \right\|_v \leqslant \sum_{n=N+1}^{\infty} \|\varphi_n\|_v$$

for every integer N. Thus $\varphi = \sum_{n=1}^{\infty} \varphi_n$ in the norm of $BV[0,1]$ and the proof is complete. ∎

Recall that for φ of bounded variation on $[0,1]$, we let $\hat{\varphi}$ be the linear functional defined $\hat{\varphi}(f) = \int_0^1 f \, d\varphi$ for f in $C([0,1])$.

1.37   Theorem (Riesz)   The mapping $\varphi \to \hat{\varphi}$ is an isometrical isomorphism between $BV[0,1]$ and $C([0,1])^*$.

Proof   The fact that $\hat{\varphi}$ is in $C([0,1])^*$ follows immediately from Proposition 1.34 and we have, moreover, that $\|\hat{\varphi}\| \leqslant \|\varphi\|_v$. To complete the proof we must, given an L in $C([0,1])^*$, produce a function ψ in $BV[0,1]$ such that $\hat{\psi} = L$ and $\|\psi\|_v \leqslant \|L\|$. To do this we first use the Hahn–Banach theorem to extend L to a larger Banach space.

Let $B[0,1]$ be the space of *all* bounded complex functions on $[0,1]$. It is by now routine to verify that $B[0,1]$ is a Banach space with respect to pointwise addition and scalar multiplication, and the norm $\|f\|_s = \sup\{|f(t)| : 0 \leqslant t \leqslant 1\}$. For E a subset of $[0,1]$, let I_E denote the characteristic (or indicator) function on E, that is, $I_E(t)$ is 1 if t is in E and 0 otherwise. Then for every E the function I_E is in $B[0,1]$.

Since $C([0,1])$ is a subspace of $B[0,1]$ and since L is a bounded linear functional on $C([0,1])$, we can extend it (but not necessarily uniquely) to a bounded linear functional L' on $B[0,1]$ such that $\|L'\| = \|L\|$. Moreover, L'

can be chosen to satisfy $L'(I_{\{0\}}) = 0$, since we may first extend it in this manner to the linear span of $I_{\{0\}}$ and $C([0, 1])$ in view of the inequality

$$|L'(f + \lambda I_{\{0\}})| = |L(f)| \leqslant \|L\| \|f\|_\infty \leqslant \|L\| \|f + \lambda I_{\{0\}}\|_s$$

which holds for f in $C([0, 1])$ and λ in $\mathbb{C}$.

Now for $0 < t \leqslant 1$ define $\varphi(t) = L'(I_{(0, t]})$, where $(0, t]$ is the half open interval $\{s : 0 < s \leqslant t\}$ and set $\varphi(0) = 0$. We want first to show that φ is of bounded variation and that $\|\varphi\|_v \leqslant \|L\|$.

Let $0 = t_0 < t_1 < \cdots < t_n < t_{n+1} = 1$ be a partition of $[0, 1]$ and set

$$\lambda_k = [\varphi(t_{k+1}) - \varphi(t_k)] / |\varphi(t_{k+1}) - \varphi(t_k)|$$

if $\varphi(t_{k+1}) \neq \varphi(t_k)$ and 0 otherwise. Then the function

$$f = \sum_{k=0}^{n} \lambda_k I_{(t_k, t_{k+1}]}$$

is in $B[0, 1]$ and $\|f\|_s \leqslant 1$. Moreover, we have

$$\sum_{k=0}^{n} |\varphi(t_{k+1}) - \varphi(t_k)| = \sum_{k=0}^{n} \lambda_k (\varphi(t_{k+1}) - \varphi(t_k))$$

$$= \sum_{k=0}^{n} \lambda_k L'(I_{(t_k, t_{k+1}]})$$

$$= L'(f) \leqslant \|L'\| = \|L\|,$$

and hence φ is of bounded variation and $\|\varphi\|_v \leqslant \|L\|$.

We next want to show that $L(g) = \int_0^1 g \, d\varphi$ for every function g in $C([0, 1])$. To that end, let g be in $C([0, 1])$ and $\varepsilon > 0$; choose a partition $0 = t_0 < t_1 < \cdots < t_n < t_{n+1} = 1$ such that

$$|g(s) - g(s')| < \frac{\varepsilon}{2\|L'\|}$$

for s and s' in each subinterval $[t_k, t_{k+1}]$ and such that

$$\left| \int_0^1 g \, d\varphi - \sum_{k=0}^{n} g(t_k)(\varphi(t_{k+1}) - \varphi(t_k)) \right| < \frac{\varepsilon}{2}.$$

Then we have for $f = \sum_{k=0}^{n} g(t_k) I_{(t_k, t_{k+1}]} + g(0) I_{\{0\}}$ the inequality

$$\left| L(g) - \int_0^1 g \, d\varphi \right| \leqslant |L(g) - L'(f)| + \left| L'(f) - \int_0^1 g \, d\varphi \right|$$

$$\leqslant \|L'\| \|g - f\|_s + \left| \sum_{k=0}^{n} g(t_k)(\varphi(t_{k+1}) - \varphi(t_k)) - \int_0^1 g \, d\varphi \right|$$

$$\leqslant \frac{\varepsilon}{2} + \frac{\varepsilon}{2} = \varepsilon.$$

Thus $L(g) = \int_0^1 g \, d\varphi$ for g in $C([0,1])$.

Now the φ obtained need not be continuous from the left on $(0,1)$. However, appealing to Proposition 1.35, we obtain ψ in $BV[0,1]$ such that $\|\psi\|_v \leqslant \|\varphi\|_v \leqslant \|L\|$ and

$$\hat{\psi}(g) = \int_0^1 g \, d\psi = \int_0^1 g \, d\varphi = L(g)$$

for g in $C([0,1])$. Thus $\hat{\psi} = L$ and combining the inequality obtained from the first paragraph of the proof with the one just above, we obtain $\|\psi\|_v = \|L\|$. Thus $BV[0,1] = C([0,1])^*$. ∎

1.38   The Conjugate Space of $C(X)$ If X is an arbitrary compact Hausdorff space, then the notion of a function of bounded variation on X makes no sense. Thus one must search for a different realization of the elements of $C(X)^*$. It can be shown with little difficulty that each countably additive measure defined on the Borel sets of X gives rise to a bounded linear functional on $C(X)$. Moreover, just as in the preceding proof we can extend a bounded linear functional on $C(X)$ to the Banach space of bounded Borel functions by the Hahn–Banach theorem and then obtain a Borel measure by evaluating the extended functional at the indicator functions for Borel sets. This representation of a bounded linear functional as a Borel measure is not unique. If one restricts attention to the regular Borel measures on X, then the pairing is unique and one can identify $C(X)^*$ with the space $M(X)$ of complex regular Borel measures on X. We do not prove this in this book but refer the reader to [65]. This result is usually called the Riesz–Markov representation theorem. We shall need it at least for X a compact subset of the plane.

1.39   Quotient Spaces   Let $\mathscr{X}$ be a Banach space and $\mathscr{M}$ be a closed subspace of $\mathscr{X}$. We want to show that there is a natural norm on the quotient space $\mathscr{X}/\mathscr{M}$ making it into a Banach space. Let $\mathscr{X}/\mathscr{M}$ denote the linear space of equivalence classes $\{[f] : f \in \mathscr{X}\}$, where $[f] = \{f + g : g \in \mathscr{M}\}$, and define a norm on $\mathscr{X}/\mathscr{M}$ by

$$\|[f]\| = \inf_{g \in \mathscr{M}} \|f + g\| = \inf_{h \in [f]} \|h\|.$$

Then $\|[f]\| = 0$ implies there exists a sequence $\{g_n\}_{n=1}^{\infty}$ in $\mathscr{M}$ with $\lim_{n \to \infty} \|f + g_n\| = 0$. Since $\mathscr{M}$ is closed, it follows that f is in $\mathscr{M}$ so that $[f] = [0]$. Conversely, if $[f] = [0]$, then f is in $\mathscr{M}$ and $0 \leqslant \|[f]\| \leqslant \|f - f\| = 0$. Thus, $\|[f]\| = 0$ if and only if $[f] = [0]$. Further, if f_1 and f_2 are in $\mathscr{X}$ and

λ is in $\mathbb{C}$, then

$$\|\lambda[f_1]\| = \|[\lambda f_1]\| = \inf_{g \in \mathcal{M}} \|\lambda f_1 + g\| = |\lambda| \inf_{h \in \mathcal{M}} \|f_1 + h\| = |\lambda| \|[f_1]\|$$

and

$$\|[f_1] + [f_2]\| = \|[f_1 + f_2]\| = \inf_{g \in \mathcal{M}} \|f_1 + f_2 + g\|$$

$$= \inf_{g_1, g_2 \in \mathcal{M}} \|f_1 + g_1 + f_2 + g_2\| \leqslant \inf_{g_1 \in \mathcal{M}} \|f_1 + g_1\| + \inf_{g_2 \in \mathcal{M}} \|f_2 + g_2\|$$

$$\leqslant \|[f_1]\| + \|[f_2]\|.$$

Therefore, $\|\cdot\|$ is a norm on $\mathcal{X}/\mathcal{M}$ and it remains only to prove that the space is complete.

If $\{[f_n]\}_{n=1}^{\infty}$ is a Cauchy sequence in $\mathcal{X}/\mathcal{M}$, then there exists a subsequence $\{f_{n_k}\}_{k=1}^{\infty}$ such that $\|[f_{n_{k+1}}] - [f_{n_k}]\| < 1/2^k$. If we choose h_k in $[f_{n_{k+1}} - f_{n_k}]$ such that $\|h_k\| < 1/2^k$, then $\sum_{k=1}^{\infty} \|h_k\| < 1$ and hence the sequence $\{h_k\}$ is absolutely summable. Therefore, $h = \sum_{k=1}^{\infty} h_k$ exists by Proposition 1.9. Since

$$[f_{n_k} - f_{n_1}] = \sum_{i=1}^{k-1} [f_{n_{i+1}} - f_{n_i}] = \sum_{i=1}^{k-1} [h_i],$$

we have $\lim_{k \to \infty} [f_{n_k} - f_{n_1}] = [h]$. Therefore $\lim_{k \to \infty} [f_{n_k}] = [h + f_{n_1}]$ and $\mathcal{X}/\mathcal{M}$ is seen to be a Banach space.

We conclude by pointing out that the natural map $f \to [f]$ from $\mathcal{X}$ to $\mathcal{X}/\mathcal{M}$ is a contraction and is an open map. For suppose f is in $\mathcal{X}$, $\varepsilon > 0$, and $N_\varepsilon(f) = \{g \in \mathcal{X} : \|f - g\| < \varepsilon\}$. If $[h]$ is in

$$N_\varepsilon([f]) = \{[k] \in \mathcal{X}/\mathcal{M} : \|[f] - [k]\| < \varepsilon\},$$

then there exists h_0 in $[h]$ such that $\|f - h_0\| < \varepsilon$. Hence, $[h]$ and, in fact, all of $N_\varepsilon([f])$ is in the image of $N_\varepsilon(f)$ under the natural map. Therefore, the natural map is open.

1.40 Definition Let $\mathcal{X}$ and $\mathcal{Y}$ be Banach spaces. A linear transformation T from $\mathcal{X}$ to $\mathcal{Y}$ is said to be bounded if

$$\|T\| = \sup_{f \neq 0} \frac{\|Tf\|}{\|f\|} < \infty.$$

The set of bounded linear transformations of $\mathcal{X}$ to $\mathcal{Y}$ is denoted $\mathfrak{L}(\mathcal{X}, \mathcal{Y})$ with $\mathfrak{L}(\mathcal{X}, \mathcal{X})$ abbreviated $\mathfrak{L}(\mathcal{X})$. A linear transformation is bounded if and only if it is continuous.

1.41 Proposition The space $\mathfrak{L}(\mathcal{X}, \mathcal{Y})$ is a Banach space.

Proof The only thing that needs proof is the completeness of $\mathfrak{L}(\mathscr{X}, \mathscr{Y})$ and that is left as an exercise. ∎

Although an essential feature of a Banach space is that it is complete in the metric induced by the norm, we have not yet made any real use of this property. The importance of completeness is due mainly to the applicability of the Baire category theorem. We now present one of the principal applications, namely the *open mapping theorem*. The equally important *uniform boundedness theorem* will be given in the exercises.

1.42   Theorem   If $\mathscr{X}$ and $\mathscr{Y}$ are Banach spaces and T in $\mathfrak{L}(\mathscr{X}, \mathscr{Y})$ is one-to-one and onto, then T^{-1} exists and is bounded.

Proof The transformation T^{-1} is well defined and we must show it to be bounded. For $r > 0$ let $(\mathscr{X})_r = \{f \in \mathscr{X} : \|f\| \leqslant r\}$. To show that T^{-1} is bounded it is sufficient to establish $T^{-1}(\mathscr{Y})_1 \subset (\mathscr{X})_r$ for some $r > 0$ or equivalently, that $(\mathscr{Y})_1 \subset T(\mathscr{X})_N$ for some integer N.

Since T is onto, we have $\bigcup_{n=1}^{\infty} T[(\mathscr{X}_n)] = \mathscr{Y}$. Further, since $\mathscr{Y}$ is a complete metric space, the Baire category theorem states that $\mathscr{Y}$ is not the countable union of nowhere dense sets. Thus, for some N the closure $\mathrm{clos}\{T[(\mathscr{X})_N]\}$ of $T[(\mathscr{X})_N]$ contains a nonempty open set. It follows that there is an h in $(\mathscr{X})_N$ and an $\varepsilon > 0$ such that

$$Th + (\mathscr{Y})_\varepsilon = \{f \in \mathscr{Y} : \|f - Th\| < \varepsilon\} \subset \mathrm{clos}\{T[(\mathscr{X})_N]\}.$$

Therefore, $(\mathscr{Y})_\varepsilon \subset -Th + \mathrm{clos}\{T[(\mathscr{X})_N]\} \subset \mathrm{clos}\{T[(\mathscr{X})_{2N}]\}$ so that $(\mathscr{Y})_1 \subset \mathrm{clos}\{T[(\mathscr{X})_r]\}$, where $r = 2N/\varepsilon$. Except for the fact that this is the closure, this is what we need to prove. Thus we want to remove the closure.

Let f be in $(\mathscr{Y})_1$. There exists g_1 in $(\mathscr{X})_r$ with $\|f - Tg_1\| < \frac{1}{2}$. Since $f - Tg_1$ is in $(\mathscr{Y})_{1/2}$, there exists g_2 in $(\mathscr{X})_{r/2}$ with $\|f - Tg_1 - Tg_2\| < \frac{1}{4}$. Since $f - Tg_1 - Tg_2$ is in $(\mathscr{Y})_{1/4}$, there exists g_3 in $(\mathscr{X})_{r/4}$ with $\|f - Tg_1 - Tg_2 - Tg_3\| < \frac{1}{8}$. Continuing by induction, we obtain a sequence $\{g_n\}_{n=1}^{\infty}$ such that $\|g_n\| \leqslant r/2^{n-1}$ and $\|f - \sum_{i=1}^{n} Tg_i\| < 1/2^n$. Since

$$\sum_{n=1}^{\infty} \|g_n\| \leqslant \sum_{n=1}^{\infty} \frac{r}{2^{n-1}} = 2r,$$

the series $\sum_{n=1}^{\infty} g_n$ converges to an element g in $(\mathscr{X})_{2r}$. Further,

$$Tg = T\left(\lim_{n \to \infty} \sum_{k=1}^{n} g_k\right) = \lim_{n \to \infty} \sum_{k=1}^{n} Tg_k = f.$$

Therefore, $(\mathscr{Y})_1 \subset T[(\mathscr{X})_{2r}]$ which completes the proof. ∎

1.43   Corollary (Open Mapping Theorem)   If $\mathscr{X}$ and $\mathscr{Y}$ are Banach spaces and T is an onto operator in $\mathfrak{L}(\mathscr{X}, \mathscr{Y})$, then T is an open map.

Proof   Since T is continuous, the set $\mathscr{M} = \{f \in \mathscr{X} : Tf = 0\}$ is a closed subspace of $\mathscr{X}$. We want to define a transformation S (see accompanying figure) from the quotient space $\mathscr{X}/\mathscr{M}$ to $\mathscr{Y}$ as follows: for $[f]$ in $\mathscr{X}/\mathscr{M}$ set

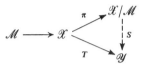

$S[f] = Tg$ for g in $[f]$. Since g_1 and g_2 in $[f]$ imply that $g_1 - g_2$ is in $\mathscr{M}$, we have $Tg_1 = Tg_2$ and hence S is well defined. Obviously, S is linear and the inequality

$$\|S[f]\| = \inf_{g \in [f]} \|Tg\| \leqslant \|T\| \inf_{g \in [f]} \|g\| = \|T\| \|[f]\|$$

which holds for $[f]$ in $\mathscr{X}/\mathscr{M}$, shows that S is bounded. Moreover, if $S[f] = 0$, then $Tf = 0$, which implies that f is in $\mathscr{M}$ and $[f] = [0]$. Therefore, S is one-to-one. Lastly, S is onto, since T is, and hence the preceding theorem demonstrates that S is an open map. Since the natural homomorphism π from $\mathscr{X}$ to $\mathscr{X}/\mathscr{M}$ is open and $T = S\pi$, we obtain that T is open.   ∎

We conclude this chapter with some classical examples of Banach spaces due to Lebesgue and Hardy. (It is assumed in what follows that the reader is familiar with standard measure theory.)

1.44   The Lebesgue Spaces   Let μ be a probability measure on a σ-algebra $\mathscr{S}$ of subsets of a set X. Let $\mathscr{L}^1$ denote the linear space of integrable complex functions on X with pointwise addition and scalar multiplication, and let $\mathscr{N}$ be the subspace of null functions. Hence, a measurable function f on X is in $\mathscr{L}^1$ if $\int_X |f| \, d\mu < \infty$ and is in $\mathscr{N}$ if $\int_X |f| \, du = 0$. We let L^1 denote the quotient space $\mathscr{L}^1/\mathscr{N}$ with the norm $\|[f]\|_1 = \int_X |f| \, d\mu$. That this satisfies the properties of a norm (that is, (1)–(3) of Section 1.1) is easy; the completeness is only slightly more difficult.

Let $\{[f_n]\}_{n=1}^\infty$ be a sequence in L^1 such that $\sum_{n=1}^\infty \|[f_n]\|_1 \leqslant M < \infty$. Choose representatives f_n from each $[f_n]$; then the sequence $\{\sum_{n=1}^N |f_n|\}_{N=1}^\infty$ is an increasing sequence of nonnegative measurable functions having the property that the integrals

$$\int_X \left(\sum_{n=1}^N |f_n| \right) d\mu = \sum_{n=1}^N \|[f_n]\|_1 \leqslant M$$

are uniformly bounded. Thus, it follows from Fatou's lemma that the function $h = \sum_{n=1}^{\infty} |f_n|$ is integrable. Therefore, the sequence $\{\sum_{n=1}^{N} f_n\}_{N=1}^{\infty}$ converges almost everywhere to an integrable function k in $\mathscr{L}^1$. Finally, we have that

$$\left\| [k] - \sum_{n=1}^{N} [f_n] \right\|_1 = \int_X \left| \sum_{n=1}^{\infty} f_n - \sum_{n=1}^{N} f_n \right| d\mu$$

$$\leqslant \sum_{n=N+1}^{\infty} \int_X |f_n| \, d\mu \leqslant \sum_{n=N+1}^{\infty} \| [f_n] \|_1 ,$$

and hence $\sum_{n=1}^{\infty} [f_n] = [k]$. Thus, L^1 is a Banach space.

For $1 < p < \infty$ let $\mathscr{L}^p$ denote the collection of functions f in $\mathscr{L}^1$ which satisfy $\int_X |f|^p \, d\mu < \infty$ and set $\mathcal{N}^p = \mathcal{N} \cap \mathscr{L}^p$. Then it can be shown that $\mathscr{L}^p$ is a linear subspace of $\mathscr{L}^1$ and that the quotient space $L^p = \mathscr{L}^p/\mathcal{N}^p$ is a Banach space for the norm

$$\| [f] \|_p = \left(\int_X |f|^p \, d\mu \right)^{1/p}$$

The details of this will be carried out for the case $p = 2$ in Chapter 3; we refer the reader to [65] for details concerning the other cases.

Now let $\mathscr{L}^\infty$ denote the subspace of $\mathscr{L}^1$ consisting of the essentially bounded functions, that is, the functions f for which the set

$$\{x \in X : |f(x)| > M\}$$

has measure zero for M sufficiently large, and let $\| f \|_\infty$ denote the smallest such M. If we set $\mathcal{N}^\infty = \mathcal{N} \cap \mathscr{L}^\infty$, then we can easily show that for f in $\mathscr{L}^\infty$ we have $\| f \|_\infty = 0$ if and only if f is in $\mathcal{N}^\infty$. Thus $\| \ \|_\infty$ defines a norm on the quotient space $L^\infty = \mathscr{L}^\infty/\mathcal{N}^\infty$. To show that L^∞ is a Banach space we need only verify completeness and we do this using Corollary 1.10. Let $\{[f_n]\}_{n=1}^{\infty}$ be a sequence of elements of L^∞ such that $\sum_{n=1}^{\infty} \| [f_n] \|_\infty \leqslant M < \infty$. Choose representatives f_n for each $[f_n]$ such that $|f_n|$ is bounded everywhere by $\| [f_n] \|_\infty$. Then for x in $[0, 1]$ we have

$$\sum_{n=1}^{\infty} |f_n(x)| \leqslant \sum_{n=1}^{\infty} \| [f_n] \|_\infty \leqslant M.$$

Therefore the function $h(x) = \sum_{n=1}^{\infty} f_n(x)$ is well defined, measurable, and bounded since

$$|h(x)| = \left| \sum_{n=1}^{\infty} f_n(x) \right| \leqslant \sum_{n=1}^{\infty} |f_n(x)| \leqslant M.$$

Thus, h is in $\mathscr{L}^\infty$ and we omit the verification that $\lim_{N \to \infty} \| [h] - \sum_{n=1}^{N} [f_n] \| = 0$. Hence, L^∞ is a Banach space.

Although the elements of an L^p space are actually equivalence classes of functions, one normally treats them as functions. Thus when we write f in L^1, we mean f is a function in $\mathscr{L}^1$ and f denotes the equivalence class in L^1 containing f. Hereafter we adopt this abuse of notation.

We conclude by showing that $(L^1)^*$ can be identified as L^∞. This result should be compared with that of Example 1.15. We indicate a different proof of this result not using the Radon–Nikodym theorem in Problem 3.22.

For φ in L^∞ we let $\hat{\varphi}$ denote the linear functional defined as

$$\hat{\varphi}(f) = \int_X f\varphi \, d\mu \qquad \text{for } f \text{ in } L^1.$$

1.45   Theorem   The map $\varphi \to \hat{\varphi}$ is an isometrical isomorphism of L^∞ onto $(L^1)^*$.

Proof   If φ is in L^∞, then for f in L^1 we have $|(\varphi f)(x)| \leqslant \|\varphi\|_\infty |f(x)|$ for almost every x in X. Thus $f\varphi$ is integrable, $\hat{\varphi}$ is well defined and linear, and

$$|\hat{\varphi}(f)| = \left| \int_X f\varphi \, d\mu \right| \leqslant \|\varphi\|_\infty \int_X |f| \, d\mu \leqslant \|\varphi\|_\infty \|f\|_1 \, .$$

Therefore, $\hat{\varphi}$ is in $(L^1)^*$ and $\|\hat{\varphi}\| \leqslant \|\varphi\|_\infty$.

Now let L be an element of $(L^1)^*$. For E a measurable subset of X the indicator function I_E is in L^1 and $\|I_E\|_1 = \int_X I_E \, d\mu = \mu(E)$. If we set $\lambda(E) = L(I_E)$, then it is easily verified that λ is a finitely additive set function and that $|\lambda(E)| \leqslant \mu(E)\|L\|$. Moreover, if $\{E_n\}_{n=1}^\infty$ is a nested sequence of measurable sets such that $\bigcap_{n=1}^\infty E_n = \varnothing$, then

$$\left| \lim_{n \to \infty} \lambda(E_n) \right| \leqslant \lim_{n \to \infty} |\lambda(E_n)| \leqslant \|L\| \lim_{n \to \infty} \mu(E_n) = 0.$$

Therefore, λ is a complex measure on X dominated by μ. Hence by the Radon–Nikodym theorem there exists an integrable function φ on X such that $\lambda(E) = \int_X I_E \varphi \, d\mu$ for all measurable E. It remains to prove that φ is essentially bounded by $\|L\|$ and that $L(f) = \int_X f\varphi \, d\mu$ for f in L^1.

For N an integer, set

$$E_N = \left\{ x \in X : \|L\| + \frac{1}{N} \leqslant |\varphi(x)| \leqslant N \right\}.$$

Then E_N is measurable and $I_{E_N}\varphi$ is bounded. If $f = \sum_{i=1}^k c_i I_{E_i}$ is a simple step function, then it is easy to see that $L(f) = \int_X f\varphi \, d\mu$. Moreover, a simple approximation argument shows that if f is in L^1 and supported on

E_N, then again $L(f) = \int_X f\varphi \, d\mu$. Let g be the function defined to be $\overline{\varphi(x)}/|\varphi(x)|$ if x is in E_N and $\varphi(x) \neq 0$ and 0 otherwise. Then g is in L^1, is supported on E_N, and $\|g\|_1 = \mu(E_N)$. Therefore, we have

$$\mu(E_N)\|L\| \geqslant |L(g)| = \left| \int_X g\varphi \, d\mu \right| = \int_X |\varphi| \, I_{E_N} \, d\mu \geqslant \left(\|L\| + \frac{1}{N} \right) \mu(E_N)$$

which implies $\mu(E_N) = 0$. Hence, we obtain $\mu(\bigcup_{N=1}^{\infty} E_N) = 0$, which implies φ is essentially bounded and $\|\varphi\|_\infty \leqslant \|L\|$. Moreover, the above argument can be used to show that

$$L(f) = \int_X f\varphi \, d\mu \qquad \text{for all } f \text{ in } L^1,$$

which completes the proof.   ∎

We now consider the Banach spaces first studied by Hardy. Although these spaces can be viewed as subspaces of the L^p spaces, this point of view is quite different from that of Hardy, who considered them as spaces of analytic functions on the unit disk. Moreover, although we study these spaces in some detail in later chapters, here we do little more than give the definition and make a few elementary observations concerning them.

1.46   The Hardy Spaces   If $\mathbb{T}$ denotes the unit circle in the complex plane and μ is the Lebesgue measure on $\mathbb{T}$ normalized so that $\mu(\mathbb{T}) = 1$, then we can define the Lebesgue spaces $L^p(\mathbb{T})$ with respect to μ. The Hardy space H^p will be defined as a closed subspace of $L^p(\mathbb{T})$. As in the previous section we consider only the cases $p = 1$ or ∞.

For n in $\mathbb{Z}$ let χ_n denote the function on $\mathbb{T}$ defined $\chi_n(z) = z^n$. If we define

$$H^1 = \left\{ f \in L^1(\mathbb{T}) : \frac{1}{2\pi} \int_0^{2\pi} f\chi_n \, dt = 0 \qquad \text{for } n = 1, 2, 3, \cdots, \right\},$$

then H^1 is obviously a linear subspace of $L^1(\mathbb{T})$. Moreover, since the set

$$\left\{ f \in L^1(\mathbb{T}) : \frac{1}{2\pi} \int_0^{2\pi} f\chi_n \, dt = 0 \right\}$$

is the kernel of a bounded linear functional on $L^1(\mathbb{T})$, we see that H^1 is a closed subspace of $L^1(\mathbb{T})$ and hence a Banach space.

For precisely the same reasons, the set

$$H^\infty = \left\{ \varphi \in L^\infty(\mathbb{T}) : \frac{1}{2\pi} \int_0^{2\pi} \varphi\chi_n \, dt = 0 \qquad \text{for } n = 1, 2, 3, \cdots, \right\}$$

is a closed subspace of $L^\infty(\mathbb{T})$. Moreover, in this case

$$\left\{ \varphi \in L^\infty(\mathbb{T}) : \frac{1}{2\pi} \int_0^{2\pi} \varphi \chi_n \, dt = 0 \right\}$$

is the zero set or kernel of the w^*-continuous function

$$\hat{\chi}_n(\varphi) = \frac{1}{2\pi} \int_0^{2\pi} \varphi \chi_n \, dt$$

and hence is w^*-closed. Therefore, H^∞ is a w^*-closed subspace of $L^\infty(\mathbb{T})$. If we let H_0^∞ denote the closed subspace

$$\left\{ \varphi \in H^\infty : \frac{1}{2\pi} \int_0^{2\pi} \varphi \, dt = 0 \right\},$$

then the conjugate space of H^1 can be shown to be naturally isometrically isomorphic to $L^\infty(\mathbb{T})/H_0^\infty$. We do not prove this here but consider this question in Chapter 6.

Notes

The basic theory of Banach spaces is covered in considerable detail in most textbooks on functional analysis. Accounts are contained in Bourbaki [7], Goffman and Pedrick [44], Naimark [80], Riesz and Sz.-Nagy [92], Rudin [95], and Yoshida [117]. The reader may also find it of interest to consult Banach [5].

Exercises

Assume in the following that X is a compact Hausdorff space and that $\mathscr{X}$ is a Banach space.

1.1   Show that the space $C(X)$ is finite dimensional if and only if X is finite.

1.2   Show that every linear functional on $\mathscr{X}$ is continuous if and only if $\mathscr{X}$ is finite dimensional.

1.3   If $\mathscr{M}$ is a normed linear space, then there exists a unique (up to isomorphism) Banach space $\mathscr{X}$ containing $\mathscr{M}$ such that clos $\mathscr{M} = \mathscr{X}$.

1.4   Complete the proof begun in Section 1.15 that $l^1(\mathbb{Z}^+)^* = l^\infty(\mathbb{Z}^+)$.

1.5   Determine whether each of the following spaces is separable in the norm topology: $c_0(\mathbb{Z}^+)$, $l^1(\mathbb{Z}^+)$, $l^\infty(\mathbb{Z}^+)$, and $l^\infty(\mathbb{Z}^+)^*$.

Definition   An element f of the convex subset K of $\mathscr{X}$ is said to be an extreme point of K if for no distinct pair f_1 and f_2 in K is $f = \frac{1}{2}(f_1 + f_2)$.

1.6   Show that an element f of $C(X)$ is an extreme point of the unit ball if and only if $|f(x)| = 1$ for each x in X.

1.7   Show that the linear span of the extreme points of the unit ball of $C(X)$ is $C(X)$.

1.8   Show that the smallest closed convex set containing the extreme points of the unit ball of $C([0, 1])$ is the unit ball but that the same is not true for $C([0, 1] \times [0, 1])$.*

1.9   Show that the unit ball of $c_0(\mathbb{Z}^+)$ has no extreme points. Determine the extreme points of the unit ball of $l^1(\mathbb{Z}^+)$. What about the extreme points of the unit ball of $L^1([0, 1])$?

1.10   If K is a bounded w^*-closed convex subset of $\mathscr{X}^*$, then $\{\varphi(f) : \varphi \in K\}$ is a compact convex subset of $\mathbb{C}$ for each f in $\mathscr{X}$. Moreover, if λ_0 is an extreme point of $\{\varphi(f_0) : \varphi \in K\}$, then any extreme point of the set $\{\varphi \in K : \varphi(f_0) = \lambda_0\}$ is an extreme point of K.

1.11   If K is a bounded w^*-closed convex subset of $\mathscr{X}^*$, then K contains an extreme point.* (Hint: If $\{f_\alpha\}_{\alpha \in A}$ is a well-ordering of $\mathscr{X}$, define nested subsets $\{K_\alpha\}_{\alpha \in A}$ such that

$$K_\alpha = \left\{\varphi \in \bigcap_{\beta < \alpha} K_\beta : \varphi(f_\alpha) = \lambda_\alpha\right\},$$

where λ_α is an extreme point of the set $\{\varphi(f_\alpha) : \varphi \in \bigcap_{\beta < \alpha} K_\beta\}$. Show that $\bigcap_{\alpha \in A} K_\alpha$ consists of a single point which is an extreme point of K.)

1.12   (*Kreĭn–Mil'man*)   A bounded w^*-closed convex subset of $\mathscr{X}^*$ is the w^*-closed convex hull of its extreme points.*

1.13   Prove that the relative w^*-topology on the unit ball of $\mathscr{X}^*$ is metrizable if and only if $\mathscr{X}$ is separable.

1.14   Let $\mathscr{N}$ be a subspace of $\mathscr{X}$, x be in $\mathscr{X}$, and set

$$d = \inf\{\|x - y\| : y \in \mathscr{N}\}.$$

If $d > 0$, then show that there exists φ in $\mathscr{X}^*$ such that $\varphi(y) = 0$ for y in $\mathscr{N}$, $\varphi(x) = 1$, and $\|\varphi\| = 1/d$.

1.15   Show that if we define the function $\hat{f}(\varphi) = \varphi(f)$ for f in $\mathscr{X}$ and φ in $\mathscr{X}^*$, then $\hat{f}$ is in $\mathscr{X}^{**}$ and that the mapping $f \to \hat{f}$ is an isometrical isomorphism of $\mathscr{X}$ into $\mathscr{X}^{**}$.

Definition A Banach space is said to be reflexive if the image of $\mathscr{X}$ is all of $\mathscr{X}^{**}$.

1.16 Show that $\mathscr{X}$ is reflexive for $\mathscr{X}$ finite dimensional but that none of the spaces $c_0(\mathbb{Z}^+)$, $l^1(\mathbb{Z}^+)$, $l^\infty(\mathbb{Z}^+)$, $C([0,1])$, and $L^1([0,1])$ is reflexive.

1.17 Let $\mathscr{X}$ and $\mathscr{Y}$ be Banach spaces. Define the 1-norm $\|f \oplus g\|_1 = \|f\| + \|g\|$ and the ∞-norm $\|f \oplus g\|_\infty = \sup\{\|f\|, \|g\|\}$ on the algebraic direct sum $\mathscr{X} \oplus \mathscr{Y}$. Show that $\mathscr{X} \oplus \mathscr{Y}$ is a Banach space with respect to both norms and that the conjugate space of $\mathscr{X} \oplus \mathscr{Y}$ with the 1-norm is $\mathscr{X}^* \oplus \mathscr{Y}^*$ with the ∞-norm.

1.18 Let $\mathscr{X}$ and $\mathscr{Y}$ be Banach spaces and $\|\ \|$ be a norm on $\mathscr{X} \oplus \mathscr{Y}$ making it into a Banach space such that the projections $\pi_1 : \mathscr{X} \oplus \mathscr{Y} \to \mathscr{X}$ and $\pi_2 : \mathscr{X} \oplus \mathscr{Y} \to \mathscr{Y}$ are continuous. Show that the identity map between $\mathscr{X} \oplus \mathscr{Y}$ in the given norm and $\mathscr{X} \oplus \mathscr{Y}$ with the 1-norm is a homeomorphism. Thus the norm topology on $\mathscr{X} \oplus \mathscr{Y}$ is independent of the norm chosen.

1.19 (*Closed Graph Theorem*) If T is a linear transformation from the Banach space $\mathscr{X}$ to the Banach space $\mathscr{Y}$ such that the graph $\{(f, Tf) : f \in \mathscr{X}\}$ of T is a closed subspace of $\mathscr{X} \oplus \mathscr{Y}$, then T is bounded. (Hint: Consider the map $f \to (f, Tf)$.)

1.20 (*Uniform Boundedness Theorem*) Let $\mathscr{X}$ be a Banach space and $\{\varphi_n\}_{n=1}^\infty$ be a sequence in $\mathscr{X}^*$ such that $\sup\{|\varphi_n(f)| : n \in \mathbb{Z}^+\} < \infty$ for f in $\mathscr{X}$. Show that $\sup\{\|\varphi_n\| : n \in \mathbb{Z}^+\} < \infty$.* (Hint: Baire category.)

1.21 If $\mathscr{X}$ is a Banach space and $\{\varphi_n\}_{n=1}^\infty$ is a sequence in $\mathscr{X}^*$ such that $\{\varphi_n(f)\}_{n=1}^\infty$ is a Cauchy sequence for each f in $\mathscr{X}$, then $\lim_{n \to \infty} \varphi_n$ exists in the w^*-topology. Moreover, the corresponding result for nets is false.

1.22 Show that if $\mathscr{X}$ is a Banach space and φ is a (not necessarily continuous) linear functional on $\mathscr{X}$, then there exists a net $\{\varphi_\alpha\}_{\alpha \in A}$ in $\mathscr{X}^*$ such that $\lim_{\alpha \in A} \varphi_\alpha(f) = \varphi(f)$ for f in $\mathscr{X}$.

1.23 Let $\mathscr{X}$ and $\mathscr{Y}$ be Banach spaces and T be a bounded linear transformation from $\mathscr{X}$ *onto* $\mathscr{Y}$. Show that if $\mathscr{M} = \ker T$, then $\mathscr{X}/\mathscr{M}$ is topologically isomorphic to $\mathscr{Y}$.

Definition If $\mathscr{X}$ is a Banach space, then the collection of functions $\{\varphi \in \mathscr{X}^*\}$ defines a weak topology on $\mathscr{X}$ called the *w*-topology.

1.24 Show that a subspace $\mathscr{M}$ of the Banach space $\mathscr{X}$ is norm closed if and only if it is *w*-closed. Show that the unit sphere in $\mathscr{X}$ is *w*-closed if and only if $\mathscr{X}$ is finite dimensional.*

1.25 Show that if $\mathscr{X}$ is a Banach space, then $\mathscr{X}$ is w^*-dense in $\mathscr{X}^{**}$.

1.26 Show that a Banach space $\mathscr{X}$ is reflexive if and only if the w- and w^*-topologies coincide on $\mathscr{X}^*$.

1.27 Let $\mathscr{X}$ and $\mathscr{Y}$ be Banach spaces and T be in $\mathfrak{L}(\mathscr{X}, \mathscr{Y})$. Show that if φ is in $\mathscr{Y}^*$ then $f \to \varphi(Tf)$ defines an element ψ of $\mathscr{X}^*$. Show that the map $T^*\varphi = \psi$ is in $\mathfrak{L}(\mathscr{Y}^*, \mathscr{X}^*)$. (The operator T^* is called the adjoint of T.)

1.28 If $\mathscr{X}$ and $\mathscr{Y}$ are Banach spaces and T is in $\mathfrak{L}(\mathscr{X}, \mathscr{Y})$, then T is one-to-one if and only if T^* has dense range.

1.29 If $\mathscr{X}$ and $\mathscr{Y}$ are Banach spaces and T is in $\mathfrak{L}(\mathscr{X}, \mathscr{Y})$, then T has a closed range if and only if T^* has a closed range. (Hint: Consider first the case when T is one-to-one and onto.)

Definition If $\mathscr{M}$ is a subspace of the Banach space $\mathscr{X}$, then the annihilator $\mathscr{M}^{\perp}$ of $\mathscr{M}$ is defined as $\mathscr{M}^{\perp} = \{\varphi \in \mathscr{X}^* : \varphi(x) = 0 \text{ for } x \in \mathscr{M}\}$.

1.30 If $\mathscr{X}$ is a Banach space and $\mathscr{M}$ is a closed subspace of $\mathscr{X}$, then $\mathscr{M}^*$ is naturally isometrically isomorphic to $\mathscr{X}^*/\mathscr{M}^{\perp}$.

1.31 If $\mathscr{X}$ is a Banach space and $\mathscr{N}$ is a subspace of $\mathscr{X}^*$, then there exists a subspace $\mathscr{M}$ of $\mathscr{X}$ such that $\mathscr{M}^{\perp} = \mathscr{N}$ if and only if $\mathscr{N}$ is w^*-closed.

1.32 If the restriction of a linear functional φ on the Banach space $M(X)$ of complex regular Borel measures on X to the unit ball of $M(X)$ is continuous in the relative w^*-topology, then there exists a function f in $C(X)$ such that

$$\varphi(\mu) = \int_X f \, d\mu \qquad \text{for} \quad \mu \text{ in } M(X).^*$$

(Hint: Obtain f by evaluating φ at the point measure δ_x at x and use the fact that measures of the form $\sum_{i=1}^{n} \alpha_i \delta_{x_i}$ for $\sum_{i=1}^{n} |\alpha_i| \leqslant 1$ are w^*-dense in the unit ball of $M(X)$.)

1.33 (*Grothendieck*) A linear functional φ in $\mathscr{X}^{**}$ is w^*-continuous if and only if the restriction of φ to $(\mathscr{X}^*)_1$ is continuous in the relative w^*-topology.* (Hint: Embed $\mathscr{X}$ in $C(X)$, extend φ to $M(X)$ via the homomorphism from $M(X)$ to $M(X)/\mathscr{X}^{\perp} = \mathscr{X}^*$, and show that the function f obtained from the preceding problem is in $\mathscr{X}$.)

1.34 (*Kreĭn–Smul'yan*) If $\mathscr{X}$ is a Banach space and $\mathscr{M}$ is a subspace of $\mathscr{X}^*$, then $\mathscr{M}$ is w^*-closed if and only if $\mathscr{M} \cap (\mathscr{X}^*)_1$ is w^*-closed.* (Hint: Show that $\mathscr{M}$ is the intersection of the zero sets of a collection of w^*-continuous linear functionals on $\mathscr{X}^*$.)

1.35 (*Banach*) If $\mathscr{X}$ is a separable Banach space and $\mathscr{M}$ is a subspace of $\mathscr{X}^*$, then $\mathscr{M}$ is w^*-closed if and only if $\mathscr{M}$ is w^*-sequentially complete.

1.36 Let $\mathscr{X}$ and $\mathscr{Y}$ be Banach spaces and $\mathscr{X} \underset{a}{\otimes} \mathscr{Y}$ be the algebraic tensor product of $\mathscr{X}$ and $\mathscr{Y}$ as linear spaces over C. Show that if for w in $\mathscr{X} \underset{a}{\otimes} \mathscr{Y}$ we define

$$\|w\|_\pi = \inf\left\{ \sum_{i=1}^n \|x_i\| \|y_i\| : x_1,\ldots,x_n \in \mathscr{X}; y_1,\ldots,y_n \in \mathscr{Y}, \quad w = \sum_{i=1}^n x_i \otimes y_i \right\},$$

then $\|\cdot\|_\pi$ is a norm on $\mathscr{X} \underset{a}{\otimes} \mathscr{Y}$. The completion of $\mathscr{X} \underset{a}{\otimes} \mathscr{Y}$ is the projective tensor product of $\mathscr{X}$ and $\mathscr{Y}$ and is denoted $\mathscr{X} \hat{\otimes} \mathscr{Y}$.

1.37 Let $\mathscr{X}$ and $\mathscr{Y}$ be Banach spaces and $\mathscr{X} \underset{a}{\otimes} \mathscr{Y}$ be the algebraic tensor product of $\mathscr{X}$ and $\mathscr{Y}$ as linear spaces over C. Show that if for w in $\mathscr{X} \underset{a}{\otimes} \mathscr{Y}$ we define

$$\|w\|_i = \sup\left\{ \left| \sum_{i=1}^n \varphi(x_i) \psi(y_i) \right| : x_1,\ldots,x_n \in \mathscr{X}; y_1,\ldots,y_n \in \mathscr{Y}; \right.$$

$$\left. \varphi \in (\mathscr{X}^*)_1; \psi \in (\mathscr{Y}^*)_1; w = \sum_{i=1}^n x_i \otimes y_i \right\},$$

then $\|\cdot\|_i$ is a norm on $\mathscr{X} \underset{a}{\otimes} \mathscr{Y}$. The completion of $\mathscr{X} \underset{a}{\otimes} \mathscr{Y}$ is the inductive tensor product of $\mathscr{X}$ and $\mathscr{Y}$ and is denoted $\mathscr{X} \overline{\otimes} \mathscr{Y}$.

1.38 For $\mathscr{X}$ and $\mathscr{Y}$ Banach spaces, show that the identity mapping extends to a contractive transformation from $\mathscr{X} \hat{\otimes} \mathscr{Y}$ to $\mathscr{X} \overline{\otimes} \mathscr{Y}$.

1.39 For X and Y compact Hausdorff spaces show that $C(X) \overline{\otimes} C(Y) = C(X \times Y)$. (Hint: Show that it is sufficient in defining $\|\cdot\|_i$ to take φ and ψ to be extreme points of the unit ball of $\mathscr{X}^*$ and $\mathscr{Y}^*$.)

1.40 For X and Y compact Hausdorff spaces show that

$$C(X) \hat{\otimes} C(Y) = C(X \times Y)$$

if and only if X or Y is finite. (Hint: show that there are functions $h(x,y) = \sum_{i=1}^n f_i(x) g_i(y)$ for which $\|h\|_\infty = 1$ but $\|h\|_\pi$ is arbitrarily large.) Thus the tensor product of two Banach spaces is not unique.

2 *Banach Algebras*

2.1 In Chapter 1 we showed that $C(X)$ is a Banach space and that every Banach space is, in fact, isomorphic to a subspace of some $C(X)$. In addition to being a linear space, $C(X)$ is also an algebra and multiplication is continuous in the norm topology. In this chapter we study $C(X)$ as a Banach algebra and show that $C(X)$ is a "universal" commutative Banach algebra in a sense which we will later make precise. We shall indicate the usefulness and power of this result in some examples.

2.2 Recall that in Section 1.1 we observed that $C(X)$ is an algebra over $\mathbb{C}$ with pointwise multiplication and that the supremum norm satisfies $\|fg\|_\infty \leqslant \|f\|_\infty \|g\|_\infty$ for f and g in $C(X)$. These properties make $C(X)$ into what we will call a Banach algebra.

In the study of Banach spaces the notion of bounded linear functional is important. For Banach algebras and, in particular, for $C(X)$ the important idea is that of a multiplicative linear functional. (We do not assume the functional to be continuous because we show later that such a functional is necessarily continuous.) Except for the zero functional, which is obviously both multiplicative and linear, every multiplicative linear functional φ satisfies $\varphi(1) = 1$ since $\varphi \not\equiv 0$ means there exists an f in $C(X)$ with $\varphi(f) \neq 0$, and then the equation $\varphi(1)\varphi(f) = \varphi(f)$ implies $\varphi(1) = 1$. Thus we restrict our attention to the set $M_{C(X)}$ of complex multiplicative linear functionals φ on $C(X)$ which satisfy $\varphi(1) = 1$. For each x in X we define the complex function

φ_x on $C(X)$ such that $\varphi_x(f) = f(x)$ for f in $C(X)$. It is immediate that φ_x is in $M_{C(X)}$, and we let ψ denote the mapping from X to $M_{C(X)}$ defined $\psi(x) = \varphi_x$. The following proposition shows that ψ maps onto $M_{C(X)}$.

2.3 Proposition The map ψ defines a homeomorphism from X onto $M_{C(X)}$, where $M_{C(X)}$ is given the relative w^*-topology on $C(X)^*$.

Proof Let φ be in $M_{C(X)}$ and set

$$\Re = \ker \varphi = \{f \in C(X) : \varphi(f) = 0\}.$$

We show first that there exists x_0 in X such that $f(x_0) = 0$ for each f in $\Re$. If that were not the case, then for each x in X, there would exist f_x in $\Re$ such that $f_x(x) \neq 0$. Since f_x is continuous, there exists a neighborhood U_x of x on which $f_x \neq 0$. Since X is compact and $\{U_x\}_{x \in X}$ is an open cover of X, there exists $U_{x_1}, \ldots, U_{x_N}$ with $X = \bigcup_{n=1}^{N} U_{x_n}$. If we set $g = \sum_{n=1}^{N} \bar{f}_{x_n} f_{x_n}$, then $\varphi(g) = \sum_{n=1}^{N} \varphi(\bar{f}_{x_n}) \varphi(f_{x_n}) = 0$, implying that g is in $\Re$. But $g \neq 0$ on X and hence is invertible in $C(X)$. This in turn implies $\varphi(1) = \varphi(g) \cdot \varphi(1/g) = 0$, which is a contradiction. Thus there exists x_0 in X such that $f(x_0) = 0$ for f in $\Re$.

If f is in $C(X)$, then $f - \varphi(f) \cdot 1$ is in $\Re$ since $\varphi(f - \varphi(f) \cdot 1) = \varphi(f) - \varphi(f) = 0$. Thus

$$f(x_0) - \varphi(f) = (f - \varphi(f) \cdot 1)(x_0) = 0,$$

since $\varphi - \varphi(f) \cdot 1$ is in $\Re$ and therefore $\varphi = \varphi_{x_0}$.

Since each φ in $M_{C(X)}$ is bounded (in fact, of norm one), we can give $M_{C(X)}$ the relative w^*-topology on $C(X)^*$ and consider the map $\psi : X \to M_{C(X)}$. If x and y are distinct points of X, then by Urysohn's lemma there exists f in $C(X)$ such that $f(x) \neq f(y)$. Thus

$$\psi(x)(f) = \varphi_x(f) = f(x) \neq f(y) = \varphi_y(f) = \psi(y)(f),$$

which implies that ψ is one-to-one.

To show that ψ is continuous, let $\{x_\alpha\}_{\alpha \in A}$ be a net in X converging to x. Then $\lim_{\alpha \in A} f(x_\alpha) = f(x)$ for f in $C(X)$ or equivalently $\lim_{\alpha \in A} \psi(x_\alpha)(f) = \psi(x)(f)$ for each f in $C(X)$. Thus the net $\{\psi(x_\alpha)\}_{\alpha \in A}$ converges in the w^*-topology to $\psi(x)$ so that ψ is seen to be continuous. Since ψ is a one-to-one continuous map from a compact space onto a Hausdorff space, it follows that ψ is a homeomorphism. This completes the proof. ∎

We next state the definition of Banach algebra and proceed to show that the collection of multiplicative linear functionals on a general Banach algebra can always be made into a compact Hausdorff space in a natural way.

2.4   Definition   A Banach algebra $\mathfrak{B}$ is an algebra over $\mathbb{C}$ with identity 1 which has a norm making it into a Banach space and satisfying $\|1\| = 1$ and the inequality $\|fg\| \leqslant \|f\| \|g\|$ for f and g in $\mathfrak{B}$.

We let λ denote the element of $\mathfrak{B}$ obtained upon multiplying the identity by the complex number λ.

The following fundamental proposition will be used to show that the collection of invertible elements in $\mathfrak{B}$ is an open set and that inversion is continuous in the norm topology.

2.5   Proposition   If f is in the Banach algebra $\mathfrak{B}$ and $\|1-f\| < 1$, then f is invertible and

$$\|f^{-1}\| \leqslant \frac{1}{1 - \|1-f\|}.$$

Proof   If we set $\eta = \|1-f\| < 1$, then for $N \geqslant M$ we have

$$\left\| \sum_{n=0}^{N} (1-f)^n - \sum_{n=0}^{M} (1-f)^n \right\| = \left\| \sum_{n=M+1}^{N} (1-f)^n \right\| \leqslant \sum_{n=M+1}^{N} \|1-f\|^n$$

$$= \sum_{n=M+1}^{N} \eta^n \leqslant \frac{\eta^{M+1}}{1-\eta}$$

and the sequence of partial sums $\{\sum_{n=0}^{N} (1-f)^n\}_{N=0}^{\infty}$ is seen to be a Cauchy sequence. If $g = \sum_{n=0}^{\infty} (1-f)^n$, then

$$fg = [1-(1-f)]\left(\sum_{n=0}^{\infty} (1-f)^n \right) = \lim_{N \to \infty} \left([1-(1-f)] \sum_{n=0}^{N} (1-f)^n \right)$$

$$= \lim_{N \to \infty} (1-(1-f)^{N+1}) = 1,$$

since $\lim_{N \to \infty} \|(1-f)^{N+1}\| = 0$. Similarly, $gf = 1$ so that f is invertible with $f^{-1} = g$. Further,

$$\|g\| = \lim_{N \to \infty} \left\| \sum_{n=0}^{N} (1-f)^n \right\| \leqslant \lim_{N \to \infty} \sum_{n=0}^{N} \|1-f\|^n = \frac{1}{1 - \|1-f\|}. \quad \blacksquare$$

2.6   Definition   For $\mathfrak{B}$ a Banach algebra, let $\mathscr{G}$ denote the collection of invertible elements in $\mathfrak{B}$ and let $\mathscr{G}_l$, respectively, $\mathscr{G}_r$ denote the collection of left, respectively, right invertible elements in $\mathfrak{B}$ that are not invertible.

The following result will be of interest in this chapter only as it concerns $\mathscr{G}$ but we will need the results about $\mathscr{G}_l$ and $\mathscr{G}_r$ in Chapter 5 when we study index theory.

2.7 Proposition If $\mathfrak{B}$ is a Banach algebra, then each of the sets $\mathscr{G}$, $\mathscr{G}_1$, and $\mathscr{G}_r$ is open in $\mathfrak{B}$.

Proof If f is in $\mathscr{G}$ and $\|f-g\| < 1/\|f^{-1}\|$, then $1 > \|f^{-1}\| \, \|f-g\| \geqslant \|1-f^{-1}g\|$. Thus the preceding proposition implies that $f^{-1}g$ is in $\mathscr{G}$ and hence $g = f(f^{-1}g)$ is in $\mathscr{G}$. Therefore $\mathscr{G}$ contains the open ball of radius $1/\|f^{-1}\|$ about each element of f in $\mathscr{G}$. Thus $\mathscr{G}$ is an open set in $\mathfrak{B}$.

If f is in $\mathscr{G}_1$, then there exists h in $\mathfrak{B}$ such that $hf = 1$. If $\|f-g\| < 1/\|h\|$, then $1 > \|h\| \, \|f-g\| \geqslant \|hf-hg\| = \|1-hg\|$. Again the proposition implies that $k = hg$ is invertible and the identity $(k^{-1}h)g = 1$ implies that g is left invertible. Moreover, if g is invertible, then $h = kg^{-1}$ is invertible which in turn implies that f is invertible. This contradiction shows that g is in $\mathscr{G}_1$ so that $\mathscr{G}_1$ is seen to contain the open ball of radius $1/\|h\|$ about f. Thus $\mathscr{G}_1$ is open. The proof that $\mathscr{G}_r$ is open proceeds in the same manner. ■

2.8 Corollary If $\mathfrak{B}$ is a Banach algebra, then the map on $\mathscr{G}$ defined $f \to f^{-1}$ is continuous. Thus, $\mathscr{G}$ is a topological group.

Proof If f is in $\mathscr{G}$, then the inequality $\|f-g\| < 1/2\|f^{-1}\|$ implies that $\|1-f^{-1}g\| < \frac{1}{2}$ and hence

$$\|g^{-1}\| \leqslant \|g^{-1}f\| \, \|f^{-1}\| = \|(f^{-1}g)^{-1}\| \, \|f^{-1}\| \leqslant 2\|f^{-1}\|.$$

Thus the inequality

$$\|f^{-1} - g^{-1}\| = \|f^{-1}(f-g)g^{-1}\| \leqslant 2\|f^{-1}\|^2 \, \|f-g\|$$

shows that the map $f \to f^{-1}$ is continuous. ■

There is another group which is important in some problems.

2.9 Proposition Let $\mathfrak{B}$ be a Banach algebra, $\mathscr{G}$ be the group of invertible elements in $\mathfrak{B}$, and $\mathscr{G}_0$ be the connected component in $\mathscr{G}$ which contains the identity. Then $\mathscr{G}_0$ is an open and closed normal subgroup of $\mathscr{G}$, the cosets of $\mathscr{G}_0$ are the components of $\mathscr{G}$, and $\mathscr{G}/\mathscr{G}_0$ is a discrete group.

Proof Since $\mathscr{G}$ is an open subset of a locally connected space, its components are open and closed subsets of $\mathscr{G}$. Further, if f and g are in $\mathscr{G}_0$, then $f\mathscr{G}_0$ is a connected subset of $\mathscr{G}$ which contains fg and f. Therefore, $\mathscr{G}_0 \cup f\mathscr{G}_0$ is connected and hence is contained in $\mathscr{G}_0$. Thus fg is in $\mathscr{G}_0$ so that $\mathscr{G}_0$ is a semigroup. Similarly, $f^{-1}\mathscr{G}_0 \cup \mathscr{G}_0$ is connected, hence contained in $\mathscr{G}_0$, and therefore $\mathscr{G}_0$ is a subgroup of $\mathscr{G}$. Lastly, if f is in $\mathscr{G}$, then the conjugate group $f\mathscr{G}_0 f^{-1}$ is a connected subset containing the identity and therefore $f\mathscr{G}_0 f^{-1} = \mathscr{G}_0$. Thus, $\mathscr{G}_0$ is a normal subgroup of $\mathscr{G}$ and $\mathscr{G}/\mathscr{G}_0$ is a group.

Further, since $f\mathscr{G}_0$ is an open and closed connected subset of $\mathscr{G}$ for each f in $\mathscr{G}$, the cosets of $\mathscr{G}_0$ are the components of $\mathscr{G}$. Lastly, $\mathscr{G}/\mathscr{G}_0$ is discrete since $\mathscr{G}_0$ is an open and closed subset of $\mathscr{G}$.   ■

2.10 Definition If $\mathfrak{B}$ is a Banach algebra, then the abstract index group for $\mathfrak{B}$, denoted $\Lambda_{\mathfrak{B}}$, is the discrete quotient group $\mathscr{G}/\mathscr{G}_0$. Moreover, the abstract index is the natural homomorphism γ from $\mathscr{G}$ to $\Lambda_{\mathfrak{B}}$.

We next consider the abstract index group for a Banach algebra in a little more detail.

2.11 Definition If $\mathfrak{B}$ is a Banach algebra, then the exponential map on $\mathfrak{B}$, denoted exp, is defined

$$\exp f = \sum_{n=0}^{\infty} \frac{1}{n!} f^n.$$

The absolute convergence of this series is established just as in the scalar case from whence follows the continuity of exp. If $\mathfrak{B}$ is not commutative, then many of the familiar properties of the exponential function do not hold. The following key formula is valid, however, with the additional hypothesis of commutativity.

2.12 Lemma If $\mathfrak{B}$ is a Banach algebra and f and g are elements of $\mathfrak{B}$ which commute, then $\exp(f+g) = \exp f \exp g$.

Proof Multiply the series defining $\exp f$ and $\exp g$ and rearrange.   ■

In a general Banach algebra it is difficult to determine the elements in the range of the exponential map, that is, the elements which have a "logarithm." The following lemma gives a sufficient condition.

2.13 Lemma If $\mathfrak{B}$ is a Banach algebra and f is an element of $\mathfrak{B}$ such that $\|1-f\| < 1$, then f is in $\exp \mathfrak{B}$.

Proof If we set $g = \sum_{n=1}^{\infty} (1/n)(1-f)^n$, then the series converges absolutely and, as in the scalar case, substituting this series into the series expansion for $\exp g$ yields $\exp g = f$.   ■

Although it is difficult to characterize $\exp \mathfrak{B}$ for an arbitrary Banach algebra, the collection of finite products of elements in $\exp \mathfrak{B}$ is a familiar object.

2.14 Theorem If $\mathfrak{B}$ is a Banach algebra, then the collection of finite products of elements in $\exp \mathfrak{B}$ is $\mathscr{G}_0$.

Proof If $f = \exp g$, then $f \cdot \exp(-g) = \exp(g-g) = 1 = \exp(-g)f$, and hence f is in $\mathscr{G}$. Moreover, the map φ from $[0,1]$ to $\exp \mathfrak{B}$ defined $\varphi(\lambda) = \exp(\lambda g)$ is an arc connecting 1 to f and hence f is in $\mathscr{G}_0$. Thus $\exp \mathfrak{B}$ is contained in $\mathscr{G}_0$. Further, if $\mathscr{F}$ denotes the collection of finite products of elements of $\exp \mathfrak{B}$, then $\mathscr{F}$ is a subgroup contained in $\mathscr{G}_0$. Moreover, by the previous lemma $\mathscr{F}$ contains an open set and hence $\mathscr{F}$ being a subgroup is an open set. Lastly, since each of the left cosets of $\mathscr{F}$ is an open set, it follows that $\mathscr{F}$ is an open and closed subset of $\mathscr{G}_0$. Since $\mathscr{G}_0$ is connected we conclude that $\mathscr{G}_0 = \mathscr{F}$, which completes the proof. ∎

The following corollary shows that the problem of identifying the elements of a commutative Banach algebra which have a logarithm is much easier.

2.15 Corollary If $\mathfrak{B}$ is a commutative Banach algebra, then $\exp \mathfrak{B} = \mathscr{G}_0$.

Proof By Lemma 2.12 if $\mathfrak{B}$ is commutative, then $\exp \mathfrak{B}$ is a subgroup. ∎

Before continuing, we identify the abstract index group for $C(X)$ with a more familiar object from algebraic topology. This identification is actually valid for arbitrary commutative Banach algebras but we will not pursue this any further (see [40]).

2.16 Let X be a compact Hausdorff space and let $\mathscr{G}$ denote the invertible elements of $C(X)$. Hence a function f in $C(X)$ is in $\mathscr{G}$ if and only if $f(x) \neq 0$ for all x in X, that is, $\mathscr{G}$ consists of the continuous functions from X to $\mathbb{C}^* = \mathbb{C}\backslash\{0\}$. Since $\mathscr{G}$ is locally arcwise connected, a function f is in $\mathscr{G}_0$ if there exists a continuous arc $\{f_\lambda\}_{\lambda \in [0,1]}$ of functions in $\mathscr{G}$ such that $f_0 = 1$ and $f_1 = f$. If we define the function F from $X \times [0,1]$ to $\mathbb{C}^*$ such that $F(x, \lambda) = f_\lambda(x)$, then F is continuous, $F(x, 0) = 1$ and $F(x, 1) = f(x)$ for x in X. Hence f is homotopic to the constant function 1. Conversely, if g is a function in $\mathscr{G}$ which is homotopic to 1, then g is in $\mathscr{G}_0$. Similarly, two functions g_1 and g_2 in $\mathscr{G}$ represent the same element of $\Lambda = \mathscr{G}/\mathscr{G}_0$ if and only if g_1 is homotopic to g_2. Thus Λ is the group of homotopy classes of maps from X to $\mathbb{C}^*$.

2.17 Definition If X is a compact Hausdorff space, then the first co-homotopy group $\pi^1(X)$ of X is the group of homotopy classes of continuous maps from X to the circle group $\mathbb{T}$ with pointwise multiplication.

2.18 Theorem If X is a compact Hausdorff space, then the abstract index group Λ for $C(X)$ and $\pi^1(X)$ are naturally isomorphic.

Proof We define the mapping Φ from $\pi^1(X)$ to Λ as follows: A continuous function f from X to $\mathbb{T}$ determines first an element $\{f\}$ of $\pi^1(X)$ and second, viewed as an invertible function on X, determines a coset $f+\mathscr{G}_0$ of Λ. We define $\Phi(\{f\}) = f+\mathscr{G}_0$. To show, however, that Φ is well defined we need to observe that if g is a continuous function from X to $\mathbb{T}$ such that $\{f\} = \{g\}$, then f is homotopic to g and hence $f+\mathscr{G}_0 = g+\mathscr{G}_0$. Moreover, since multiplication in both $\pi^1(X)$ and $\mathscr{G}$ is defined pointwise, the mapping Φ is obviously a homomorphism. It remains only to show that Φ is one to-one and onto.

To show Φ is onto let f be an invertible element of $C(X)$. Define the function F from $X \times [0, 1]$ to $\mathbb{C}^*$ such that $F(x,t) = f(x)/|f(x)|^t$. Then F is continuous, $F(x,0) = f(x)$ for x in X, and $g(x) = F(x,1)$ has modulus one for x in X. Hence, $f+\mathscr{G}_0 = g+\mathscr{G}_0$ so that $\Phi(\{g\}) = f+\mathscr{G}_0$ and therefore Φ is onto.

If f and g are continuous functions from X to $\mathbb{T}$ such that $\Phi(\{f\}) = \Phi(\{g\})$, then f is homotopic to g in the functions in $\mathscr{G}$, that is, there exists a continuous function G from $X \times [0,1]$ to $\mathbb{C}^*$ such that $G(x,0) = f(x)$ and $G(x,1) = g(x)$ for x in X. If, however, we define $F(x,t) = G(x,t)/|G(x,t)|$, then F is continuous and establishes that f and g are homotopic in the class of continuous functions from X to $\mathbb{T}$. Thus $\{f\} = \{g\}$ and therefore Φ is one-to-one, which completes the proof. ■

The preceding result is usually stated in a slightly different way.

2.19 Corollary If X is a compact Hausdorff space, then Λ is naturally isomorphic to the first Čech cohomology group $H^1(X,\mathbb{Z})$ with integer coefficients.

Proof It is proved in algebraic topology (see [67]) that $\pi^1(X)$ and $H^1(X,\mathbb{Z})$ are naturally isomorphic. ■

These results enable us to determine the abstract index group for simple commutative Banach algebras.

2.20   Corollary   The abstract index group of $C(\mathbb{T})$ is isomorphic to $\mathbb{Z}$.

Proof   The first cohomotopy group of $\mathbb{T}$ is the same as the first homotopy group of $\mathbb{T}$ and hence is $\mathbb{Z}$.   ■

We now return to the basic structure theory for Banach algebras.

2.21   Definition   Let $\mathfrak{B}$ be a Banach algebra. A complex linear functional φ on $\mathfrak{B}$ is said to be *multiplicative* if:

(1) $\varphi(fg) = \varphi(f)\varphi(g)$ for f and g in $\mathfrak{B}$; and
(2) $\varphi(1) = 1$.

The set of all multiplicative linear functionals on $\mathfrak{B}$ is denoted by $M = M_{\mathfrak{B}}$.

We will show that the elements of M are bounded and that M is a w^*-compact subset of the unit ball of the conjugate space of $\mathfrak{B}$. We show later that M is nonempty if we further assume that $\mathfrak{B}$ is commutative.

2.22   Proposition   If $\mathfrak{B}$ is a Banach algebra and φ is in M, then $\|\varphi\| = 1$.

Proof   Let $\mathfrak{K} = \ker\varphi = \{f \in \mathfrak{B} : \varphi(f) = 0\}$. Since $\varphi(f - \varphi(f)\cdot 1) = 0$, it follows that every element in $\mathfrak{B}$ can be written in the form $\lambda + f$ for some λ in $\mathbb{C}$ and f in $\mathfrak{K}$. Thus

$$\|\varphi\| = \sup_{g \neq 0}\frac{|\varphi(g)|}{\|g\|} = \sup_{\substack{f \in \mathfrak{K} \\ \lambda \neq 0}}\frac{|\varphi(\lambda + f)|}{\|\lambda + f\|} = \sup_{\substack{f \in \mathfrak{K} \\ \lambda \neq 0}}\frac{|\lambda|}{\|\lambda + f\|} = \sup_{h \in \mathfrak{K}}\frac{1}{\|1 + h\|} = 1$$

because $\|1 + h\| < 1$ implies that h is invertible by Proposition 2.5, which implies in turn that h is not in $\mathfrak{K}$. Therefore $\|\varphi\| = 1$ and the proof is complete.   ■

Whenever we deduce topological properties from algebraic hypotheses, completeness is usually crucial; the use of completeness in the proof of Theorem 1.42 was obvious. Less obvious is the role played by completeness in the preceding proposition.

2.23   Proposition   If $\mathfrak{B}$ is a Banach algebra, then M is a w^*-compact subset of $(\mathfrak{B}^*)_1$.

Proof   Let $\{\varphi_\alpha\}_{\alpha \in A}$ be a net of multiplicative linear functionals in M that converge in the w^*-topology on $(\mathfrak{B}^*)_1$ to a φ in $(\mathfrak{B}^*)_1$. To show that M is w^*-compact it is sufficient in view of Theorem 1.23 to prove that φ is multiplicative and $\varphi(1) = 1$. To this end we have $\varphi(1) = \lim_{\alpha \in A} \varphi_\alpha(1) = \lim_{\alpha \in A} 1 = 1$. Further, for f and g in $\mathfrak{B}$, we have

$$\varphi(fg) = \lim_{\alpha \in A} \varphi_\alpha(fg) = \lim_{\alpha \in A} \varphi_\alpha(f) \varphi_\alpha(g)$$

$$= \lim_{\alpha \in A} \varphi_\alpha(f) \lim_{\alpha \in A} \varphi_\alpha(g) = \varphi(f) \cdot \varphi(g).$$

Thus φ is in M and the proof is complete. ∎

Thus M is a compact Hausdorff space in the relative w^*-topology. Recall that for each f in $\mathfrak{B}$ there is a w^*-continuous function $\hat{f} : (\mathfrak{B}^*)_1 \to \mathbb{C}$ given by $\hat{f}(\varphi) = \varphi(f)$. Since M is contained in $(\mathfrak{B}^*)_1$, then $\hat{f}|M$ is also continuous. We formalize this in the following:

2.24   Definition   For the Banach algebra $\mathfrak{B}$, the Gelfand transform is the function $\Gamma : \mathfrak{B} \to C(M)$ given by $\Gamma(f) = \hat{f}|M$, that is, $\Gamma(f)(\varphi) = \varphi(f)$ for φ in M.

2.25   Elementary Properties of the Gelfand Transform   If $\mathfrak{B}$ is a Banach algebra and Γ is the Gelfand transform on $\mathfrak{B}$, then:

(1) Γ is an algebra homomorphism; and
(2) $\|\Gamma f\|_\infty \leqslant \|f\|$ for f in $\mathfrak{B}$.

Proof   The only nonobvious property needed to conclude that Γ is an algebra homomorphism is that Γ is multiplicative and that argument goes as follows: For f and g in $\mathfrak{B}$ we have

$$\Gamma(fg)(\varphi) = \varphi(fg) = \varphi(f)\varphi(g) = \Gamma(f)(\varphi) \cdot \Gamma(g)(\varphi) = [\Gamma(f) \cdot \Gamma(g)](\varphi),$$

and hence Γ is multiplicative. To show that Γ is a contractive mapping we let f be in $\mathfrak{B}$ and then

$$\|\Gamma f\|_\infty = \|\hat{f}|M\|_\infty \leqslant \|\hat{f}\|_\infty = \|f\|.$$

Thus Γ is a contractive algebra homomorphism and the proof is complete. ∎

2.26   Before proceeding we want to make a few remarks about the Gelfand transform. Note first that Γ sends all elements of the form $fg - gf$ to 0. Thus, if $\mathfrak{B}$ is not commutative, then the subalgebra of $C(M)$ that is the range of Γ may fail to reflect the properties of $\mathfrak{B}$. (In particular, we indicate in the problems at the end of this chapter an example of a Banach algebra for which M is empty.) In the commutative case, however, M is not only not empty but is sufficiently large that the invertibility of an element f in $\mathfrak{B}$ is determined by the invertibility of Γf in $C(M)$. This fact alone makes the Gelfand transform a powerful tool for the study of commutative Banach algebras.

To establish this further property of the Gelfand transform in the commutative case, we must first consider the basic facts of spectral theory. We will not assume, in what follows, that $\mathfrak{B}$ is commutative until this assumption is actually needed.

2.27   Definition   For $\mathfrak{B}$ a Banach algebra and f an element of $\mathfrak{B}$ we define the spectrum of f to be the set

$$\sigma_{\mathfrak{B}}(f) = \{\lambda \in \mathbb{C} : f - \lambda \text{ is not invertible in } \mathfrak{B}\},$$

and the resolvent set of f to be the set

$$\rho_{\mathfrak{B}}(f) = \mathbb{C} \backslash \sigma_{\mathfrak{B}}(f).$$

Further, the spectral radius of f is defined

$$r_{\mathfrak{B}}(f) = \sup\{|\lambda| : \lambda \in \sigma_{\mathfrak{B}}(f)\}.$$

When no confusion will result we omit the subscript $\mathfrak{B}$ and write only $\sigma(f)$, $\rho(f)$, and $r(f)$.

The following elementary proposition shows that $\sigma(f)$ is compact. The fact that $\sigma(f)$ is nonempty lies deeper and is the content of the next theorem.

2.28   Proposition   If $\mathfrak{B}$ is a Banach algebra and f is in $\mathfrak{B}$, then $\sigma(f)$ is compact and $r(f) \leqslant \|f\|$.

Proof   If we define the function $\varphi : \mathbb{C} \to \mathfrak{B}$ by $\varphi(\lambda) = f - \lambda$, then φ is continuous and $\rho(f) = \varphi^{-1}(\mathscr{G})$ is open since $\mathscr{G}$ is open. Thus the set $\sigma(f)$ is closed.

If $|\lambda| > \|f\|$, then

$$1 > \frac{\|f\|}{|\lambda|} = \left\|\frac{f}{\lambda}\right\| = \left\|1 - \left(1 - \frac{f}{\lambda}\right)\right\|,$$

so that $1 - f/\lambda$ is invertible by Proposition 2.5. Thus $f - \lambda$ is invertible. Therefore, λ is in $\rho(f)$, $\sigma(f)$ is bounded and hence compact, and $r(f) \leqslant \|f\|$.   ∎

2.29   Theorem   If $\mathfrak{B}$ is a Banach algebra and f is in $\mathfrak{B}$, then $\sigma(f)$ is nonempty.

Proof   Consider the function $F : \rho(f) \to \mathfrak{B}$ defined $F(\lambda) = (f - \lambda)^{-1}$. We show that F is an analytic $\mathfrak{B}$-valued function on $\rho(f)$ that is bounded at infinity and use the Liouville theorem to obtain a contradiction.

First, since inversion is continuous we have for λ_0 in $\rho(f)$ that

$$\lim_{\lambda \to \lambda_0} \left\{ \frac{F(\lambda) - F(\lambda_0)}{\lambda - \lambda_0} \right\} = \lim_{\lambda \to \lambda_0} \left\{ \frac{(f-\lambda_0)^{-1}[(f-\lambda_0) - (f-\lambda)](f-\lambda)^{-1}}{\lambda - \lambda_0} \right\}$$

$$= \lim_{\lambda \to \lambda_0} (f-\lambda_0)^{-1}(f-\lambda)^{-1} = (f-\lambda_0)^{-2}.$$

In particular, for φ in the conjugate space $\mathfrak{B}^*$, the function $\varphi(F)$ is a complex analytic function on $\rho(f)$.

Further, for $|\lambda| > \|f\|$ we have, using Proposition 2.5, that $1 - f/\lambda$ is invertible and

$$\left\| \left(1 - \frac{f}{\lambda}\right)^{-1} \right\| \leqslant \frac{1}{1 - \|f/\lambda\|} .$$

Thus it follows that

$$\lim_{\lambda \to \infty} \|F(\lambda)\| = \lim_{\lambda \to \infty} \left\| \frac{1}{\lambda} \left(\frac{f}{\lambda} - 1 \right)^{-1} \right\|$$

$$\leqslant \lim_{|\lambda| \to \infty} \sup \frac{1}{|\lambda|} \frac{1}{1 - \|f/\lambda\|} = 0.$$

Therefore for φ in $\mathfrak{B}^*$ we have $\lim_{\lambda \to \infty} \varphi(F(\lambda)) = 0$.

If we now assume that $\sigma(f)$ is empty, then $\rho(f) = \mathbb{C}$. Thus for φ in $\mathfrak{B}^*$ it follows that $\varphi(F)$ is an entire function that vanishes at infinity. By Liouville's theorem we have $\varphi(F) \equiv 0$. In particular, since for a fixed λ in $\mathbb{C}$ we have $\varphi(F(\lambda)) = 0$ for each φ in $\mathfrak{B}^*$, it follows from Corollary 1.28 that $F(\lambda) = 0$. This, however, is a contradiction, since $F(\lambda)$ is by definition an invertible element of $\mathfrak{B}$. Therefore $\sigma(f)$ is nonempty. ∎

Note that although $\mathfrak{B}$ is not assumed to be commutative, the subalgebra of $\mathfrak{B}$ spanned by $1, f$, and elements of form $(f-\lambda)^{-1}$ is commutative, and the result really concerns only this subalgebra.

2.30   The following theorem is an immediate corollary to the preceding and is crucial in establishing the desired properties of the Gelfand transform. Recall that a division algebra is an algebra in which each nonzero element is invertible.

2.31 Theorem (Gelfand–Mazur)   If $\mathfrak{B}$ is a Banach algebra which is a division algebra, then there is a unique isometric isomorphism of $\mathfrak{B}$ onto $\mathbb{C}$.

Proof If f is in $\mathfrak{B}$, then $\sigma(f)$ is nonempty by the preceding theorem. If λ_f is in $\sigma(f)$, then $f - \lambda_f$ is not invertible by definition. Since $\mathfrak{B}$ is a division algebra, then $f - \lambda_f = 0$. Moreover, for $\lambda \neq \lambda_f$ we have $f - \lambda = \lambda_f - \lambda$ which is invertible. Thus $\sigma(f)$ consists of exactly one complex number λ_f for each f in $\mathfrak{B}$. The map $\psi : \mathfrak{B} \to \mathbb{C}$ defined $\psi(f) = \lambda_f$ is obviously an isometric isomorphism of $\mathfrak{B}$ onto $\mathbb{C}$. Moreover, if ψ' were any other, then $\psi'(f)$ would be in $\sigma(f)$ implying that $\psi(f) = \psi'(f)$. This completes the proof. ∎

2.32 Quotient Algebras We now consider the notion of a quotient algebra. Let $\mathfrak{B}$ be a Banach algebra and suppose that $\mathfrak{M}$ is a closed two-sided ideal of $\mathfrak{B}$. Since $\mathfrak{M}$ is a closed subspace of $\mathfrak{B}$ we can define a norm on $\mathfrak{B}/\mathfrak{M}$ following Section 1.39 making it into a Banach space. Further, since $\mathfrak{M}$ is a two-sided ideal in $\mathfrak{B}$, we also know that $\mathfrak{B}/\mathfrak{M}$ is an algebra. There remain two facts to verify before we can assert that $\mathfrak{B}/\mathfrak{M}$ is a Banach algebra.

First, we must show that $\|[1]\| = 1$, and this proof proceeds as follows: $\|[1]\| = \inf_{g \in \mathfrak{M}} \|1 - g\| = 1$, for if $\|1 - g\| < 1$, then g is invertible by Proposition 2.5.

Secondly, for f and g in $\mathfrak{B}$ we have

$$
\begin{aligned}
\|[f][g]\| = \|[fg]\| &= \inf_{h \in \mathfrak{M}} \|fg - h\| \\
&\leqslant \inf_{h_1, h_2 \in \mathfrak{M}} \|(f - h_1)(g - h_2)\| \leqslant \inf_{h_1 \in \mathfrak{M}} \|f - h_1\| \inf_{h_2 \in \mathfrak{M}} \|g - h_2\| \\
&= \|[f]\| \, \|[g]\|
\end{aligned}
$$

so that $\|[f][g]\| \leqslant \|[f]\| \, \|[g]\|$. Thus $\mathfrak{B}/\mathfrak{M}$ is a Banach algebra. Moreover, the natural map $f \to [f]$ is a contractive homomorphism.

2.33 Proposition If $\mathfrak{B}$ is a commutative Banach algebra, then the set M of multiplicative linear functionals on $\mathfrak{B}$ is in one-to-one correspondence with the set of maximal two-sided ideals in $\mathfrak{B}$.

Proof Let φ be a multiplicative linear functional on $\mathfrak{B}$ and let $\mathfrak{K} = \ker \varphi = \{f \in \mathfrak{B} : \varphi(f) = 0\}$. The kernel $\mathfrak{K}$ of a homomorphism is a proper two-sided ideal and if f is not in $\mathfrak{K}$, then

$$
1 = \left(1 - \frac{f}{\varphi(f)}\right) + \frac{f}{\varphi(f)}.
$$

Since $(1 - f/\varphi(f))$ is in $\mathfrak{K}$, the linear span of f with $\mathfrak{K}$ contains the identity 1. Thus an ideal containing both $\mathfrak{K}$ and f would have to be all of $\mathfrak{B}$ so that $\mathfrak{K}$ is seen to be a maximal two-sided ideal.

Suppose $\mathfrak{M}$ is a maximal proper two-sided ideal in $\mathfrak{B}$. Since each element f of $\mathfrak{M}$ is not invertible, then $\|1 - f\| \geqslant 1$ by Proposition 2.5. Thus 1 is not in the closure of $\mathfrak{M}$. Moreover, since the closure $\overline{\mathfrak{M}}$ of $\mathfrak{M}$ is obviously a two-sided ideal and $\mathfrak{M} \subset \overline{\mathfrak{M}} \subsetneqq \mathfrak{B}$, then $\mathfrak{M} = \overline{\mathfrak{M}}$ and $\mathfrak{M}$ is closed. The quotient algebra $\mathfrak{B}/\mathfrak{M}$ is a Banach algebra which because $\mathfrak{M}$ is maximal and $\mathfrak{B}$ is commutative, is a division algebra. Thus by Theorem 2.31, there is a natural isometrical isomorphism ψ of $\mathfrak{B}/\mathfrak{M}$ onto $\mathbb{C}$. If π denotes the natural homomorphism of $\mathfrak{B}$ onto $\mathfrak{B}/\mathfrak{M}$, then the composition $\varphi = \psi\pi$ is a nonzero multiplicative linear functional on $\mathfrak{B}$. Thus φ is in M and $\mathfrak{M} = \ker \varphi$.

Lastly, we want to show that the correspondence $\varphi \leftrightarrow \ker \varphi$ is one-to-one. If φ_1 and φ_2 are in M with $\ker \varphi_1 = \ker \varphi_2 = \mathfrak{M}$, then

$$\varphi_1(f) - \varphi_2(f) = (f - \varphi_2(f)) - (f - \varphi_1(f))$$

is both in $\mathfrak{M}$ and a scalar multiple of the identity for each f in $\mathfrak{B}$ and hence must be 0. Therefore $\ker \varphi_1 = \ker \varphi_2$ implies $\varphi_1 = \varphi_2$ and this completes the proof. ∎

This last proposition is the only place in the preceding development where the assumption that $\mathfrak{B}$ is commutative is required.

Hereafter, we refer to $M_\mathfrak{B}$ as the maximal ideal space for $\mathfrak{B}$.

2.34 Proposition If $\mathfrak{B}$ is a commutative Banach algebra and f is in $\mathfrak{B}$, then f is invertible in $\mathfrak{B}$ if and only if $\Gamma(f)$ is invertible in $C(M)$.

Proof If f is invertible in $\mathfrak{B}$, the $\Gamma(f^{-1})$ is the inverse of $\Gamma(f)$. If f is not invertible in $\mathfrak{B}$, then $\mathfrak{M}_0 = \{gf : g \in \mathfrak{B}\}$ is a proper ideal in $\mathfrak{B}$ since 1 is not in $\mathfrak{M}_0$. Since $\mathfrak{B}$ is commutative, $\mathfrak{M}_0$ is contained in some maximal ideal $\mathfrak{M}$. By the preceding proposition there exists φ in M such that $\ker \varphi = \mathfrak{M}$. Thus $\Gamma(f)(\varphi) = \varphi(f) = 0$ so that $\Gamma(f)$ is not invertible in $C(M)$. ∎

We summarize the results for the commutative case.

2.35 Theorem (Gelfand) If $\mathfrak{B}$ is a commutative Banach algebra, M is its maximal ideal space, and $\Gamma : \mathfrak{B} \to C(M)$ is the Gelfand transform, then:

(1) M is not empty;
(2) Γ is an algebra homomorphism;
(3) $\|\Gamma f\|_\infty \leqslant \|f\|$ for f in $\mathfrak{B}$; and
(4) f is invertible in $\mathfrak{B}$ if and only if $\Gamma(f)$ is invertible in $C(M)$.

The crucial fact about statement (4) is that it refers to $\Gamma(f)$ being invertible in $C(M)$ rather than in the range of Γ.

We obtain two corollaries before proceeding to a result concerning the spectral radius.

2.36 Corollary If $\mathfrak{B}$ is a commutative Banach algebra and f is in $\mathfrak{B}$, then $\sigma(f) = \text{range } \Gamma f$ and $r(f) = \|\Gamma f\|_\infty$.

Proof If λ is not in $\sigma(f)$, then $f - \lambda$ is invertible in $\mathfrak{B}$ by definition. This implies that $\Gamma(f) - \lambda$ is invertible in $C(M)$, which in turn implies that $(\Gamma f - \lambda)(\varphi) \neq 0$ for φ in M. Thus $(\Gamma f)(\varphi) \neq \lambda$ for φ in M. If λ is not in the range of Γf, then $\Gamma f - \lambda$ is invertible in $C(M)$ and hence, by the preceding theorem, $f - \lambda$ is invertible in $\mathfrak{B}$. Therefore, λ is not in $\sigma(f)$ and the proof is complete. ■

If $\varphi(z) = \sum_{n=0}^\infty a_n z^n$ is an entire function with complex coefficients and f is an element of the Banach algebra $\mathfrak{B}$, then we let $\varphi(f)$ denote the element $\sum_{n=0}^\infty a_n f^n$ of $\mathfrak{B}$.

2.37 Corollary (Spectral Mapping Theorem) If $\mathfrak{B}$ is a Banach algebra, f is in $\mathfrak{B}$, and φ is an entire function on $\mathbb{C}$, then

$$\sigma(\varphi(f)) = \varphi(\sigma(f)) = \{\varphi(\lambda) : \lambda \in \sigma(f)\}.$$

Proof If $\varphi(z) = \sum_{n=0}^\infty a_n z^n$ is the Taylor series expansion for φ, then $\varphi(f) = \sum_{n=0}^\infty a_n f^n$ can be seen to converge to an element of $\mathfrak{B}$. If $\mathfrak{B}_0$ is the subalgebra of $\mathfrak{B}$ generated by $1, f$, and elements of the form $(f - \lambda)^{-1}$ for λ in $\rho(f)$ and $(\varphi(f) - \mu)^{-1}$ for μ in $\rho(\varphi(f))$, then $\mathfrak{B}_0$ is commutative and $\sigma_{\mathfrak{B}}(f) = \sigma_{\mathfrak{B}_0}(f)$ and $\sigma_{\mathfrak{B}}(\varphi(f)) = \sigma_{\mathfrak{B}_0}(\varphi(f))$. Thus, we can assume that $\mathfrak{B}$ is commutative and use the Gelfand transform.

Using the preceding corollary we obtain

$$\sigma(\varphi(f)) = \text{range } \Gamma(\varphi(f)) = \text{range } \varphi(\Gamma f)$$

$$= \varphi(\text{range } \Gamma f) = \varphi(\sigma(f)),$$

since $\Gamma(\varphi(f)) = \varphi(\Gamma f)$ by continuity; thus the proof is complete. ■

We next prove a basic result due to Beurling and Gelfand relating the spectral radius to the norm.

2.38 Theorem If $\mathfrak{B}$ is a Banach algebra and f is in $\mathfrak{B}$, then $r_{\mathfrak{B}}(f) = \lim_{n \to \infty} \|f^n\|^{1/n}$.

Proof If $\mathfrak{B}_0$ denotes the closed subalgebra of $\mathfrak{B}$ generated by the identity, f, and $\{(f^n - \lambda)^{-1} : \lambda \in \rho_{\mathfrak{B}}(f^n), n \in \mathbb{Z}^+\}$, then $\mathfrak{B}_0$ is commutative and $\sigma_{\mathfrak{B}_0}(f^n) = \sigma_{\mathfrak{B}}(f^n)$ for all positive integers n. From the preceding corollary, we have $\sigma_{\mathfrak{B}_0}(f^n) = \sigma_{\mathfrak{B}_0}(f)^n$ and hence $r_{\mathfrak{B}}(f)^n = r_{\mathfrak{B}}(f^n) \leqslant \|f^n\|$; thus the inequality $r_{\mathfrak{B}}(f) \leqslant \liminf_{n\to\infty} \|f^n\|^{1/n}$ follows.

Next consider the analytic function

$$G(\lambda) = -\lambda \sum_{n=0}^{\infty} \frac{f^n}{\lambda^n}$$

which converges for $|\lambda| > \limsup_{n\to\infty} \|f^n\|^{1/n}$ by Proposition 1.9. For $|\lambda| > \|f\|$ we have $G(\lambda) = (f - \lambda)^{-1}$ and therefore

$$(f - \lambda)\, G(\lambda) = G(\lambda)(f - \lambda) = 1 \qquad \text{for} \quad |\lambda| > \limsup_{n\to\infty} \|f^n\|^{1/n}.$$

Thus

$$r_{\mathfrak{B}}(f) \geqslant \limsup_{n\to\infty} \|f^n\|^{1/n} \geqslant \liminf_{n\to\infty} \|f^n\|^{1/n} \geqslant r_{\mathfrak{B}}(f),$$

from which the result follows. ∎

2.39 Corollary If $\mathfrak{B}$ is a commutative Banach algebra, then the Gelfand transform is an isometry if and only if $\|f^2\| = \|f\|^2$ for every f in $\mathfrak{B}$.

Proof Since $r(f) = \|\Gamma_{\mathfrak{B}} f\|_\infty$ for f in $\mathfrak{B}$ by Corollary 2.36, we see that $\Gamma_{\mathfrak{B}}$ is an isometry if and only if $r(f) = \|f\|$ for f in $\mathfrak{B}$. Moreover, since $r(f^2) = r(f)^2$ by Corollary 2.37, the result now follows from the theorem. ∎

We now study the self-adjoint subalgebras of $C(X)$ for X a compact Hausdorff space. We begin with the generalization due to Stone of the classical theorem of Weierstrass on the density of polynomials. A subset $\mathfrak{A}$ of $C(X)$ is said to be self-adjoint if f in $\mathfrak{A}$ implies $\bar{f}$ is in $\mathfrak{A}$.

2.40 Theorem (Stone–Weierstrass) Let X be a compact Hausdorff space. If $\mathfrak{A}$ is a closed self-adjoint subalgebra of $C(X)$ which separates the points of X and contains the constant function 1, then $\mathfrak{A} = C(X)$.

Proof If $\mathfrak{A}_r$ denotes the set of real functions in $\mathfrak{A}$, then $\mathfrak{A}_r$ is a closed subalgebra of the real algebra $C_r(X)$ of continuous functions on X which separates points and contains the function 1. Moreover, the theorem reduces to showing that $\mathfrak{A}_r = C_r(X)$.

We begin by showing that f in $\mathfrak{A}_r$ implies that $|f|$ is in $\mathfrak{A}_r$. Recall that the binomial series for the function $\varphi(t) = (1-t)^{1/2}$ is $\sum_{n=0}^{\infty} \alpha_n t^n$, where $\alpha_n = (-1)^n \binom{1/2}{n}$. It is an easy consequence of the comparison theorem that the sequence $\{\sum_{n=0}^{N} \alpha_n t^n\}_{N=1}^{\infty}$ converges uniformly to φ on the closed interval $[0, 1-\delta]$ for $\delta > 0$. (The sequence actually converges uniformly to φ on $[-1, 1]$.) Let f be in $\mathfrak{A}_r$ such that $\|f\|_\infty \leqslant 1$ and set $g_\delta = \delta + (1-\delta)f^2$ for δ in $(0, 1]$; then $0 \leqslant 1 - g_\delta \leqslant 1 - \delta$. For fixed $\delta > 0$, set $h_N = \sum_{n=0}^{N} \alpha_n (1 - g_\delta)^n$. Then h_N is in $\mathfrak{A}_r$ and

$$\|h_N - (g_\delta)^{1/2}\|_\infty = \sup_{x \in X} \left| \sum_{n=0}^{N} \alpha_n (1 - g_\delta(x))^n - \varphi(1 - g_\delta(x)) \right|$$

$$\leqslant \sup_{t \in [0, 1-\delta]} \left| \sum_{n=0}^{N} \alpha_n t^n - \varphi(t) \right|.$$

Therefore, $\lim_{N \to \infty} \|h_N - (g_\delta)^{1/2}\|_\infty = 0$, implying that $(g_\delta)^{1/2}$ is in $\mathfrak{A}_r$. Now since the square root function is uniformly continuous on $[0, 1]$, we have $\lim_{\delta \to 0} \||f| - (g_\delta)^{1/2}\|_\infty = 0$, and thus $|f|$ is in $\mathfrak{A}_r$.

We next show that $\mathfrak{A}_r$ is a lattice, that is, for f and g in $\mathfrak{A}_r$ the functions $f \vee g$ and $f \wedge g$ are in $\mathfrak{A}_r$, where $(f \vee g)(x) = \max\{f(x), g(x)\}$, and $(f \wedge g)(x) = \min\{f(x), g(x)\}$. This follows from the identities

$$f \vee g = \tfrac{1}{2}\{f + g + |f - g|\}, \qquad \text{and} \qquad f \wedge g = \tfrac{1}{2}\{f + g - |f - g|\}$$

which can be verified pointwise.

Further, if x and y are distinct points in X and a and b arbitrary real numbers, and f is a function in $\mathfrak{A}_r$ such that $f(x) \neq f(y)$, then the function g defined by

$$g(z) = a + (b-a)\frac{f(z) - f(x)}{f(y) - f(x)}$$

is in $\mathfrak{A}_r$ and has the property that $g(x) = a$ and $g(y) = b$. Thus there exist functions in $\mathfrak{A}_r$ taking prescribed values at two points.

We now complete the proof. Take f in $C_r(X)$ and $\varepsilon > 0$. Fix x_0 in X. For each x in X, we can find a g_x in $\mathfrak{A}_r$ such that $g_x(x_0) = f(x_0)$ and $g_x(x) = f(x)$. Since f and g are continuous, there exists an open set U_x of x such that $g_x(y) \leqslant f(y) + \varepsilon$ for all y in U_x. The open sets $\{U_x\}_{x \in X}$ cover X and hence by compactness, there is a finite subcover $U_{x_1}, U_{x_2}, \ldots, U_{x_n}$. Let $h_{x_0} = g_{x_1} \wedge g_{x_2} \wedge \cdots \wedge g_{x_n}$. Then h_{x_0} is in $\mathfrak{A}_r$, $h_{x_0}(x_0) = f(x_0)$, and $h_{x_0}(y) \leqslant f(y) + \varepsilon$ for y in X.

Thus for each x_0 in X there exists h_{x_0} in $\mathfrak{A}_r$ such that $h_{x_0}(x_0) = f(x_0)$ and $h_{x_0}(y) \leqslant f(y) + \varepsilon$ for y in X. Since h_{x_0} and f are continuous, there exists an open set V_{x_0} of x_0 such that $h_{x_0}(y) \geqslant f(y) - \varepsilon$ for y in V_{x_0}. Again, the family $\{V_{x_0}\}_{x_0 \in X}$ covers X, and hence there exists a finite subcover $V_{x_1}, V_{x_2}, ..., V_{x_m}$. If we set $k = h_{x_1} \vee h_{x_2} \vee \cdots \vee h_{x_m}$, then k is in $\mathfrak{A}_r$ and $f(y) - \varepsilon \leqslant k(y) \leqslant f(y) + \varepsilon$ for y in X. Therefore, $\|f - k\|_\infty \leqslant \varepsilon$ and the proof is complete. ∎

2.41 If $[a, b]$ is a closed interval of $\mathbb{R}$, then the collection of polynomials $\{\sum_{n=0}^N \alpha_n t^n\}$ with complex coefficients is a self-adjoint subalgebra of $C([a,b])$ which separates points and contains the constant function 1. Thus its closure must be $C([a, b])$, and this is the statement of the Weierstrass theorem.

We now consider the closed self-adjoint subalgebras of $C(X)$ containing the constant function 1 that do not separate points and show that they can be identified as $C(Y)$ for some compact Hausdorff space Y.

Let X be a compact Hausdorff space and $\mathfrak{A}$ be a closed subalgebra of $C(X)$ which contains the constant functions. For x in X we let φ_x denote the multiplicative linear functional in $M_{\mathfrak{A}}$ defined $\varphi_x(f) = f(x)$. The following proposition is of interest even in the nonself-adjoint case.

2.42 Proposition If η is the map defined from X to $M_{\mathfrak{A}}$ so that $\eta(x) = \varphi_x$, then η is continuous.

Proof If $\{x_\alpha\}_{\alpha \in A}$ is a net in X which converges to x, then $\lim_{\alpha \in A} f(x_\alpha) = f(x)$ for f in $\mathfrak{A}$. Therefore, $\lim_{\alpha \in A} \varphi_{x_\alpha}(f) = \varphi_x(f)$ and hence $\lim_{\alpha \in A} \eta(x_\alpha) = \eta(x)$ in the topology of $M_{\mathfrak{A}}$. Thus, η is continuous. ∎

In general, η is neither one-to-one nor onto. The latter property, however, holds if $\mathfrak{A}$ is self-adjoint.

2.43 Proposition If $\mathfrak{A}$ is self-adjoint, then η maps X onto $M_{\mathfrak{A}}$.

Proof Fix φ in $M_{\mathfrak{A}}$ and set $K_f = \{x \in X : f(x) = \varphi(f)\}$. First of all, each K_f is a closed subset of X, since f is continuous. Secondly, we want to show that not only is each K_f nonempty but that the collection of sets $\{K_f : f \in \mathfrak{A}\}$ has the finite intersection property. Suppose

$$K_{f_1} \cap K_{f_2} \cap \cdots \cap K_{f_n} = \varnothing$$

for some functions $f_1, f_2, ..., f_n$ in $\mathfrak{A}$. Then the function

$$g(x) = \sum_{i=1}^n |f_i(x) - \varphi(f_i)|^2$$

does not vanish on X. Moreover, g is in $\mathfrak{A}$ since the latter is a self-adjoint algebra. But $g(x) > 0$ for x in X and the fact that X is compact implies that there exists $\varepsilon > 0$ such that $1 \geqslant g(x)/\|g\|_\infty \geqslant \varepsilon$, and hence such that $\|1 - (g/\|g\|_\infty)\|_\infty < 1$. But then g^{-1} is in $\mathfrak{A}$ by Proposition 2.5 which implies $\varphi(g) \neq 0$. However,

$$\varphi(g) = \sum_{i=1}^n (\varphi(f_i) - \varphi(f_i))(\varphi(\bar{f_i}) - \overline{\varphi(\bar{f_i})}) = 0,$$

which is a contradiction. Thus the collection $\{K_f : f \in \mathfrak{A}\}$ has the finite intersection property. If x is in $\bigcap_{f \in \mathfrak{A}} K_f$, then $\eta(x) = \varphi$ and the proof is complete. ∎

The reader should consider carefully how the self-adjointness of $\mathfrak{A}$ was used in the preceding proof. We give an example in this chapter of a subalgebra for which η is not onto. Even for examples where η is onto, the Gelfand transform Γ need not be onto. It is, however, for self-adjoint subalgebras.

2.44 Proposition If $\mathfrak{A}$ is a closed self-adjoint subalgebra of $C(X)$ containing the constant function 1, then the Gelfand transform Γ is an isometric isomorphism from $\mathfrak{A}$ onto $C(M_\mathfrak{A})$.

Proof For f in $\mathfrak{A}$ there exists x_0 in X such that $f(x_0) = \|f\|_\infty$, since X is compact. Therefore,

$$\|f\|_\infty \geqslant \|\Gamma f\|_\infty = \sup_{\varphi \in M_\mathfrak{A}} |(\Gamma f)(\varphi)| \geqslant |(\Gamma f)(\eta x_0)| = f(x_0) = \|f\|_\infty,$$

and hence Γ is an isometry. Since Γ is known to be an algebraic homomorphism, it remains only to prove that Γ is onto. The range of Γ is a subalgebra of $C(M_\mathfrak{A})$ that contains 1, since $\Gamma 1 = 1$, is uniformly closed since Γ is an isometry, and separates points. Moreover, since by the preceding proposition for φ in $M_\mathfrak{A}$ there exists x in X such that $\eta x = \varphi$, we have for f in $\mathfrak{A}$ that

$$\overline{(\Gamma f)(\varphi)} = \overline{(\Gamma f)(\varphi)} = \overline{(\Gamma f)(\eta x)} = \overline{f(x)} = \bar{f}(x) = \Gamma(\bar{f})(\eta x) = \Gamma(\bar{f})(\varphi).$$

Therefore, $\overline{\Gamma f} = \Gamma(\bar{f})$ and $\Gamma \mathfrak{A}$ is self-adjoint because $\mathfrak{A}$ is. By the Stone–Weierstrass theorem, we have $\Gamma \mathfrak{A} = C(M_\mathfrak{A})$ and the proof is complete. ∎

2.45 Lemma Let X and Y be compact Hausdorff spaces and θ be a continuous map from X onto Y. The map θ^* defined by $\theta^* f = f \circ \theta$ from $C(Y)$ into $C(X)$ is an isometrical isomorphism onto the subalgebra of continuous functions on X which are constant on the closed partition $\{\theta^{-1}(y) : y \in Y\}$ of X.

Proof That θ^* is an isometrical isomorphism of $C(Y)$ into $C(X)$ is obvious. Moreover, it is clear that a function of the form $f \circ \theta$ is constant on the partition $\{\theta^{-1}(y) : y \in Y\}$. Now suppose g is continuous on X and constant on each of the sets $\theta^{-1}(y)$ for y in Y. We can unambiguously define a function f on Y such that $f \circ \theta = g$; the only question is whether this f is continuous. Suppose $\{y_\alpha\}_{\alpha \in A}$ is a net of points in Y and y is in Y such that $\lim_{\alpha \in A} y_\alpha = y$. Choose x_α in $\theta^{-1}(y_\alpha)$ for each α in A and consider the net $\{x_\alpha\}_{\alpha \in A}$. In general, $\lim_{\alpha \in A} x_\alpha$ does not exist; however, since X is compact there exists a subnet $\{x_{\alpha_\beta}\}_{\beta \in B}$ and an x in X such that $\lim_{\beta \in B} x_{\alpha_\beta} = x$. Since θ is continuous, we have $\theta(x) = y$ and $\lim_{\beta \in B} g(x_{\alpha_\beta}) = g(x) = f(y)$; thus f is continuous and the proof is complete. ∎

2.46   Proposition   If $\mathfrak{A}$ is a closed self-adjoint subalgebra of $C(X)$ containing the constant function 1, and η a continuous map from X onto $M_\mathfrak{A}$, then η^* is an isometrical isomorphism of $C(M_\mathfrak{A})$ onto $\mathfrak{A}$ which is the left inverse of the Gelfand transform, that is, $\eta^* \circ \Gamma = 1$.

Proof   For f in $\mathfrak{A}$ and x in X, we have $((\eta^* \circ \Gamma)f)(x) = (\Gamma f)(\eta x) = f(x)$. Therefore η^* is the left inverse of the Gelfand transform. Since Γ maps $\mathfrak{A}$ onto $C(M_\mathfrak{A})$ by Proposition 2.44, we have that η^* maps $C(M_\mathfrak{A})$ onto $\mathfrak{A}$. ∎

We state and prove the generalized Stone–Weierstrass theorem after introducing the following terminology. For X a set and $\mathfrak{A}$ a collection of functions on X define the equivalence relation on X such that two points x_1 and x_2 are related if $f(x_1) = f(x_2)$ for every f in $\mathfrak{A}$. This relation partitions X into the sets on which the functions in $\mathfrak{A}$ are constant. Let $\Pi_\mathfrak{A}$ denote this collection of subsets of X.

2.47   Theorem   Let X be a compact Hausdorff space and $\mathfrak{A}$ be a closed self-adjoint subalgebra of $C(X)$ which contains the constants. Then $\mathfrak{A}$ is the collection of continuous functions on X which are constant on the sets of $\Pi_\mathfrak{A}$.

Proof   This follows by combining Lemma 2.45 and Proposition 2.46. ∎

If $\mathfrak{A}$ separates the points of X, then $\Pi_\mathfrak{A}$ consists of one-point sets and the usual Stone–Weierstrass theorem follows.

2.48   As we have just seen, the self-adjoint subalgebras of $C(X)$ are all of the form $C(Y)$ for some compact Hausdorff space Y. This is far from true, however, for the nonself-adjoint subalgebras. Let $\mathfrak{A}$ be a closed subalgebra of $C(X)$ which contains the constant functions. If we let $\mathfrak{B}$ denote the smallest closed self-adjoint subalgebra of $C(X)$ that contains $\mathfrak{A}$, then $\mathfrak{B}$ is isometrically isomorphic to $C(Y)$, where Y is the maximal ideal space of $\mathfrak{B}$, and the Gelfand transform Γ implements the isomorphism. Then $\Gamma\mathfrak{A}$ is a closed subalgebra of $C(Y)$ that contains the constant functions and, more importantly, separates the points of Y. Therefore, rather than study $\mathfrak{A}$ as a subalgebra of $C(X)$, we choose to study $\Gamma\mathfrak{A}$ as a subalgebra of $C(Y)$. Thus we make the following definition.

2.49   Definition   Let X be a compact Hausdorff space and $\mathfrak{A}$ be a subset of $C(X)$. Then $\mathfrak{A}$ is said to be a function algebra if $\mathfrak{A}$ is a closed subalgebra of $C(X)$ which separates points and contains the constant functions.

The theory of function algebras is very extensive and draws on the techniques of approximation theory and complex function theory as well as those of functional analysis. In this book we will be limited to considering only a few important examples of function algebras.

2.50   Example   Let $\mathbb{T}$ denote the circle group $\{z \in \mathbb{C} : |z| = 1\}$. For n in $\mathbb{Z}$ let χ_n be the function on $\mathbb{T}$ defined $\chi_n(z) = z^n$. Then $\chi_0 = 1$, $\chi_{-n} = \bar{\chi}_n$, and $\chi_m \chi_n = \chi_{m+n}$ for n and m in $\mathbb{Z}$. The functions in the set

$$\mathscr{P} = \left\{ \sum_{n=-N}^{N} \alpha_n \chi_n : \alpha_n \in \mathbb{C} \right\}$$

are called the trigonometric polynomials. Since $\mathscr{P}$ is a self-adjoint subalgebra of $C(\mathbb{T})$ which contains the constant functions and separates points, then the uniform closure of $\mathscr{P}$ is $C(\mathbb{T})$ by the Stone–Weierstrass theorem.

Let $\mathscr{P}_+ = \{\sum_{n=0}^{N} \alpha_n \chi_n : \alpha_n \in \mathbb{C}\}$; the functions in $\mathscr{P}_+$ are called analytic trigonometric polynomials. If we let A denote the uniform closure of $\mathscr{P}_+$ in $C(\mathbb{T})$, then A is a function algebra, but at this point it is not obvious that $A \neq C(\mathbb{T})$. We prove this by showing that the maximal ideal space of A is not $\mathbb{T}$. For this we need a lemma.

2.51   Lemma   If $\sum_{n=0}^{N} \alpha_n \chi_n$ is in $\mathscr{P}_+$ and w is in $\mathbb{C}$, $|w| < 1$, then

$$\sum_{n=0}^{N} \alpha_n w^n = \frac{1}{2\pi} \int_0^{2\pi} \left(\sum_{n=0}^{N} \alpha_n \chi_n \right)(e^{it}) \frac{1}{1 - we^{-it}} \, dt.$$

Proof   Expand $1/(1 - we^{-it}) = \sum_{m=0}^{\infty} (we^{-it})^m$, where the series converges uniformly for t in $[0, 2\pi]$. Therefore,

$$\frac{1}{2\pi} \int_0^{2\pi} \left(\sum_{n=0}^{N} \alpha_n \chi_n \right)(e^{it}) \frac{1}{1 - we^{-it}} \, dt = \frac{1}{2\pi} \sum_{n=0}^{N} \alpha_n \sum_{m=0}^{\infty} w^m \int_0^{2\pi} e^{i(n-m)t} \, dt$$

$$= \frac{1}{2\pi} \sum_{n=0}^{N} \alpha_n w^n,$$

since $(1/2\pi) \int_0^{2\pi} e^{ikt} \, dt = 1$ for $k = 0$ and 0 otherwise.   ∎

For w in $\mathbb{C}$, $|w| < 1$, define φ_w on $\mathscr{P}_+$ such that

$$\varphi_w \left(\sum_{n=0}^{N} \alpha_n \chi_n \right) = \sum_{n=0}^{N} \alpha_n w^n.$$

It is clear that φ_w is a multiplicative linear functional on $\mathscr{P}_+$. However, since $\mathscr{P}_+$ is not a Banach algebra, that is, $\mathscr{P}_+$ is not complete, we cannot conclude apriori that φ_w is continuous. That follows, however, from the preceding lemma since

$$\left| \varphi_w \left(\sum_{n=0}^{N} \alpha_n \chi_n \right) \right| = \left| \sum_{n=0}^{N} \alpha_n w^n \right| = \left| \frac{1}{2\pi} \int_0^{2\pi} \left(\sum_{n=0}^{N} \alpha_n \chi_n \right)(e^{it}) \frac{1}{1 - we^{-it}} \, dt \right|$$

$$\leqslant \frac{1}{2\pi} \left\| \sum_{n=0}^{N} \alpha_n \chi_n \right\|_\infty \int_0^{2\pi} \frac{1}{|1 - we^{-it}|} \, dt.$$

Therefore, φ_w is bounded on $\mathscr{P}_+$ and hence can be extended to a multiplicative linear functional on A. Now for w in $\mathbb{C}$, $|w| = 1$, let φ_w denote the evaluation functional on A, that is, $\varphi_w(f) = f(w)$ for f in A. The latter is well defined, since $A \subset C(\mathbb{T})$.

Now set $\overline{\mathbb{D}} = \{z \in \mathbb{C} : |z| \leqslant 1\}$, let M denote the maximal ideal space of A, and let ψ be the function from $\overline{\mathbb{D}}$ to M defined by $\psi(z) = \varphi_z$.

2.52   Theorem   The function ψ is a homeomorphism of $\overline{\mathbb{D}}$ onto the maximal ideal space M of A.

Proof   By the remarks preceding the theorem, the function ψ is well defined. If z_1 and z_2 are in $\overline{\mathbb{D}}$, then $\psi(z_1) = \psi(z_2)$ implies that $z_1 = \varphi_{z_1}(\chi_1) = \varphi_{z_2}(\chi_1) = z_2$; thus ψ is one-to-one.

If φ is in M, then $\|\chi_1\| = 1$ implies that $z = \varphi(\chi_1)$ is in $\overline{\mathbb{D}}$. Moreover, the identity

$$\varphi \left(\sum_{n=0}^{N} \alpha_n \chi_n \right) = \sum_{n=0}^{N} \alpha_n [\varphi(\chi_1)]^n$$

$$= \sum_{n=0}^{N} \alpha_n z^n = \varphi_z \left(\sum_{n=0}^{N} \alpha_n \chi_n \right)$$

proves that φ agrees with φ_z on the dense subset $\mathscr{P}_+$ of A. Therefore, $\varphi = \varphi_z$ and ψ is seen to be onto M.

Since both $\overline{\mathbb{D}}$ and M are compact Hausdorff spaces and ψ is one-to-one and onto, to complete the proof it suffices to show that ψ is continuous. To this end suppose $\{z_\beta\}_{\beta \in B}$ is a net in $\overline{\mathbb{D}}$ such that $\lim_{\beta \in B} z_\beta = z$. Since $\sup_{\beta \in B}\{\|\varphi_{z_\beta}\|\} = 1$ and $\mathscr{P}_+$ is dense in A, and since

$$\lim_{\beta \in B} \varphi_{z_\beta}\left(\sum_{n=0}^N \alpha_n \chi_n\right) = \lim_{\beta \in B}\left(\sum_{n=0}^N \alpha_n z_\beta{}^n\right) = \sum_{n=0}^N \alpha_n z^n$$

$$= \varphi_z\left(\sum_{n=0}^N \alpha_n \chi_n\right)$$

for every function $\sum_{n=0}^N \alpha_n \chi_n$ in $\mathscr{P}_+$, it follows from Proposition 1.21 that ψ is continuous. ∎

2.53 From Proposition 2.3 we know that the maximal ideal space of $C(\mathbb{T})$ is just $\mathbb{T}$. We have just shown that the maximal ideal space of the closed subalgebra A of $C(\mathbb{T})$ is $\overline{\mathbb{D}}$. Moreover, if φ_z is a multiplicative linear functional on A, and $|z| = 1$, then φ_z is the restriction to A of the "evaluation at z" map on $C(\mathbb{T})$. Thus the maximal ideal space of $C(\mathbb{T})$ is embedded in that of A. This example also shows how the maximal ideal space of a function algebra is, at least roughly speaking, the natural domain of the functions in it. In this case although the elements of A are functions on $\mathbb{T}$, there are "hidden points" inside the circle which "ought" to be in the domain. In particular, viewing χ_1 as a function on $\mathbb{T}$, there is no reason why it should not be invertible; on $\overline{\mathbb{D}}$, however, it is obvious why it is not—it vanishes at the origin.

Let us consider this example from another viewpoint. The element χ_1 is contained in both of the algebras A and $C(\mathbb{T})$. In $C(\mathbb{T})$ we have $\sigma_{C(\mathbb{T})}(\chi_1) = \mathbb{T}$, while in A we have $\sigma_A(\chi_1) = \overline{\mathbb{D}}$. Hence not only is the "A-spectrum" of χ_1 larger, but it is obtained from the $C(\mathbb{T})$-spectrum by "filling in a hole." That this is true, in general, is a corollary to the next theorem.

2.54 Theorem (Šilov) If $\mathfrak{B}$ is a Banach algebra, $\mathfrak{A}$ is a closed subalgebra of $\mathfrak{B}$, and f is an element of $\mathfrak{A}$, then the boundary of $\sigma_{\mathfrak{A}}(f)$ is contained in the boundary of $\sigma_{\mathfrak{B}}(f)$.

Proof If $(f - \lambda)$ has an inverse in $\mathfrak{A}$, then it has an inverse in $\mathfrak{B}$. Thus $\sigma_{\mathfrak{A}}(f)$ contains $\sigma_{\mathfrak{B}}(f)$ and hence it is sufficient to show that the boundary of $\sigma_{\mathfrak{A}}(f) \subset \sigma_{\mathfrak{B}}(f)$. If λ_0 is in the boundary of $\sigma_{\mathfrak{A}}(f)$, then there exists a sequence $\{\lambda_n\}_{n=1}^\infty$ contained in $\rho_{\mathfrak{A}}(f)$ such that $\lim_{n \to \infty} \lambda_n = \lambda_0$. If for some integer

n it were true that $\|(f-\lambda_n)^{-1}\| < |1/(\lambda_0-\lambda_n)|$, then it would follow that

$$\|(f-\lambda_0) - (f-\lambda_n)\| < 1/\|(f-\lambda_n)^{-1}\|$$

and hence $f-\lambda_0$ would be invertible as in the proof of Proposition 2.7. Thus we have $\lim_{n \to \infty} \|(f-\lambda_n)^{-1}\| = \infty$.

If λ_0 were not in $\sigma_{\mathfrak{B}}(f)$, then it would follow from Corollary 2.8 that $\|(f-\lambda)^{-1}\|$ is bounded for λ in some neighborhood of λ_0, which is a contradiction. ■

2.55 Corollary If $\mathfrak{B}$ is a Banach algebra, $\mathfrak{A}$ is a closed subalgebra of $\mathfrak{B}$, and f is an element of $\mathfrak{A}$, then $\sigma_{\mathfrak{A}}(f)$ is obtained by adding to $\sigma_{\mathfrak{B}}(f)$ certain of the bounded components of $\mathbb{C} \backslash \sigma_{\mathfrak{B}}(f)$.

Proof Elementary topology and the theorem yield this result. ■

2.56 Example We next consider an example for which the Gelfand transform is not an isometry.

In Section 1.15 we showed that $l^1(\mathbb{Z}^+)$ is a Banach space. Analogously, if we let $l^1(\mathbb{Z})$ denote the collection of complex functions f on $\mathbb{Z}$ such that $\sum_{n=-\infty}^{\infty} |f(n)| < \infty$, then with pointwise addition and scalar multiplication and the norm $\|f\|_1 = \sum_{n=-\infty}^{\infty} |f(n)| < \infty$, $l^1(\mathbb{Z})$ is a Banach space. Moreover, $l^1(\mathbb{Z})$ can also be made into a Banach algebra in a nonobvious way. For f and g in $l^1(\mathbb{Z})$ define the convolution product

$$(f \circ g)(n) = \sum_{k=-\infty}^{\infty} f(n-k)\, g(k).$$

To show that this sum converges for each n in $\mathbb{Z}$ and that the resulting function is in $l^1(\mathbb{Z})$, we write

$$\sum_{n=-\infty}^{\infty} |(f \circ g)(n)| = \sum_{n=-\infty}^{\infty} \left| \sum_{k=-\infty}^{\infty} f(n-k)\, g(k) \right| \leqslant \sum_{n=-\infty}^{\infty} \sum_{k=-\infty}^{\infty} |f(n-k)|\, |g(k)|$$

$$= \sum_{k=-\infty}^{\infty} |g(k)| \sum_{n=-\infty}^{\infty} |f(n-k)| = \|f\|_1 \sum_{k=-\infty}^{\infty} |g(k)|$$

$$= \|f\|_1 \|g\|_1.$$

Therefore, $f \circ g$ is well defined and is in $l^1(\mathbb{Z})$, and $\|f \circ g\|_1 \leqslant \|f\|_1 \|g\|_1$. We leave to the reader the exercise of showing that this multiplication is associative and commutative. Assuming this, then $l^1(\mathbb{Z})$ is a commutative Banach algebra.

For n in $\mathbb{Z}$ let e_n denote the function on $\mathbb{Z}$ defined to be 1 at n and 0 otherwise. Then e_1 is the identity element of $l^1(\mathbb{Z})$ and $e_n \circ e_m = e_{n+m}$ for n and m in $\mathbb{Z}$.

Let M be the maximal ideal space of $l^1(\mathbb{Z})$. For each z in $\mathbb{T}$, let φ_z be the function defined on $l^1(\mathbb{Z})$ such that $\varphi_z(f) = \sum_{n=-\infty}^{\infty} f(n) z^n$. It is easily verified that φ_z is well defined and in M. Thus we can define a function from $\mathbb{T}$ to M by setting $\psi(z) = \varphi_z$.

2.57   Theorem   The function ψ is a homeomorphism from $\mathbb{T}$ onto the maximal ideal space M of $l^1(\mathbb{Z})$.

Proof   If z_1 and z_2 are in $\mathbb{T}$, and $\varphi_{z_1} = \varphi_{z_2}$, then $z_1 = \varphi_{z_1}(e_1) = \varphi_{z_2}(e_1) = z_2$; hence ψ is one-to-one. Suppose φ is an element of M and $z = \varphi(e_1)$; then

$$1 = \|e_1\|_1 \geqslant |\varphi(e_1)| = |z| = \frac{1}{|z^{-1}|} = \frac{1}{|\varphi(e_{-1})|} \geqslant \frac{1}{\|e_{-1}\|} = 1,$$

which implies that z is in $\mathbb{T}$. Moreover, since $\varphi(e_n) = [\varphi(e_1)]^n = z^n = \varphi_z(e_n)$ for n in $\mathbb{Z}$, it follows that $\varphi = \varphi_z = \psi(z)$. Therefore, again ψ is one-to-one and onto and it remains only to show that ψ is continuous, since both M and $\mathbb{T}$ are compact Hausdorff spaces. Thus, suppose $\{z_\beta\}_{\beta \in B}$ is a net of points in $\mathbb{T}$ such that $\lim_{\beta \in B} z_\beta = z$. Then for f in $l^1(\mathbb{Z})$ we have

$$|\varphi_{z_\beta}(f) - \varphi_z(f)| \leqslant \sum_{|n| \leqslant N} |f(n)| \, |z_\beta{}^n - z^n| + \sum_{|n| > N} |f(n)| \, |z_\beta{}^n - z^n|$$
$$\leqslant \|f\|_1 \sup_{|n| \leqslant N} |z_\beta{}^n - z^n| + 2 \sum_{|n| > N} |f(n)|.$$

Hence for $\varepsilon > 0$, if N is chosen such that $\sum_{|n| > N} |f(n)| < \varepsilon/4$ and β_0 is then chosen in B such that $\beta > \beta_0$ implies $\sup_{|n| \leqslant N} |z_\beta{}^n - z^n| < \varepsilon/2\|f\|_1$, then $|\varphi_{z_\beta}(f) - \varphi_z(f)| < \varepsilon$ for $\beta \geqslant \beta_0$. Therefore, $\lim_{\beta \in B} \varphi_{z_\beta}(f) = \varphi_z(f)$ and ψ is continuous and the proof is complete.   ∎

2.58   Using the homeomorphism ψ we identify the maximal ideal space of $l^1(\mathbb{Z})$ with $\mathbb{T}$. Thus the Gelfand transform is the operator Γ defined from $l^1(\mathbb{Z})$ to $C(\mathbb{T})$ such that $(\Gamma f)(z) = \sum_{n=-\infty}^{\infty} f(n) z^n$ for z in $\mathbb{T}$, where the series converges uniformly and absolutely on $\mathbb{T}$ to Γf. The values of f on $\mathbb{Z}$ can be recaptured from Γf, since they coincide with the Fourier coefficients of Γf. More specifically,

$$f(n) = \frac{1}{2\pi} \int_0^{2\pi} (\Gamma f)(e^{it}) e^{-int} \, dt \qquad \text{for} \quad n \text{ in } \mathbb{Z},$$

since

$$\frac{1}{2\pi} \int_0^{2\pi} (\Gamma f)(e^{it}) e^{-int} \, dt = \frac{1}{2\pi} \int_0^{2\pi} \sum_{m=-\infty}^{\infty} f(m) e^{i(m-n)t} \, dt$$

$$= \frac{1}{2\pi} \sum_{m=-\infty}^{\infty} f(m) \int_0^{2\pi} e^{i(m-n)t} \, dt = f(n),$$

where the interchange of integration and summation is justified since the series converges uniformly. In particular, φ in $C(\mathbb{T})$ is in the range of Γ if and only if the Fourier coefficients of φ are an absolutely convergent series, that is, if and only if

$$\sum_{n=-\infty}^{\infty} \left| \frac{1}{2\pi} \int_0^{2\pi} \varphi(e^{it}) e^{-int} \, dt \right| < \infty.$$

We leave to the exercises the task of showing that this is not always the case.

Since not every function φ in $C(\mathbb{T})$ has an absolutely convergent Fourier series, it is not obvious whether $1/\varphi$ does if φ does and $\varphi(z) \neq 0$. That this is the case is a nontrivial theorem due to Wiener. The proof below is due to Gelfand and indicates the power of his theory for commutative Banach algebras.

2.59   Theorem   If φ in $C(\mathbb{T})$ has an absolutely convergent Fourier series and $\varphi(z) \neq 0$ for z in $\mathbb{T}$, then $1/\varphi$ has an absolutely convergent Fourier series.

Proof   By hypothesis there exists f in $l^1(\mathbb{Z})$ such that $\Gamma f = \varphi$. Moreover, it follows from Theorem 2.35 that $\varphi(z) \neq 0$ for z in the maximal ideal space $\mathbb{T}$ of $l^1(\mathbb{Z})$ implies that f is invertible in $l^1(\mathbb{Z})$. If $g = f^{-1}$, then

$$1 = \Gamma(e_0) = \Gamma(g \circ f) = \Gamma g \cdot \varphi \qquad \text{or} \qquad \frac{1}{\varphi} = \Gamma g.$$

Therefore, $1/\varphi$ has an absolutely convergent Fourier series and the proof is complete.   ∎

2.60   Example   We conclude this chapter with an example of a commutative Banach algebra for which the Gelfand transform is as nice as possible, namely an isometric isomorphism onto the space of all continuous functions on the maximal ideal space.

In Section 1.44 we showed that L^∞ is a Banach space. If f and g are elements of L^∞, then the pointwise product is well defined, is in L^∞, and $\|fg\|_\infty \leqslant$

$\|f\|_\infty \|g\|_\infty$. (That is, $\mathcal{N}^\infty$ is an ideal in $\mathcal{L}^\infty$.) Thus L^∞ is a commutative Banach algebra. Although it is not at all obvious, L^∞ is isometrically isomorphic to $C(Y)$ for some compact Hausdorff space Y. We prove this after determining the spectrum of an element of L^∞. For this we need the following notion of range for a measurable function.

2.61 Definition If f is a measurable function on X, then the essential range $\mathcal{R}(f)$ of f is the set of all λ in $\mathbb{C}$ for which $\{x \in X : |f(x) - \lambda| < \varepsilon\}$ has positive measure for every $\varepsilon > 0$.

2.62 Lemma If f is in L^∞, then $\mathcal{R}(f)$ is a compact subset of $\mathbb{C}$ and $\|f\|_\infty = \sup\{|\lambda| : \lambda \in \mathcal{R}(f)\}$.

Proof If λ_0 is not in $\mathcal{R}(f)$, then there exists $\varepsilon > 0$ such that the set $\{x \in X : |f(x) - \lambda_0| < \varepsilon\}$ has measure zero. Clearly, then each λ in the open disk of radius ε about λ_0 fails to be in the essential range of f. Therefore, the complement of $\mathcal{R}(f)$ is open and hence $\mathcal{R}(f)$ is closed. If λ_0 in $\mathbb{C}$ is such that $\lambda_0 \geqslant |f(x)| + \delta$ for almost all x in X, then the set $\{x \in X : |f(x) - \lambda_0| < \delta/2\}$ has measure zero, and hence $\sup\{|\lambda| : \lambda \in \mathcal{R}(f)\} \leqslant \|f\|_\infty$. Thus $\mathcal{R}(f)$ is a compact subset of $\mathbb{C}$ for f in L^∞.

Now suppose f is in L^∞ and no λ satisfying $|\lambda| = \|f\|_\infty$ is in $\mathcal{R}(f)$. Then about every such λ there is an open disk D_λ of radius δ_λ such that the set $\{x \in X : |f(x) - \lambda| < \delta_\lambda\}$ has measure zero. Since the circle $\{\lambda \in \mathbb{C} : |\lambda| = \|f\|_\infty\}$ is compact, there exists a finite subcover of open disks $D_{\lambda_1}, D_{\lambda_2}, ..., D_{\lambda_n}$ such that the sets $\{x \in X : f(x) \in D_{\lambda_i}\}$ have measure zero. Then the set $\{x \in X : f(x) \in \bigcup_{i=1}^n D_{\lambda_i}\}$ has measure zero, which implies that there exists an $\varepsilon > 0$ such that the set $\{x \in X : |f(x)| > \|f\|_\infty - \varepsilon\}$ has measure zero. This contradiction completes the proof. ∎

2.63 Lemma If f is in L^∞, then $\sigma(f) = \mathcal{R}(f)$.

Proof If λ is not in $\sigma(f)$, then $1/(f - \lambda)$ is essentially bounded, which implies that the set $\{x \in X : |f(x) - \lambda| < 1/2\|f - \lambda\|_\infty\}$ has measure zero. Conversely, if $\{x \in X : |f(x) - \lambda| < \delta\}$ has measure zero for some $\delta > 0$, then $1/(f - \lambda)$ is essentially bounded by $1/\delta$, and hence λ is not in $\sigma(f)$. Therefore $\sigma(f) = \mathcal{R}(f)$. ∎

2.64 Theorem If M is the maximal ideal space of L^∞, then the Gelfand transform Γ is an isometrical isomorphism of L^∞ onto $C(M)$. Moreover, $\overline{\Gamma f} = \Gamma(\bar{f})$ for f in L^∞.

Proof We show first that Γ is an isometry. For f in L^∞ we have, combining the previous lemma and Corollary 2.36, that range $\Gamma f = \mathscr{R}(f)$, and hence

$$\|\Gamma(f)\|_\infty = \sup\{|\lambda| : \lambda \in \text{range } \Gamma f\} = \sup\{|\lambda| : \lambda \in \mathscr{R}(f)\} = \|f\|_\infty.$$

Therefore Γ is an isometry and $\Gamma(L^\infty)$ is a closed subalgebra of $C(M)$.

For f in L^∞ set $f = f_1 + if_2$, where each of f_1 and f_2 is real valued. Since the essential range of a real function is real and range $\Gamma f_1 = \mathscr{R}(f_1)$ and range $\Gamma f_2 = \mathscr{R}(f_2)$, we have

$$\overline{\Gamma f} = \overline{\Gamma f_1} + \overline{i\Gamma f_2} = \Gamma f_1 - i\Gamma f_2 = \Gamma(\bar{f}).$$

Therefore ΓL^∞ is a closed self-adjoint subalgebra of $C(M)$. Since it obviously separates points and contains the constant functions, we have by the Stone–Weierstrass theorem that $\Gamma L^\infty = C(M)$ and the theorem is proved. ∎

Whereas in preceding examples we computed the maximal ideal space, in this case the maximal ideal space is a highly pathological space having $2^{2^{\aleph_0}}$ points. We shall have reason to make use of certain properties of this space later on.

2.65 It can be easily verified that $l^\infty(\mathbb{Z}^+)$ (Section 1.15) is also a Banach algebra with respect to pointwise multiplication. It will follow from one of the problems that the Gelfand transform is an onto isometrical isomorphism in this case also. The maximal ideal space of $l^\infty(\mathbb{Z}^+)$ is denoted $\beta\mathbb{Z}^+$ and is called the Stone–Čech compactification of $\mathbb{Z}^+$.

Notes

The elementary theory of commutative Banach algebras is due to Gelfand [41] but the model provided by Wiener's theory of generalized harmonic analysis should be mentioned. Further results can be found in the treatises of Gelfand, Raikov and Šilov [42], Naimark [80], and Rickart [89]. The determination of the self-adjoint subalgebras of $C(X)$ including the generalization of the Weierstrass approximation theorem was made by Stone [105]. The literature on function algebras is quite extensive but two excellent sources are the books of Browder [10] and Gamelin [40].

Exercises

2.1 Let $\mathscr{D} = \{f \in C([0,1]) : f' \in C([0,1])\}$ and define $\|f\|_d = \|f\|_\infty + \|f'\|_\infty$. Show that $\mathscr{D}$ is a Banach algebra and that the Gelfand transform is neither isometric nor onto.

2.2   Let X be a compact Hausdorff space, K be a closed subset of X, and

$$\mathfrak{I} = \{f \in C(X) : f(x) = 0 \text{ for } x \in K\}.$$

Show that $\mathfrak{I}$ is a closed ideal in $C(X)$. Show further that every closed ideal in $C(X)$ is of this form. In particular, every closed ideal in $C(X)$ is the intersection of the maximal ideals which contain it.*

2.3   Show that every closed ideal in $\mathscr{D}$ is not the intersection of the maximal ideals which contain it.

2.4   Let $\mathscr{X}$ be a Banach space and $\mathfrak{L}(\mathscr{X})$ be the collection of bounded linear operators on $\mathscr{X}$. Show that $\mathfrak{L}(\mathscr{X})$ is a Banach algebra.

2.5   Show that if $\mathscr{X}$ is a finite (>1) dimensional Banach space, then the only multiplicative linear functional on $\mathfrak{L}(\mathscr{X})$ is the zero functional.

Definition   An element T of $\mathfrak{L}(\mathscr{X})$ is finite rank if the range of T is finite dimensional.

2.6   Show that if $\mathscr{X}$ is a Banach space, then the finite rank operators form a two-sided ideal in $\mathfrak{L}(\mathscr{X})$ which is contained in every proper two-sided ideal.

2.7   If f is a continuous function on $[0, 1]$ show that the range of f is the essential range of f.

2.8   Let f be a bounded real-valued function on $[0, 1]$ continuous except at the point $\frac{1}{2}$. Let $\mathfrak{A}$ be the uniformly closed algebra generated by f and $C([0\,1])$. Determine the maximal ideal space of $\mathfrak{A}$.*

2.9   If X is a compact Hausdorff space, then $C(X)$ is the closed linear span of the idempotent functions in $C(X)$ if and only if X is totally disconnected.

2.10   Show that the maximal ideal space of L^∞ is totally disconnected.

2.11   Let X be a completely regular Hausdorff space and $B(X)$ be the space of bounded continuous functions on X. Show that $B(X)$ is a commutative Banach algebra in the supremum norm. If βX denotes the maximal ideal space of $B(X)$, then the Gelfand transform is an isometrical isomorphism of $B(X)$ onto $C(\beta X)$ which preserves conjugation. Moreover, there exists a natural embedding β of X into βX. The space βX is the Stone–Čech compactification of X.

2.12   Let X be a completely regular Hausdorff space, Y be a compact Hausdorff space, and φ be a continuous one-to-one mapping of X onto a dense subset of Y. Show that there exists a continuous mapping ψ from βX onto Y such that $\rho = \psi \circ \varphi$. (Hint: Consider the restriction of the functions in $C(Y)$ as forming a subalgebra of $C(\beta X)$.)

2.13   Let $\mathfrak{B}$ be a commutative Banach algebra and

$$\mathfrak{R} = \{ f \in \mathfrak{B} : 1 + \lambda f \in \mathscr{G} \text{ for } \lambda \in \mathbb{C} \}.$$

Show that $\mathfrak{R}$ is a closed ideal in $\mathfrak{B}$.

Definition   If $\mathfrak{B}$ is a commutative Banach algebra, then

$$\mathfrak{R} = \{ f \in \mathfrak{B} : 1 + \lambda f \in \mathscr{G} \text{ for } \lambda \in \mathbb{C} \}$$

is the radical of $\mathfrak{B}$ and $\mathfrak{B}$ is said to be semisimple if $\mathfrak{R} = \{0\}$.

2.14   If $\mathfrak{B}$ is a commutative Banach algebra, then $\mathfrak{R}$ is the intersection of the maximal ideals in $\mathfrak{B}$.

2.15   If $\mathfrak{B}$ is a commutative Banach algebra, then $\mathfrak{B}$ is semisimple if and only if the Gelfand transform is one-to-one.

2.16   If $\mathfrak{B}$ is a commutative Banach algebra, then $\mathfrak{B}/\mathfrak{R}$ is semisimple.

2.17   Show that $L^1([0,1]) \oplus \mathbb{C}$ with the 1-norm (see Exercise 1.17) is a commutative Banach algebra for the multiplication defined by

$$[(f \oplus \lambda)(g \oplus \mu)](t) = \left\{ \mu f(t) + \lambda g(t) + \int_0^t f(t-x) g(x) \, dx \right\} \oplus \lambda \mu.$$

Show that $L^1([0,1]) \oplus \mathbb{C}$ is not semisimple.

2.18   (*Riesz Functional Calculus*)   Let $\mathfrak{B}$ be a commutative Banach algebra, x be an element of $\mathfrak{B}$, Ω be an open set in $\mathbb{C}$ containing $\sigma(x)$, and Λ be a finite collection of rectifiable simple closed curves contained in Ω such that Λ forms the boundary of an open subset of $\mathbb{C}$ which contains $\sigma(x)$. Let $A(\Omega)$ denote the algebra of complex holomorphic functions on Ω. Show that the mapping

$$\varphi \to \int_\Lambda \varphi(z)(x-z)^{-1} \, dz$$

defines a homomorphism from $A(\Omega)$ to $\mathfrak{B}$ such that $\sigma(\varphi(x)) = \varphi(\sigma(x))$ for φ in $A(\Omega)$.

2.19   If $\mathfrak{B}$ is a commutative Banach algebra, x is an element of $\mathfrak{B}$ with $\sigma(x) \subset \Omega$, and there exists a nonzero φ in $A(\Omega)$ such that $\varphi(x) = 0$, then $\sigma(x)$ is finite. Show that there exists a polynomial $p(z)$ such that $p(x) = 0$.

2.20   Show that for no constant M is it true that $\sum_{n=-N}^{N} |a_n| \leq M \|p\|_\infty$ for all trigonometric polynomials $p = \sum_{n=-N}^{N} a_n \chi_n$ on $\mathbb{T}$.

2.21 Show that the assumption that every continuous function on $\mathbb{T}$ has an absolutely convergent Fourier series implies that the Gelfand transform on $l^1(\mathbb{Z})$ is invertible, and hence conclude in view of the preceding problem that there exists a continuous function whose Fourier series does not converge absolutely.*

Definition If $\mathfrak{B}$ is a Banach algebra, then an automorphism on $\mathfrak{B}$ is a continuous isomorphism from $\mathfrak{B}$ onto $\mathfrak{B}$. The collection of all automorphisms on $\mathfrak{B}$ is denoted $\text{Aut}(\mathfrak{B})$.

2.22 If X is a compact Hausdorff space, then every isomorphism from $C(X)$ onto $C(X)$ is continuous.

2.23 If X is a compact Hausdorff space and φ is a homeomorphism on X, then $(\Phi f)(x) = f(\varphi x)$ defines an automorphism Φ in $\text{Aut}[C(X)]$. Show that the mapping $\varphi \to \Phi$ defines an isomorphism between the group $\text{Hom}(X)$ of homeomorphisms on X and $\text{Aut}[C(X)]$.

2.24 If $\mathfrak{A}$ is a function algebra with maximal ideal space M, then there is a natural isomorphism of $\text{Aut}(\mathfrak{A})$ into $\text{Hom}(M)$.

2.25 If A is the disk algebra with maximal ideal space the closed unit disk, then the range of $\text{Aut}(A)$ in $\text{Hom}(\mathbb{D})$ is the group of fractional linear transformations on $\mathbb{D}$, that is, the maps $z \to \beta(z-\alpha)/(1-\bar{\alpha}z)$ for complex numbers α and β satisfying $|\alpha| < 1$ and $|\beta| = 1$.*

Definition If $\mathfrak{A}$ is a function algebra contained in $C(X)$, then a closed subset M of X is a boundary for $\mathfrak{A}$ if $\|f\|_\infty = \sup\{|f(m)| : m \in M\}$ for f in $\mathfrak{A}$.

2.26 If $\mathfrak{A}$ is a function algebra contained in $C(X)$; M is a boundary for $\mathfrak{A}$; $f_1,\ldots,f_n$ are functions in $\mathfrak{A}$; and U is the open subset of M defined by

$$\{x \in X : |f_i(x)| < 1 \text{ for } i = 1, 2, \ldots, n\}.$$

Show either that $M\backslash U$ is a boundary for $\mathfrak{A}$ or U intersects every boundary for $\mathfrak{A}$.

2.27 (Šilov) If $\mathfrak{A}$ is a function algebra, then the intersection of all boundaries for $\mathfrak{A}$ is a boundary (called the Šilov boundary for $\mathfrak{A}$).*

2.28 Give a functional analytic proof of the maximum modulus principle for the functions in the disk algebra A. (Hint: Show that

$$\hat{f}(re^{i\theta}) = \frac{1}{2\pi} \int_0^{2\pi} k_r(\theta - t) f(e^{i\theta}) \, dt,$$

where the function $k_r(t) = \sum_{n=-\infty}^{\infty} r^n e^{int}$ is positive.)

2.29 Show that the Šilov boundary for the disk algebra A is the unit circle.

2.30 Show that the abstract index group for a commutative Banach algebra contains no elements of finite order.

2.31 If $\mathfrak{B}_1$ and $\mathfrak{B}_2$ are Banach algebras, then $\mathfrak{B}_1 \hat{\otimes} \mathfrak{B}_2$ and $\mathfrak{B}_1 \bar{\otimes} \mathfrak{B}_2$ are Banach algebras. Moreover, if $\mathfrak{B}_1$ and $\mathfrak{B}_2$ are commutative with maximal ideal spaces M_1 and M_2, respectively, then $M_1 \times M_2$ is the maximal ideal space of both $\mathfrak{B}_1 \hat{\otimes} \mathfrak{B}_2$ and $\mathfrak{B}_1 \bar{\otimes} \mathfrak{B}_2$ (see Exercises 1.36 and 1.37 for the definition of $\hat{\otimes}$ and $\bar{\otimes}$.)

2.32 If $\mathfrak{B}$ is an algebra over $\mathbb{C}$, which has a norm making it into a Banach space such that $\|fg\| \leqslant \|f\| \|g\|$ for f and g in $\mathfrak{B}$, then $\mathfrak{B} \oplus \mathbb{C}$ is a Banach algebra in the 1-norm (see Exercise 1.17) for the multiplication

$$(f \oplus \lambda)(g \oplus \mu) = (fg + \lambda g + \mu f) \oplus \lambda \mu$$

with identity $0 \oplus 1$.

2.33 If φ is a multiplicative linear functional on $\mathfrak{B}$, then φ has a unique extension to an element of $M_{\mathfrak{B} \oplus \mathbb{C}}$. Moreover, the collection of nonzero multiplicative linear functionals on $\mathfrak{B}$ is a locally compact Hausdorff space.

2.34 Show that $L^1(\mathbb{R})$ is a commutative Banach algebra without identity for the multiplication defined by

$$(f \circ g)(x) = \int_{-\infty}^{\infty} f(x-t) g(t) \, dt \qquad \text{for } f \text{ and } g \text{ in } L^1(\mathbb{R}).$$

2.35 Show that for t in $\mathbb{R}$ the linear function on $L^1(\mathbb{R})$ defined by

$$\varphi_t(f) = \int_{-\infty}^{\infty} f(x) e^{ixt} \, dx$$

is multiplicative. Conversely, every nonzero multiplicative linear functional in $L^1(\mathbb{R})$ is of this form.* (Hint: Every bounded linear functional on $L^1(\mathbb{R})$ is given by a φ in $L^\infty(\mathbb{R})$. Show for f and g in $L^1(\mathbb{R})$ that

$$\int_{-\infty}^{\infty} \int_{-\infty}^{\infty} f(x-t) g(t) \{\varphi(x-t)\varphi(t) - \varphi(x)\} \, dt \, dx = 0$$

and that implies $\varphi(x-t)\varphi(t) = \varphi(x)$ for (x,t) not in a planar set of Lebesgue measure.)

2.36 Show that the maximal ideal space of $L^1(\mathbb{R})$ is homeomorphic to $\mathbb{R}$ and that the Gelfand transform coincides with the Fourier transform.

3 *Geometry of Hilbert Space*

3.1 The notion of Banach space abstracts many of the important properties of finite-dimensional linear spaces. The geometry of a Banach space can, however, be quite different from that of Euclidean n-space; for example, the unit ball of a Banach space may have corners, and closed convex sets need not possess a unique vector of smallest norm. The most important geometrical property absent in general Banach spaces is a notion of perpendicularity or orthogonality.

In the study of analytic geometry we recall that the orthogonality of two vectors was determined analytically by considering their inner (or dot) product. In this chapter we introduce the abstract notion of an inner product and show how a linear space equipped with an inner product can be made into a normed linear space. If the linear space is complete in the metric defined by this norm, then it is said to be a Hilbert space. This chapter is devoted to studying the elementary geometry of Hilbert spaces and to showing that such spaces possess many of the more pleasant properties of Euclidean n-space. We will show, in fact, that a finite dimensional Hilbert space is isomorphic to Euclidean n-space for some integer n.

3.2 Definition An inner product on a complex linear space $\mathscr{L}$ is a function φ from $\mathscr{L} \times \mathscr{L}$ to $\mathbb{C}$ such that:

    (1) $\varphi(\alpha_1 f_1 + \alpha_2 f_2, g) = \alpha_1 \varphi(f_1, g) + \alpha_2 \varphi(f_2, g)$ for α_1, α_2 in $\mathbb{C}$ and f_1, f_2, g in $\mathscr{L}$;

(2) $\varphi(f, \beta_1 g_1 + \beta_2 g_2) = \bar{\beta}_1 \varphi(f, g_1) + \bar{\beta}_2 \varphi(f, g_2)$ for β_1, β_2 in $\mathbb{C}$ and f, g_1, g_2 in $\mathscr{L}$;

(3) $\varphi(f, g) = \overline{\varphi(g, f)}$ for f and g in $\mathscr{L}$; and

(4) $\varphi(f, f) \geqslant 0$ for f in $\mathscr{L}$ and $\varphi(f, f) = 0$ if and only if $f = 0$.

A linear space equipped with an inner product is said to be an inner product space.

The following lemma contains a useful polarization identity, the importance of which lies in the fact that the value of the inner product φ is expressed solely in terms of the values of the associated quadratic form ψ defined by $\psi(f) = \varphi(f, f)$ for f in $\mathscr{L}$.

3.3   Lemma   If $\mathscr{L}$ is an inner product space with the inner product φ, then

$$\varphi(f, g) = \tfrac{1}{4}\{\varphi(f+g, f+g) - \varphi(f-g, f-g) + i\varphi(f+ig, f+ig)$$
$$- i\varphi(f-ig, f-ig)\}$$

for f and g in $\mathscr{L}$.

Proof   Compute.   ∎

An inner product is usually denoted $(\,,\,)$, that is, $(f, g) = \varphi(f, g)$ for f and g in $\mathscr{L}$.

3.4   Definition   If $\mathscr{L}$ is an inner product space, then the norm $\|\ \|$ on $\mathscr{L}$ associated with the inner product is defined by $\|f\| = (f, f)^{1/2}$ for f in $\mathscr{L}$.

The following inequality is basic in the study of inner product spaces. We show that the norm just defined has the required properties of a norm after the proof of this inequality.

3.5   Proposition (Cauchy–Schwarz Inequality)   If f and g are in the inner product space $\mathscr{L}$, then

$$|(f, g)| \leqslant \|f\|\,\|g\|.$$

Proof   For f and g in $\mathscr{L}$ and λ in $\mathbb{C}$, we have

$$|\lambda|^2 \|g\|^2 + 2\operatorname{Re}[\bar{\lambda}(f, g)] + \|f\|^2 = (f+\lambda g, f+\lambda g)$$
$$= \|f+\lambda g\|^2 \geqslant 0.$$

Setting $\lambda = te^{i\theta}$, where t is real and $e^{i\theta}$ is chosen such that $e^{-i\theta}(f, g) \geqslant 0$, we obtain the inequality

$$\|g\|^2 t^2 + 2|(f, g)|\,t + \|f\|^2 \geqslant 0.$$

Hence the quadratic equation

$$\|g\|^2 t^2 + 2|(f,g)|t + \|f\|^2 = 0$$

in t has at most one real root, and therefore its discriminant must be non-positive. Substituting we obtain

$$[2|(f,g)|]^2 - 4\|g\|^2 \|f\|^2 \leqslant 0,$$

from which the desired inequality follows. ∎

3.6 Observe that the property $(f,f) = 0$ implies that $f = 0$ was not needed in the preceding proof.

3.7 Proposition If $\mathscr{L}$ is an inner product space, then $\|\cdot\|$ defines a norm on $\mathscr{L}$.

Proof We must verify properties (1)–(3) of Definition 1.3. The fact that $\|f\| = 0$ if and only if $f = 0$ is immediate from (4) (Definition 3.2) and thus (1) holds.

Since

$$\|\lambda f\| = (\lambda f, \lambda f)^{\frac{1}{2}} = (\lambda \bar{\lambda}(f,f))^{\frac{1}{2}} = |\lambda| \|f\| \qquad \text{for } \lambda \text{ in } \mathbb{C} \text{ and } f \text{ in } \mathscr{L},$$

we see that (2) holds. Lastly, using the Cauchy–Schwarz inequality, we have

$$\begin{aligned}
\|f+g\|^2 &= (f+g, f+g) = (f,f) + (f,g) + (g,f) + (g,g) \\
&= \|f\|^2 + \|g\|^2 + 2\operatorname{Re}(f,g) \leqslant \|f\|^2 + \|g\|^2 + 2|(f,g)| \\
&\leqslant \|f\|^2 + \|g\|^2 + 2\|f\| \|g\| \leqslant (\|f\| + \|g\|)^2
\end{aligned}$$

for f and g in $\mathscr{L}$. Thus (3) holds and $\|\cdot\|$ is a norm. ∎

3.8 Proposition In an inner product space, the inner product is continuous.

Proof Let $\mathscr{L}$ be an inner product space and $\{f_\alpha\}_{\alpha \in A}$ and $\{g_\alpha\}_{\alpha \in A}$ be nets in $\mathscr{L}$ such that $\lim_{\alpha \in A} f_\alpha = f$ and $\lim_{\alpha \in A} g_\alpha = g$. Then

$$\begin{aligned}
|(f,g) - (f_\alpha, g_\alpha)| &\leqslant |(f - f_\alpha, g)| + |(f_\alpha, g - g_\alpha)| \\
&\leqslant \|f - f_\alpha\| \|g\| + \|f_\alpha\| \|g - g_\alpha\|,
\end{aligned}$$

and hence $\lim_{\alpha \in A} (f_\alpha, g_\alpha) = (f,g)$. ∎

3.9 Definition In the inner product space $\mathscr{L}$ two vectors f and g are said to be orthogonal, denoted $f \perp g$, if $(f,g) = 0$. A subset $\mathscr{S}$ of $\mathscr{L}$ is said to be orthogonal if $f \perp g$ for f and g in $\mathscr{S}$ and orthonormal if, in addition, $\|f\| = 1$ for f in $\mathscr{S}$.

This notion of orthogonality generalizes the usual one in Euclidean space. It is now possible to extend various theorems from Euclidean geometry to inner product spaces. We give two that will be useful. The first is the familiar Pythagorean theorem, while the second is the result relating the lengths of the sides of a parallelogram to the lengths of the diagonals.

3.10   Proposition (Pythagorean Theorem)   If $\{f_1, f_2, \ldots, f_n\}$ is an orthogonal subset of the inner product space $\mathscr{L}$, then

$$\left\| \sum_{i=1}^{n} f_i \right\|^2 = \sum_{i=1}^{n} \|f_i\|^2.$$

Proof   Computing, we have

$$\left\| \sum_{i=1}^{n} f_i \right\|^2 = \left(\sum_{i=1}^{n} f_i, \sum_{j=1}^{n} f_j \right) = \sum_{i=1}^{n} (f_i, f_i) + \sum_{\substack{i,j=1 \\ i \neq j}}^{n} (f_i, f_j)$$

$$= \sum_{i=1}^{n} (f_i, f_i) = \sum_{i=1}^{n} \|f_i\|^2. \qquad \blacksquare$$

3.11   Proposition (Parallelogram Law)   If f and g are in the inner product space $\mathscr{L}$, then

$$\|f+g\|^2 + \|f-g\|^2 = 2\|f\|^2 + 2\|g\|^2$$

Proof   Expand the left-hand side in terms of inner products. $\blacksquare$

As in the case of normed linear spaces the deepest results are valid only if the space is complete in the metric induced by the norm.

3.12   Definition   A Hilbert space is a complex linear space which is complete in the metric induced by the norm.

In particular, a Hilbert space is a Banach space.

3.13   Examples   We now consider some examples of Hilbert spaces.

For n a positive integer let $\mathbb{C}^n$ denote the collection of complex ordered n-tuples $\{x : x = (x_1, x_2, \ldots, x_n), x_i \in \mathbb{C}\}$. Then $\mathbb{C}^n$ is a complex linear space for the coordinate-wise operations. Define the inner product $(\,,\,)$ on $\mathbb{C}^n$ such that $(x, y) = \sum_{i=1}^{n} x_i \bar{y}_i$. The properties of an inner product are easily verified and the associated norm is the usual Euclidean norm $\|x\|_2 = (\sum_{i=1}^{n} |x_i|^2)^{\frac{1}{2}}$. To verify completeness suppose $\{x^k\}_{k=1}^{\infty}$ is a Cauchy sequence in $\mathbb{C}^n$. Then since $|x_i^k - x_i^m| \leqslant \|x^k - x^m\|_2$, it follows that $\{x_i^k\}_{k=1}^\infty$ is a Cauchy sequence in $\mathbb{C}$ for $1 \leqslant i \leqslant n$. If we set $x = (x_1, x_2, \ldots, x_k)$, where $x_i = \lim_{k \to \infty} x_i^k$, then x is in $\mathbb{C}^n$ and $\lim_{k \to \infty} x^k = x$ in the norm of $\mathbb{C}^n$. Thus $\mathbb{C}^n$ is a Hilbert space.

The space $\mathbb{C}^n$ is the complex analog of real Euclidean n-space. We show later in this chapter, in a sense to be made precise, that the $\mathbb{C}^n$'s are the only finite-dimensional Hilbert spaces.

3.14   We next consider the "union" of the $\mathbb{C}^n$'s. Let $\mathscr{L}$ be the collection of complex functions on $\mathbb{Z}^+$ which take only finitely many nonzero values. With respect to pointwise addition and scalar multiplication, $\mathscr{L}$ is a complex linear space. Moreover, $(f, g) = \sum_{n=0}^\infty f(n) \overline{g(n)}$ defines an inner product on $\mathscr{L}$, where the sum converges, since all but finitely many terms are zero. Is $\mathscr{L}$ a Hilbert space? It is if $\mathscr{L}$ is complete with respect to the metric induced by the norm $\|f\|_2 = (\sum_{n=0}^\infty |f(n)|^2)^{1/2}$. Consider the sequence $\{f_k\}_{k=1}^\infty$ contained in $\mathscr{L}$, where

$$f_k(n) = \begin{cases} (\tfrac{1}{2})^n & n \leqslant k, \\ 0 & n > k. \end{cases}$$

One can easily show that $\{f_k\}_{k=1}^\infty$ is Cauchy but does not converge to an element of $\mathscr{L}$. We leave this as an exercise for the reader. Thus $\mathscr{L}$ is not a Hilbert space.

3.15   The space $\mathscr{L}$ is not a Hilbert space because it is not large enough. Let us enlarge it to obtain our first example of an infinite-dimensional Hilbert space. (This example should be compared to Example 1.15.)

Let $l^2(\mathbb{Z}^+)$ denote the collection of all complex functions φ on $\mathbb{Z}^+$ such that $\sum_{n=0}^\infty |\varphi(n)|^2 < \infty$. Then $l^2(\mathbb{Z}^+)$ is a complex linear space, since

$$|(f+g)(n)|^2 \leqslant 2|f(n)|^2 + 2|g(n)|^2.$$

For f and g in $l^2(\mathbb{Z}^+)$, define $(f, g) = \sum_{n=0}^\infty f(n) \overline{g(n)}$. Does this make sense, that is, does the sum converge? For each N in $\mathbb{Z}^+$, the n-tuples

$$F_N = (|f(0)|, |f(1)|, \ldots, |f(N)|) \quad \text{and} \quad G_N = (|g(0)|, |g(1)|, \ldots, |g(N)|)$$

lie in $\mathbb{C}^N$. Applying the Cauchy–Schwarz inequality, we have

$$\sum_{n=0}^N |f(n) \overline{g(n)}| = |(F_N, G_N)| \leqslant \|F_N\| \|G_N\|$$

$$= \left(\sum_{n=0}^N |f(n)|^2 \right)^{1/2} \left(\sum_{n=0}^N |g(n)|^2 \right)^{1/2} \leqslant \|f\|_2 \|g\|_2.$$

Thus the series $\sum_{n=0}^{\infty} f(n)\overline{g(n)}$ converges absolutely. That $(,)$ is an inner product follows easily.

To establish the completeness of $l^2(\mathbb{Z}^+)$ in the metric given by the norm $\|\ \|_2$, suppose $\{f^k\}_{k=1}^{\infty}$ is a Cauchy sequence in $l^2(\mathbb{Z}^+)$. Then for each n in $\mathbb{Z}^+$, we have

$$|f^k(n) - f^j(n)| \leqslant \|f^k - f^j\|_2$$

and hence $\{f^k(n)\}_{k=1}^{\infty}$ is a Cauchy sequence in $\mathbb{C}$ for each n in $\mathbb{Z}^+$. Define the function f on $\mathbb{Z}^+$ to be $f(n) = \lim_{k\to\infty} f^k(n)$. Two things must be shown: that f is in $l^2(\mathbb{Z}^+)$ and that $\lim_{k\to\infty} \|f - f^k\|_2 = 0$. Since $\{f^k\}_{k=1}^{\infty}$ is a Cauchy sequence, there exists an integer K such that for $k \geqslant K$ we have $\|f^k - f^K\|_2 \leqslant 1$. Thus we obtain

$$\left\{\sum_{n=0}^{N} |f(n)|^2\right\}^{1/2} \leqslant \left\{\sum_{n=0}^{N} |f(n) - f^K(n)|^2\right\}^{1/2} + \left\{\sum_{n=0}^{N} |f^K(n)|^2\right\}^{1/2}$$

$$\leqslant \lim_{k\to\infty}\left\{\sum_{n=0}^{N} |f^k(n) - f^K(n)|^2\right\}^{1/2} + \left\{\sum_{n=0}^{N} |f^K(n)|^2\right\}^{1/2}$$

$$\leqslant \limsup_{k\to\infty} \|f^k - f^K\|_2 + \|f^K\|_2 \leqslant 1 + \|f^K\|_2,$$

and hence f is in $l^2(\mathbb{Z}^+)$. Moreover, given $\varepsilon > 0$, choose M such that $k, j \geqslant M$ implies $\|f^k - f^j\| < \varepsilon$. Then for $k \geqslant M$ and any N, we have

$$\sum_{n=0}^{N} |f(n) - f^k(n)|^2 = \lim_{j\to\infty} \sum_{n=0}^{N} |f^j(n) - f^k(n)|^2$$

$$\leqslant \limsup_{j\to\infty} \|f^j - f^k\|_2^2 \leqslant \varepsilon^2.$$

Since N is arbitrary, this proves that $\|f - f^k\|_2 \leqslant \varepsilon$ and therefore $l^2(\mathbb{Z}^+)$ is a Hilbert space.

3.16   The Space L^2   In Section 1.44 we introduced the Banach spaces L^1 and L^∞ based on a measure space $(X, \mathscr{S}, \mu)$. We now consider the corresponding L^2 space, which happens to be a Hilbert space.

We begin by letting $\mathscr{L}^2$ denote the set of all measurable complex functions f on X which satisfy $\int_X |f|^2 \, d\mu < \infty$. Since the inequality $|f + g|^2 \leqslant 2|f|^2 + 2|g|^2$ is valid for arbitrary functions f and g on X, we see that $\mathscr{L}^2$ is a linear space for pointwise addition and scalar multiplication. Let $\mathscr{N}^2$ be the subspace of functions f in $\mathscr{L}^2$ for which $\int_X |f|^2 \, d\mu = 0$, and let L^2 denote the quotient linear space $\mathscr{L}^2/\mathscr{N}^2$.

If f and g are in $\mathscr{L}^2$, then the identity

$$|f\bar{g}| = \tfrac{1}{2}\{(|f|+|g|)^2 - |f|^2 - |g|^2\}$$

shows that the function $f\bar{g}$ is integrable. If we define $\varphi(f,g) = \int_X f\bar{g}\, d\mu$ for f and g in $\mathscr{L}^2$, then φ has all the properties of an inner product except one; namely, $\varphi(f,f) = 0$ does not necessarily imply $f = 0$. By the remark following the proof of the Cauchy–Schwarz inequality, that inequality holds for φ. Thus, if f, f', g, and g' are functions in $\mathscr{L}^2$ such that $f-f'$ and $g-g'$ belong to $\mathscr{N}^2$, then

$$|\varphi(f,g) - \varphi(f',g')| \leqslant |\varphi(f-f',g)| + |\varphi(f',g-g')|$$

$$\leqslant \varphi(f-f',f-f')\,\varphi(g,g)$$

$$+ \varphi(f',f')\,\varphi(g-g',g-g') = 0.$$

Therefore, φ is a well-defined function on L^2. Moreover, if $\varphi([f],[f]) = 0$, then $\int_X |f|^2\, d\mu = 0$ and hence $[f] = [0]$. Thus φ is an inner product on L^2 and we will denote it from now on in the usual manner. Furthermore, the associated norm on L^2 is defined by $\|[f]\|_2 = (\int_X |f|^2\, d\mu)^{\frac{1}{2}}$. The only problem remaining before we can conclude that L^2 is a Hilbert space is the question of its completeness. This is slightly trickier than in the case of L^1.

We begin with a general inequality. Take f in $\mathscr{L}^2$ and define g on X such that $g(x) = f(x)/|f(x)|$ if $f(x) \neq 0$ and $g(x) = 1$ otherwise. Then g is measurable, $|g(x)| = 1$, and $f\bar{g} = |f|$. Moreover, applying the Cauchy–Schwarz inequality, we have

$$\|f\|_1 = \int_X |f|\, d\mu = \int_X f\bar{g}\, d\mu = |(f,g)| \leqslant \|f\|_2 \|g\|_2 = \|f\|_2.$$

Therefore, $\|f\|_1 \leqslant \|f\|_2$ for f in $\mathscr{L}^2$.

We now prove that L^2 is complete using Corollary 1.10. Let $\{[f_n]\}_{n=1}^{\infty}$ be a sequence in L^2 such that $\sum_{n=1}^{\infty} \|[f_n]\|_2 \leqslant M < \infty$. By the preceding inequality $\sum_{n=1}^{\infty} \|[f_n]\|_1 \leqslant M$, and hence, by the proof in Section 1.44, there exists f in $\mathscr{L}^1$ such that $\sum_{n=1}^{\infty} f_n(x) = f(x)$ for almost all x in X. Moreover

$$\int_X \left| \sum_{n=1}^{N} f_n \right|^2 d\mu \leqslant \int_X \left(\sum_{n=1}^{N} |f_n| \right)^2 d\mu = \left\| \left[\sum_{n=1}^{N} |f_n| \right] \right\|_2^2$$

$$\leqslant \left(\sum_{n=1}^{N} \|[f_n]\|_2 \right)^2 \leqslant M^2,$$

and since $\lim_{N \to \infty} |\sum_{n=1}^{N} f_n(x)|^2 = |f(x)|^2$ for almost all x in X, it follows from Fatou's lemma that $|f|^2$ is integrable and hence f is in $\mathscr{L}^2$. Moreover, since the sequence $\{(\sum_{n=1}^{N}|f_n|)^2\}_{N=1}^{\infty}$ is monotonically increasing, it follows that $k = \lim_{N\to\infty}(\sum_{n=1}^{N}|f_n|)^2$ is an integrable function. Therefore,

$$
\lim_{N\to\infty}\left\|[f] - \sum_{n=1}^{N}[f_n]\right\|_2 = \lim_{N\to\infty}\left(\int_X \left|f - \sum_{n=1}^{N}f_n\right|^2 d\mu\right)^{1/2}
$$

$$
= \lim_{N\to\infty}\left(\int_X \left|\sum_{n=N+1}^{\infty}f_n\right|^2 d\mu\right)^{1/2} = 0
$$

by the Lebesgue dominated convergence theorem, since $|\sum_{n=N+1}^{\infty}f_n|^2 \leqslant k$ for all N and $\lim_{N\to\infty}|\sum_{n=N+1}^{\infty}f_n(x)|^2 = 0$ for almost all x in X. Thus L^2 is a Hilbert space. Lastly, we henceforth adopt the convention stated in Section 1.44 for the elements of L^2; namely, we shall treat them as functions.

3.17   The Space H^2   Let $\mathbb{T}$ denote the unit circle, μ the normalized Lebesgue measure on $\mathbb{T}$, and $L^2(\mathbb{T})$ the Hilbert space defined with respect to μ. The corresponding Hardy space H^2 is defined as the closed subspace

$$
\left\{f \in L^2(\mathbb{T}): \frac{1}{2\pi}\int_0^{2\pi} f\chi_n\, dt = 0 \quad \text{for} \quad n = 1, 2, 3, \ldots\right\},
$$

where χ_n is the function $\chi_n(e^{it}) = e^{int}$. A slight variation of this definition is

$$
\{f \in L^2(\mathbb{T}): (f, \chi_n) = 0 \quad \text{for} \quad n = -1, -2, -3, \ldots\}.
$$

3.18   Whereas in Chapter 1 after defining a Banach space we proceeded to determine the conjugate space, this is unnecessary for Hilbert spaces since we show in this chapter that the conjugate space of a Hilbert space can be identified with the space itself. This will be the main result of this chapter.

We begin by extending a result on the distance to a convex set to subsets of Hilbert spaces. Although most proofs of this result for Euclidean spaces make use of the compactness of closed and bounded subsets, completeness actually suffices.

3.19   Theorem   If $\mathcal{K}$ is a nonempty, closed, and convex subset of the Hilbert space $\mathcal{H}$, then there exists a unique vector in $\mathcal{K}$ of smallest norm.

Proof   If $\delta = \inf\{\|f\|: f \in \mathcal{K}\}$, then there exists a sequence $\{f_n\}_{n=0}^{\infty}$ in $\mathcal{K}$ such that $\lim_{n\to\infty}\|f_n\| = \delta$. Applying the parallelogram law to the vectors $f_n/2$ and $f_m/2$, we obtain

$$
\left\|\frac{f_n - f_m}{2}\right\|^2 = 2\left\|\frac{f_n}{2}\right\|^2 + 2\left\|\frac{f_m}{2}\right\|^2 - \left\|\frac{f_n + f_m}{2}\right\|^2.
$$

Since $\mathcal{K}$ is convex, $(f_n + f_m)/2$ is in $\mathcal{K}$ and hence $\|(f_n + f_m)/2\|^2 \geqslant \delta^2$. Therefore, we have $\|f_n - f_m\|^2 \leqslant 2\|f_n\|^2 + 2\|f_m\|^2 - 4\delta^2$, which implies

$$\limsup_{n,m\to\infty} \|f_n - f_m\|^2 \leqslant 2\delta^2 + 2\delta^2 - 4\delta^2 = 0.$$

Thus $\{f_n\}_{n=1}^{\infty}$ is a Cauchy sequence in $\mathcal{K}$ and from the completeness of $\mathcal{H}$ and the fact that $\mathcal{K}$ is a closed subset of $\mathcal{H}$ we obtain a vector f in $\mathcal{K}$ such that $\lim_{n\to\infty} f_n = f$. Moreover, since the norm is continuous, we have $\|f\| = \lim_{n\to\infty} \|f_n\| = \delta$.

Having proved the existence of a vector in $\mathcal{K}$ of smallest norm we now consider its uniqueness. Suppose f and g are in $\mathcal{K}$ with $\|f\| = \|g\| = \delta$. Again using the parallelogram law, we have

$$\left\|\frac{f-g}{2}\right\|^2 = 2\left\|\frac{f}{2}\right\|^2 + 2\left\|\frac{g}{2}\right\|^2 - \left\|\frac{f+g}{2}\right\|^2 \leqslant \frac{\delta^2}{2} + \frac{\delta^2}{2} - \delta^2 = 0,$$

since $\|(f+g)/2\| \geqslant \delta$. Therefore, $f = g$ and uniqueness is proved. ∎

If π is a plane and l is a line in three-space perpendicular to π and both π and l contain the origin, then each vector in the space can be written uniquely as the sum of a vector which lies in π and a vector in the direction of l. We extend this idea to subspaces of a Hilbert space in the theorem following the definition.

3.20   Definition   If $\mathcal{M}$ is a subset of the Hilbert space $\mathcal{H}$, then the orthogonal complement of $\mathcal{M}$, denoted $\mathcal{M}^\perp$, is the set of vectors in $\mathcal{H}$ orthogonal to every vector in $\mathcal{M}$.

Clearly $\mathcal{M}^\perp$ is a closed subspace of $\mathcal{H}$, possibly consisting of just the zero vector. However, if $\mathcal{M}$ is not the subspace $\{0\}$ consisting of the zero vector alone, then $\mathcal{M}^1 \neq \mathcal{H}$.

3.21   Theorem   If $\mathcal{M}$ is a closed subspace of the Hilbert space $\mathcal{H}$ and f is a vector in $\mathcal{H}$, then there exist unique vectors g in $\mathcal{M}$ and h in $\mathcal{M}^\perp$ such that $f = g + h$.

Proof   If we set $\mathcal{K} = \{f - k : k \in \mathcal{M}\}$, then $\mathcal{K}$ is a nonempty, closed and convex subset of $\mathcal{H}$. Let h be the unique element of $\mathcal{K}$ with smallest norm whose existence is given by the previous theorem. If k is a unit vector in $\mathcal{M}$, then $h - (h, k)k$ is in $\mathcal{K}$, and hence

$$\|h\|^2 \leqslant \|h - (h,k)k\|^2 = \|h\|^2 - \overline{(h,k)}(h,k) - (h,k)\overline{(h,k)} + (h,k)\overline{(h,k)}\|k\|^2.$$

Therefore, $|(h,k)|^2 \leqslant 0$, which implies $(h,k) = 0$, and hence h is in $\mathcal{M}^\perp$. Since h is in $\mathcal{K}$, there exists g in $\mathcal{M}$ such that $f = g + h$ and the existence is proved.

Suppose now that $f = g_1 + h_1 = g_2 + h_2$, where g_1 and g_2 are in $\mathcal{M}$ while h_1 and h_2 are in $\mathcal{M}^\perp$. Then $g_2 - g_1 = h_1 - h_2$ is in $\mathcal{M} \cap \mathcal{M}^\perp$, and hence $g_2 - g_1$ is orthogonal to itself. Therefore, $\|g_2 - g_1\|^2 = (g_2 - g_1, g_2 - g_1) = 0$ which implies $g_1 = g_2$. Finally, $h_1 = h_2$ and the proof is complete. ∎

3.22 Corollary If $\mathcal{M}$ is a subspace of the Hilbert space $\mathcal{H}$, then $\mathcal{M}^{\perp\perp} =$ clos $\mathcal{M}$.

Proof That clos $\mathcal{M} \subset \mathcal{M}^{\perp\perp}$ follows immediately for any subset $\mathcal{M}$ of $\mathcal{H}$. If f is in $\mathcal{M}^{\perp\perp}$, then by the theorem $f = g + h$, where g is in clos $\mathcal{M}$ and h is in $\mathcal{M}^\perp$. Since f is in $\mathcal{M}^{\perp\perp}$, we have

$$0 = (f, h) = (g + h, h) = (h, h) = \|h\|^2.$$

Therefore, $h = 0$ and hence f is in clos $\mathcal{M}$. ∎

If g is a vector in the Hilbert space $\mathcal{H}$, then the complex functional defined $\varphi_g(f) = (f, g)$ for f in $\mathcal{H}$ is clearly linear. Moreover, since $|\varphi_g(f)| \leqslant \|f\| \|g\|$ for all f in $\mathcal{H}$, it follows that φ_g is bounded and that $\|\varphi_g\| \leqslant \|g\|$. Since $\|g\|^2 = \varphi_g(g) \leqslant \|\varphi_g\| \|g\|$ we have $\|g\| \leqslant \|\varphi_g\|$ and hence $\|\varphi_g\| = \|g\|$. The following theorem states that every bounded linear functional on $\mathcal{H}$ is of this form.

3.23 Theorem (Riesz Representation Theorem) If φ is a bounded linear functional on $\mathcal{H}$, then there exists a unique g in $\mathcal{H}$ such that $\varphi(f) = (f, g)$ for f in $\mathcal{H}$.

Proof Let $\mathcal{K}$ be the kernel of φ, that is, $\mathcal{K} = \{f \in \mathcal{H} : \varphi(f) = 0\}$. Since φ is continuous, $\mathcal{K}$ is a closed subspace of $\mathcal{H}$. If $\mathcal{K} = \mathcal{H}$ then $\varphi(f) = (f, 0)$ for f in $\mathcal{H}$ and the theorem is proved. If $\mathcal{K} \neq \mathcal{H}$, then there exists a unit vector h orthogonal to $\mathcal{K}$ by the remark following Definition 3.20. Since h is not in $\mathcal{K}$, then $\varphi(h) \neq 0$. For f in $\mathcal{H}$ the vector $f - (\varphi(f)/\varphi(h))h$ is in $\mathcal{K}$ since $\varphi(f - (\varphi(f)/\varphi(h))h) = 0$. Therefore, we have

$$\varphi(f) = \varphi(f)(h, h) = \left(\frac{\varphi(f)}{\varphi(h)} h, \overline{\varphi(h)} h \right)$$

$$= \left(f - \frac{\varphi(f)}{\varphi(h)} h, \overline{\varphi(h)} h \right) + \left(\frac{\varphi(f)}{\varphi(h)} h, \overline{\varphi(h)} h \right)$$

$$= (f, \overline{\varphi(h)} h)$$

for f in $\mathcal{H}$, and hence $\varphi(f) = (f, g)$ for $g = \overline{\varphi(h)} h$.

If $(f, g_1) = (f, g_2)$ for f in $\mathscr{H}$, then, in particular, $(g_1 - g_2, g_1 - g_2) = 0$ and hence $g_1 = g_2$. Therefore, $\varphi(f) = (f, g)$ for a unique g in $\mathscr{H}$. ∎

Thus we see that the mapping from $\mathscr{H}$ to $\mathscr{H}^*$ defined $g \to \varphi_g$ is not only norm preserving but onto. Moreover, a straightforward verification shows that this map is conjugate linear, that is, $\varphi_{\alpha_1 g_1 + \alpha_2 g_2} = \bar{\alpha}_1 \varphi_{g_1} + \bar{\alpha}_2 \varphi_{g_2}$ for α_1 and α_2 in $\mathbb{C}$ and g_1 and g_2 in $\mathscr{H}$. Thus for most purposes it is possible to identify $\mathscr{H}^*$ with $\mathscr{H}$ by means of this map.

3.24 In the theory of complex linear spaces, a linear space is characterized up to an isomorphism by its algebraic dimension. While this is not true for Banach spaces, it is true for Hilbert spaces with an appropriate and different definition of dimension. Before giving this definition we need an extension of the Pythagorean theorem to infinite orthogonal sets.

3.25 Theorem If $\{f_\alpha\}_{\alpha \in A}$ is an orthogonal subset of the Hilbert space $\mathscr{H}$, then $\sum_{\alpha \in A} f_\alpha$ converges in $\mathscr{H}$ if and only if $\sum_{\alpha \in A} \|f_\alpha\|^2 < \infty$ and in this case $\|\sum_{\alpha \in A} f_\alpha\|^2 = \sum_{\alpha \in A} \|f_\alpha\|^2$.

Proof Let $\mathscr{F}$ denote the collection of finite subsets of A. If $\sum_{\alpha \in A} f_\alpha$ converges, then by Definition 1.8, the continuity of the norm, and the Pythagorean theorem we have

$$\left\| \sum_{\alpha \in A} f_\alpha \right\|^2 = \left\| \lim_{F \in \mathscr{F}} \sum_{\alpha \in F} f_\alpha \right\|^2 = \lim_{F \in \mathscr{F}} \left\| \sum_{\alpha \in F} f_\alpha \right\|^2$$

$$= \lim_{F \in \mathscr{F}} \sum_{\alpha \in F} \|f_\alpha\|^2 = \sum_{\alpha \in A} \|f_\alpha\|^2.$$

Therefore, if $\sum_{\alpha \in A} f_\alpha$ converges, then $\sum_{\alpha \in A} \|f_\alpha\|^2 < \infty$.

Conversely, suppose $\sum_{\alpha \in A} \|f_\alpha\|^2 < \infty$. Given $\varepsilon > 0$, there exists F_0 in $\mathscr{F}$ such that $F \geqslant F_0$ implies $\sum_{\alpha \in F} \|f_\alpha\|^2 - \sum_{\alpha \in F_0} \|f_\alpha\|^2 < \varepsilon^2$. Thus, for F_1 and F_2 in $\mathscr{F}$ such that $F_1, F_2 \geqslant F_0$, we have

$$\left\| \sum_{\alpha \in F_1} f_\alpha - \sum_{\alpha \in F_2} f_\alpha \right\|^2 = \sum_{\alpha \in F_1 \backslash F_2} \|f_\alpha\|^2 + \sum_{\alpha \in F_2 \backslash F_1} \|f_\alpha\|^2$$

$$\leqslant \sum_{\alpha \in F_1 \cup F_2} \|f_\alpha\|^2 - \sum_{\alpha \in F_0} \|f_\alpha\|^2 < \varepsilon^2,$$

where the first equality follows from the Pythagorean theorem. Therefore, the net $\{\sum_{\alpha \in F} f_\alpha\}_{F \in \mathscr{F}}$ is Cauchy, and hence $\sum_{\alpha \in A} f_\alpha$ converges by definition. ∎

3.26 Corollary If $\{e_\alpha\}_{\alpha \in A}$ is an orthonormal subset of the Hilbert space

$\mathcal{H}$ and $\mathcal{M}$ is the smallest closed subspace of $\mathcal{H}$ containing the set $\{e_\alpha : \alpha \in A\}$, then

$$\mathcal{M} = \left\{\sum_{\alpha \in A} \lambda_\alpha e_\alpha : \lambda_\alpha \in \mathbb{C}, \sum_{\alpha \in A} |\lambda_\alpha|^2 < \infty\right\}.$$

Proof Let $\mathscr{F}$ denote the directed set of all finite subsets of A. If $\{\lambda_\alpha\}_{\alpha \in A}$ is a choice of complex numbers such that $\sum_{\alpha \in A} |\lambda_\alpha|^2 < \infty$, then $\{\lambda_\alpha e_\alpha\}_{\alpha \in A}$ is an orthogonal set and $\sum_{\alpha \in A} \|\lambda_\alpha e_\alpha\|^2 < \infty$. Thus $\sum_{\alpha \in A} \lambda_\alpha e_\alpha$ converges to a vector f in $\mathcal{H}$ by the theorem and since $f = \lim_{F \in \mathscr{F}} \sum_{\alpha \in F} \lambda_\alpha e_\alpha$, the vector is seen to lie in

$$\mathcal{N} = \left\{\sum_{\alpha \in A} \lambda_\alpha e_\alpha : \lambda_\alpha \in \mathbb{C}, \sum_{\alpha \in A} |\lambda_\alpha|^2 < \infty\right\}.$$

Since $\mathcal{N}$ contains $\{e_\alpha : \alpha \in A\}$, the proof will be complete once we show that $\mathcal{N}$ is a closed subspace of $\mathcal{H}$. If $\{\lambda_\alpha\}_{\alpha \in A}$ and $\{\mu_\alpha\}_{\alpha \in A}$ satisfy $\sum_{\alpha \in A} |\lambda_\alpha|^2 < \infty$ and $\sum_{\alpha \in A} |\mu_\alpha|^2 < \infty$, then

$$\sum_{\alpha \in A} |\lambda_\alpha + \mu_\alpha|^2 \leqslant 2 \sum_{\alpha \in A} |\lambda_\alpha|^2 + 2 \sum_{\alpha \in A} |\mu_\alpha|^2 < \infty.$$

Hence $\mathcal{N}$ is a linear subspace of $\mathcal{H}$.

Now suppose $\{f^n\}_{n=1}^\infty$ is a Cauchy sequence contained in $\mathcal{N}$ and that $f^n = \sum_{\alpha \in A} \lambda_\alpha^{(n)} e_\alpha$. Then for each α in A we have

$$|\lambda_\alpha^{(n)} - \lambda_\alpha^{(m)}| \leqslant \left(\sum_{\alpha \in A} |\lambda_\alpha^{(n)} - \lambda_\alpha^{(m)}|^2\right)^{1/2} = \|f^n - f^m\|,$$

and hence $\lambda_\alpha = \lim_{n \to \infty} \lambda_\alpha^{(n)}$ exists. Moreover, for F in $\mathscr{F}$ we have

$$\sum_{\alpha \in F} |\lambda_\alpha|^2 = \lim_{n \to \infty} \sum_{\alpha \in F} |\lambda_\alpha^{(n)}|^2 \leqslant \lim_{n \to \infty} \sum_{\alpha \in A} |\lambda_\alpha^{(n)}|^2 = \lim_{n \to \infty} \|f^n\|^2 < \infty.$$

Hence, $f = \sum_{\alpha \in A} \lambda_\alpha e_\alpha$ is well defined and an element of $\mathcal{N}$. Now given $\varepsilon > 0$, if we choose N such that $n, m \geqslant N$ implies $\|f^n - f^m\| < \varepsilon$, then for F in $\mathscr{F}$, we have

$$\sum_{\alpha \in F} |\lambda_\alpha - \lambda_\alpha^{(n)}|^2 = \lim_{m \to \infty} \sum_{\alpha \in F} |\lambda_\alpha^{(m)} - \lambda_\alpha^{(n)}|^2 \leqslant \limsup_{m \to \infty} \|f^m - f^n\|^2 \leqslant \varepsilon^2.$$

Therefore, for $n > N$, we have

$$\|f - f^n\|^2 = \sum_{\alpha \in A} |\lambda_\alpha - \lambda_\alpha^{(n)}|^2 = \lim_{F \in \mathscr{F}} \sum_{\alpha \in F} |\lambda_\alpha - \lambda_\alpha^{(n)}|^2 \leqslant \varepsilon^2,$$

and hence $\mathcal{N}$ is closed. ∎

3.27 Definition A subset $\{e_\alpha\}_{\alpha \in A}$ of the Hilbert space $\mathcal{H}$ is said to be an orthonormal basis if it is orthonormal and the smallest closed subspace containing it is $\mathcal{H}$.

An orthonormal basis has especially pleasant properties with respect to representing the elements of the space.

3.28 Corollary If $\{e_\alpha\}_{\alpha \in A}$ is an orthonormal basis for the Hilbert space $\mathscr{H}$ and f is a vector in $\mathscr{H}$, then there exist unique Fourier coefficients $\{\lambda_\alpha\}_{\alpha \in A}$ contained in $\mathbb{C}$ such that $f = \sum_{\alpha \in A} \lambda_\alpha e_\alpha$. Moreover, $\lambda_\alpha = (f, e_\alpha)$ for α in A.

Proof That $\{\lambda_\alpha\}_{\alpha \in A}$ exists such that $f = \sum_{\alpha \in A} \lambda_\alpha e_\alpha$ follows from the preceding corollary and definition. Moreover, if $\mathscr{F}$ denotes the collection of finite subsets of A and β is in A, then

$$(f, e_\beta) = \left(\sum_{\alpha \in A} \lambda_\alpha e_\alpha, e_\beta \right) = \lim_{F \in \mathscr{F}} \left(\sum_{\alpha \in F} \lambda_\alpha e_\alpha, e_\beta \right)$$

$$= \lim_{F \in \mathscr{F}} \sum_{\alpha \in F} \lambda_\alpha (e_\alpha, e_\beta) = \lim_{\substack{F \in \mathscr{F} \\ \beta \in F}} \lambda_\beta = \lambda_\beta.$$

(The limit is unaffected since the subsets of A containing β are cofinal in $\mathscr{F}$.) Therefore the set $\{\lambda_\alpha\}_{\alpha \in A}$ is unique, where $\lambda_\alpha = (f, e_\alpha)$ for α in A. ∎

3.29 Theorem Every Hilbert space ($\neq \{0\}$) possesses an orthonormal basis.

Proof Let $\mathscr{E}$ be the collection of orthonormal subsets E of $\mathscr{H}$ with the partial ordering $E_1 \leqslant E_2$ if $E_1 \subset E_2$. We want to use Zorn's lemma to assert the existence of a maximal orthonormal subset and then show that it is a basis. To this end let $\{E_\lambda\}_{\lambda \in \Lambda}$ be an increasing chain of orthonormal subsets of $\mathscr{H}$. Then clearly $\bigcup_{\lambda \in \Lambda} E_\lambda$ is an orthonormal subset of $\mathscr{H}$ and hence is in $\mathscr{E}$. Therefore, each chain has a maximal element and hence $\mathscr{E}$ itself has a maximal element E_M. Let $\mathscr{M}$ be the smallest closed subspace of $\mathscr{H}$ containing E_M. If $\mathscr{M} = \mathscr{H}$, then E_M is an orthonormal basis. If $\mathscr{M} \neq \mathscr{H}$, then for e a unit vector in $\mathscr{M}^\perp$, the set $E_M \cup \{e\}$ is an orthonormal subset of $\mathscr{H}$ greater than E_M. This contradiction shows that $\mathscr{M} = \mathscr{H}$ and E_M is the desired orthonormal basis. ∎

Although there is nothing unique about an orthonormal basis, that is, there always exist infinitely many if $\mathscr{H} \neq \{0\}$, the cardinality of an orthonormal basis is well defined.

3.30 Theorem If $\{e_\alpha\}_{\alpha \in A}$ and $\{f_\beta\}_{\beta \in B}$ are orthonormal bases for the Hilbert space $\mathscr{H}$, then card $A = $ card B.

Proof If either of A and B is finite, then the result follows from the theory of linear algebra. Assume, therefore, that card $A \geqslant \aleph_0$ and card $B \geqslant \aleph_0$.

For α in A set $B_\alpha = \{\beta \in B : (e_\alpha, f_\beta) \neq 0\}$. Since $e_\alpha = \sum_{\beta \in B} (e_\alpha, f_\beta) f_\beta$ by Corollary 3.28 and $1 = \|e_\alpha\|^2 = \sum_{\beta \in B} |(e_\alpha, f_\beta)|^2$ by Theorem 3.25, it follows that card $B_\alpha \leqslant \aleph_0$. Moreover, since $f_\beta = \sum_{\alpha \in A} (e_\alpha, f_\beta) e_\alpha$, it follows that $(e_\alpha, f_\beta) \neq 0$ for some α in A. Therefore, $B = \bigcup_{\alpha \in A} B_\alpha$ and hence card $B \leqslant \sum_{\alpha \in A}$ card $B_\alpha \leqslant \sum_{\alpha \in A} \aleph_0 = \text{card } A$ since card $A \geqslant \aleph_0$. From symmetry we obtain the reverse inequality and hence card $A = $ card B. $\blacksquare$

3.31 Definition   If $\mathscr{H}$ is a Hilbert space, then the dimension of $\mathscr{H}$, denoted dim $\mathscr{H}$, is the cardinality of any orthonormal basis for $\mathscr{H}$.

The dimension of a Hilbert space is well defined by the previous two theorems. We now show that two Hilbert spaces $\mathscr{H}$ and $\mathscr{K}$ of the same dimension are isomorphic, that is, there exists an isometrical isomorphism from $\mathscr{H}$ onto $\mathscr{K}$ which preserves the inner product.

3.32 Theorem   Two Hilbert spaces are isomorphic if and only if their dimensions are equal.

Proof   If $\mathscr{H}$ and $\mathscr{K}$ are Hilbert spaces such that dim $\mathscr{H} = $ dim $\mathscr{K}$, then there exist orthonormal bases $\{e_\alpha\}_{\alpha \in A}$ and $\{f_\alpha\}_{\alpha \in A}$ for $\mathscr{H}$ and $\mathscr{K}$, respectively. Define the map Φ from $\mathscr{H}$ to $\mathscr{K}$ such that for g in $\mathscr{H}$, we set $\Phi g = \sum_{\alpha \in A} (g, e_\alpha) f_\alpha$. Since $g = \sum_{\alpha \in A} (g, e_\alpha) e_\alpha$ by Corollary 3.28, it follows from Theorem 3.25 that $\sum_{\alpha \in A} |(g, e_\alpha)|^2 = \|g\|^2$. Therefore, Φg is well defined and

$$\|\Phi g\|^2 = \sum_{\alpha \in A} |(g, e_\alpha)|^2 = \|g\|^2.$$

That Φ is linear is obvious. Hence, Φ is an isometrical isomorphism of $\mathscr{H}$ to $\mathscr{K}$. Thus, $\Phi \mathscr{H}$ is a closed subspace of $\mathscr{K}$ which contains $\{f_\alpha : \alpha \in A\}$ and by the definition of basis, must therefore be all of $\mathscr{K}$. Lastly, since $(g, g) = \|g\|^2 = \|\Phi g\|^2 = (\Phi g, \Phi g)$ for g in $\mathscr{H}$, it follows from the polarization identity that $(g, h) = (\Phi g, \Phi h)$ for g and h in $\mathscr{H}$. $\blacksquare$

3.33   We conclude this chapter by computing the dimension of Examples 3.13, 3.15, 3.16, and 3.17. For $\mathbb{C}^n$ it is clear that the n-tuples

$$\{(1, 0, \ldots, 0), (0, 1, \ldots, 0), \ldots, (0, 0, \ldots, 1)\}$$

form an orthonormal basis, and therefore dim $\mathbb{C}^n = n$. Similarly, it is easy to see that the functions $\{e_n\}_{n=0}^\infty$ in $l^2(\mathbb{Z}^+)$ defined by $e_n(m) = 1$ if $n = m$ and 0 otherwise, form an orthonormal subset of $l^2(\mathbb{Z}^+)$. Moreover, since f in $l^2(\mathbb{Z}^+)$ can be written $f = \sum_{n=0}^\infty f(n) e_n$, it follows that $\{e_n\}_{n=0}^\infty$ is an orthonormal basis for $l^2(\mathbb{Z}^+)$ and hence that dim$[l^2(\mathbb{Z}^+)] = \aleph_0$.

In the Hilbert space $L^2([0,1])$, the set $\{e^{2\pi inx}\}_{n \in \mathbb{Z}}$ is orthonormal since $\int_0^1 e^{2\pi inx}\,dx = 1$ if $n = 0$ and 0 otherwise. Moreover, from the Stone–Weierstrass theorem it follows that $C([0,1])$ is contained in the uniform closure of the subspace spanned by the set $\{e^{2\pi inx} : n \in \mathbb{Z}\}$ and hence $C([0,1])$ is contained in the smallest closed subspace of $L^2([0,1])$ containing them. For f in $L^2([0,1])$ it follows from the Lebesgue dominated convergence theorem that $\lim_{k \to \infty} \|f - f_k\|_2 = 0$, where

$$
f_k(x) = \begin{cases} f(x), & |f(x)| \leqslant k, \\ 0, & |f(x)| > k. \end{cases}
$$

Since $C([0,1])$ is dense in $L^1([0,1])$ in the L^1-norm, there exists for each k in $\mathbb{Z}^+$, a function φ_k in $C([0,1])$ such that $|\varphi_k(x)| \leqslant k$ for x in $[0,1]$ and $\|f_k - \varphi_k\|_1 \leqslant 1/k^2$. Hence

$$
\limsup_{k \to \infty} \left(\int_0^1 |f_k - \varphi_k|^2 \, dx \right)^{1/2} \leqslant \limsup_{k \to \infty} \left(k \int_0^1 |f_k - \varphi_k| \, dx \right)
$$

$$
\leqslant \limsup_{k \to \infty} \int_0^1 \frac{1}{k} \, dx = 0.
$$

Thus, $C([0,1])$ is a dense subspace of $L^2([0,1])$ and hence the smallest closed subspace of $L^2([0,1])$ containing the functions $\{e^{2\pi inx} : n \in \mathbb{Z}\}$ is $L^2([0,1])$. Therefore, $\{e^{2\pi inx}\}_{n \in \mathbb{Z}}$ is an orthonormal basis for $L^2([0,1])$. Hence, $\dim\{L^2([0,1])\} = \aleph_0$ and therefore despite their apparent difference, $l^2(\mathbb{Z}^+)$ and $L^2([0,1])$ are isomorphic Hilbert spaces.

Similarly, since a change of variables shows that $\{\chi_n\}_{n \in \mathbb{Z}}$ is an orthonormal basis for $L^2(\mathbb{T})$, we see that $\{\chi_n\}_{n \in \mathbb{Z}^+}$ is an orthonormal basis for H^2 and hence $\dim H^2 = \aleph_0$ also.

We indicate in the exercises how to construct an example of a Hilbert space for all dimensions.

Notes

The definition of a Hilbert space is due to von Neumann and he along with Hilbert, Riesz, Stone, and others set forth the foundations of the subject. An introduction to the geometry of Hilbert space can be found in many textbooks on functional analysis and, in particular, in Stone [104], Halmos [55], Riesz and Sz.-Nagy [92], and Akhieser and Glazman [2].

Exercises

3.1   Let A be a nonempty set and let

$$l^2(A) = \left\{ f: A \to \mathbb{C} : \sum_{\alpha \in A} |f(\alpha)|^2 < \infty \right\}.$$

Show that $l^2(A)$ is a Hilbert space with the pointwise operations and with the inner product $(f, g) = \sum_{\alpha \in A} f(\alpha) \overline{g(\alpha)}$. Show that $\dim l^2(A) = \operatorname{card} A$.

3.2   Let $\mathscr{L}$ be a normed linear space for which the conclusion of the parallelogram law is valid. Show that an inner product can be defined on $\mathscr{L}$ for which the associated norm is the given norm.

3.3   Show that the completion of an inner product space is a Hilbert space.

3.4   Show that $C([0, 1])$ is not a Hilbert space, that is, there is no inner product on $C([0, 1])$ for which the associated norm is the supremum norm.

3.5   Show that $C([0, 1])$ is not homeomorphically isomorphic to a Hilbert space.*

3.6   Complete the proof began in Section 3.14 that the space $\mathscr{L}$ defined there is not complete.

3.7   Give an example of a finite dimensional space containing a closed convex set which contains more than one point of smallest norm.*

3.8   Give an example of an infinite dimensional Banach space and a closed convex set having no point of smallest norm.*

3.9   Let φ be a bounded linear functional on the subspace $\mathscr{M}$ of the Hilbert space $\mathscr{H}$. Show that there exists a unique extension of φ to $\mathscr{H}$ having the same norm.

3.10   Let $\mathscr{H}$ and $\mathscr{K}$ be Hilbert spaces and $\mathscr{H} \oplus \mathscr{K}$ denote the algebraic direct sum. Show that

$$(\langle h_1, k_1 \rangle, \langle h_2, k_2 \rangle) = (h_1, h_2) + (k_1, k_2)$$

defines an inner product on $\mathscr{H} \oplus \mathscr{K}$, that $\mathscr{H} \oplus \mathscr{K}$ is complete with respect to this inner product, and that the subspaces $\mathscr{H} \oplus \{0\}$ and $\{0\} \oplus \mathscr{K}$ are closed and orthogonal in $\mathscr{H} \oplus \mathscr{K}$.

3.11   Show that each vector of norm one is an extreme point of the unit ball of a Hilbert space.

3.12   Show that the w^*-closure of the unit sphere in an infinite-dimensional Hilbert space is the entire unit ball.

3.13 Show that every orthonormal subset of the Hilbert space $\mathcal{H}$ is contained in an orthonormal basis for $\mathcal{H}$.

3.14 Show that if $\mathcal{M}$ is a closed subspace of the Hilbert space $\mathcal{H}$, then $\dim \mathcal{M} \leqslant \dim \mathcal{H}$.

3.15 (*Gram–Schmidt*) Let $\{f_n\}_{n=1}^{\infty}$ be a subset of the Hilbert space $\mathcal{H}$ whose closed linear span is $\mathcal{H}$. Set $e_1 = f_1/\|f_1\|$ and assuming $\{e_k\}_{k=1}^{n}$ to have been defined, set

$$e_{n+1} = \left(f_{n+1} - \sum_{k=1}^{n} (f_{n+1}, e_k) e_k\right) \bigg/ \left\|f_{n+1} - \sum_{k=1}^{n} (f_{n+1}, e_k) e_k\right\|,$$

where e_{n+1} is taken to be the zero vector if

$$f_{n+1} = \sum_{k=1}^{n} (f_{n+1}, e_k) e_k.$$

Show that $\{e_n\}_{n=1}^{\infty}$ is an orthonormal basis for $\mathcal{H}$.

3.16 Show that $L^2([0,1])$ has an orthonormal basis $\{e_n\}_{n=0}^{\infty}$ such that e_n is a polynomial of degree n.

3.17 Let $\mathcal{L}$ be a dense linear subspace of the separable Hilbert space $\mathcal{H}$. Show that $\mathcal{L}$ contains an orthonormal basis for $\mathcal{H}$. Consider the same question for nonseparable $\mathcal{H}$.*

3.18 Give an example of two closed subspace $\mathcal{M}$ and $\mathcal{N}$ of the Hilbert space $\mathcal{H}$ for which the linear span

$$\mathcal{M} + \mathcal{N} = \{f + g : f \in \mathcal{M}, g \in \mathcal{N}\}$$

fails to be closed.* (Hint: Take $\mathcal{M}$ to be the graph of an appropriately chosen bounded linear transformation from $\mathcal{K}$ to $\mathcal{K}$ and $\mathcal{N}$ to be $\mathcal{K} \oplus \{0\}$, where $\mathcal{H} = \mathcal{K} \oplus \mathcal{K}$.)

3.19 Show that no Hilbert space has linear dimension $\aleph_0$. (Hint: Use the Baire category theorem.)

3.20 If $\mathcal{H}$ is an infinite-dimensional Hilbert space, then $\dim \mathcal{H}$ coincides with the smallest cardinal of a dense subset of $\mathcal{H}$.

3.21 Let $\mathcal{H}$ and $\mathcal{K}$ be Hilbert spaces and let $\mathcal{H} \underset{a}{\otimes} \mathcal{K}$ denote the algebraic tensor product of $\mathcal{H}$ and $\mathcal{K}$ considered as linear spaces over $\mathbb{C}$. Show that

$$\left(\sum_{i=1}^{m} h_i \otimes k_i, \sum_{j=1}^{n} h_j' \otimes k_j'\right) = \sum_{i=1}^{m} \sum_{j=1}^{n} (h_i, h_j')(k_i, k_j')$$

defines an inner product on $\mathcal{H} \otimes \mathcal{K}$. Denote the completion of this inner product space by $\mathcal{H} \overset{a}{\otimes} \mathcal{K}$. Show that if $\{e_\alpha\}_{\alpha \in A}$ and $\{f_\beta\}_{\beta \in B}$ are orthonormal bases for $\mathcal{H}$ and $\mathcal{K}$, respectively, then $\{e_\alpha \otimes f_\beta\}_{(\alpha, \beta) \in A \times B}$ is an orthonormal basis for $\mathcal{H} \otimes \mathcal{K}$.

3.22   Let $(X, \mathcal{S}, \mu)$ be a measure space with μ finite and φ be a bounded linear functional on $L^1(X)$. Show that the restriction of φ to $L^2(X)$ is a bounded linear functional on $L^2(X)$ and hence there exists g in $L^2(X)$ such that $\varphi(f) = \int_X f \bar{g} \, d\mu$ for f in $L^2(X)$. Show further that g is in $L^\infty(X)$ and hence obtain the characterization of $L^1(X)^*$ as $L^\infty(X)$. (Neither the result obtained in Chapter 1 nor the Radon–Nikodym theorem is to be used in this problem.)

3.23   (*von Neumann*)   Let μ and ν be positive finite measures on $(X, \mathcal{S})$ such that ν is absolutely continuous with respect to μ. Show that $f \to \int_X f \, d\mu$ is well defined and a bounded linear functional on $L^2(\mu + \nu)$. If φ is the function in $L^2(\mu + \nu)$ satisfying $\int_X f\varphi \, d(\mu + \nu) = \int_X f \, d\mu$, then $(1 - \varphi)/\varphi$ is in $L^1(\mu)$ and

$$\nu(E) = \int_E \frac{1 - \varphi}{\varphi} \, d\mu$$

for E in $\mathcal{S}$.

3.24   Interpret the results of Exercises 1.30 and 1.31 under the assumption that $\mathcal{X}$ is a Hilbert space.

4 Operators on Hilbert Space and C*-Algebras

4.1 Most of linear algebra involves the study of transformations between linear spaces which preserve the linear structure, that is, linear transformations. Such is also the case in the study of Hilbert spaces. In the remainder of the book we shall be mainly concerned with bounded linear transformations acting on Hilbert spaces. Despite the importance of certain classes of unbounded linear transformations, we consider them only in the problems.

We begin by adopting the word operator to mean bounded linear transformation. The following proposition asserts the existence and uniqueness of what we shall call the "adjoint operator."

4.2 Proposition If T is an operator on the Hilbert space $\mathscr{H}$, then there exists a unique operator S on $\mathscr{H}$ such that

$$(Tf, g) = (f, Sg) \text{ for } f \text{ and } g \text{ in } \mathscr{H}.$$

Proof For a fixed g in $\mathscr{H}$ consider the functional φ defined $\varphi(f) = (Tf, g)$ for f in $\mathscr{H}$. It is easy to verify that φ is a bounded linear functional on $\mathscr{H}$, and hence there exists by the Riesz representation theorem, a unique h in $\mathscr{H}$ such that $\varphi(f) = (f, h)$ for f in $\mathscr{H}$. Define $Sg = h$.

Obviously S is linear and $(Tf, g) = (f, Sg)$ for f and g in $\mathscr{H}$. Setting $f = Sg$

we obtain the inequality

$$\|Sg\|^2 = (Sg, Sg) = (TSg, g) \leqslant \|T\| \|Sg\| \|g\| \qquad \text{for} \quad g \text{ in } \mathscr{H}.$$

Therefore, $\|S\| \leqslant \|T\|$ and S is an operator on $\mathscr{H}$.

To show that S is unique, suppose S_0 is another operator on $\mathscr{H}$ satisfying $(f, S_0 g) = (Tf, g)$ for f and g in $\mathscr{H}$. Then $(f, Sg - S_0 g) = 0$ for f in $\mathscr{H}$ which implies $Sg - S_0 g = 0$. Hence $S = S_0$ and the proof is complete. ∎

4.3   Definition   If T is an operator on the Hilbert space $\mathscr{H}$, then the adjoint of T, denoted T^*, is the unique operator on $\mathscr{H}$ satisfying $(Tf, g) = (f, T^*g)$ for f and g in $\mathscr{H}$.

The following proposition summarizes some of the properties of the involution $T \to T^*$. In many situations this involution plays a role analogous to that of the conjugation of complex numbers.

4.4   Proposition   If $\mathscr{H}$ is a Hilbert space, then:

(1) $T^{**} = (T^*)^* = T$ for T in $\mathfrak{L}(\mathscr{H})$;

(2) $\|T\| = \|T^*\|$ for T in $\mathfrak{L}(\mathscr{H})$;

(3) $(\alpha S + \beta T)^* = \bar{\alpha} S^* + \bar{\beta} T^*$ and $(ST)^* = T^* S^*$ for α, β in $\mathbb{C}$ and S, T in $\mathfrak{L}(\mathscr{H})$;

(4) $(T^*)^{-1} = (T^{-1})^*$ for an invertible T in $\mathfrak{L}(\mathscr{H})$; and

(5) $\|T\|^2 = \|T^* T\|$ for T in $\mathfrak{L}(\mathscr{H})$.

Proof   (1) If f and g are in $\mathscr{H}$, then

$$(f, T^{**}g) = (T^*f, g) = \overline{(g, T^*f)} = \overline{(Tg, f)} = (f, Tg),$$

and hence $T^{**} = T$.

(2) In the proof of Proposition 4.2 we showed $\|T^{**}\| \leqslant \|T^*\| \leqslant \|T\|$. Combining this with (1), we have $\|T\| = \|T^*\|$.

(3) Compute.

(4) Since $T^*(T^{-1})^* = (T^{-1}T)^* = I = (TT^{-1})^* = (T^{-1})^*T^*$ by (3), it follows that T^* is invertible and $(T^*)^{-1} = (T^{-1})^*$.

(5) Since $\|T^*T\| \leqslant \|T^*\| \|T\| = \|T\|^2$ by (2), we need verify only that $\|T^*T\| \geqslant \|T\|^2$. Let $\{f_n\}_{n=1}^{\infty}$ be a sequence of unit vectors in $\mathscr{H}$ such that $\lim_{n \to \infty} \|Tf_n\| = \|T\|$. Then we have

$$\|T^*T\| \geqslant \limsup_{n \to \infty} \|T^* Tf_n\| \geqslant \limsup_{n \to \infty} (T^* Tf_n, f_n) = \lim_{n \to \infty} \|Tf_n\|^2 = \|T\|^2,$$

which completes the proof. ∎

4.5   Definition   If T is an operator on the Hilbert space $\mathcal{H}$, then the kernel of T, denoted ker T, is the closed subspace $\{f \in \mathcal{H} : Tf = 0\}$, and the range of T, denoted ran T, is the subspace $\{Tf : f \in \mathcal{H}\}$.

4.6   Proposition   If T is an operator on the Hilbert space $\mathcal{H}$, then ker $T = (\operatorname{ran} T^*)^{\perp}$ and ker $T^* = (\operatorname{ran} T)^{\perp}$.

Proof   It is sufficient to prove the first relation in view of (1) of the last proposition. To that end, if f is in ker T, then $(T^*g, f) = (g, Tf) = 0$ for g in $\mathcal{H}$, and hence f is orthogonal to ran T^*. Conversely, if f is orthogonal to ran T^*, then $(Tf, g) = (f, T^*g) = 0$ for g in $\mathcal{H}$, which implies $Tf = 0$. Therefore, f is in ker T and the proof is complete.   ■

We next derive useful criteria for the invertibility of an operator.

4.7   Definition   An operator T on the Hilbert space $\mathcal{H}$ is bounded below if there exists $\varepsilon > 0$ such that $\|Tf\| \geqslant \varepsilon \|f\|$ for f in $\mathcal{H}$.

4.8   Proposition   If T is an operator on the Hilbert space $\mathcal{H}$, then T is invertible if and only if T is bounded below and has dense range.

Proof   If T is invertible, then ran $T = \mathcal{H}$ and hence is dense. Moreover,

$$\|Tf\| \geqslant \frac{1}{\|T^{-1}\|} \|T^{-1}Tf\| = \frac{1}{\|T^{-1}\|} \|f\| \qquad \text{for } f \text{ in } \mathcal{H}$$

and therefore T is bounded below.

Conversely, if T is bounded below, there exists $\varepsilon > 0$ such that $\|Tf\| \geqslant \varepsilon \|f\|$ for f in $\mathcal{H}$. Hence, if $\{Tf_n\}_{n=1}^{\infty}$ is a Cauchy sequence in ran T, then the inequality

$$\|f_n - f_m\| \leqslant \frac{1}{\varepsilon} \|Tf_n - Tf_m\|$$

implies $\{f_n\}_{n=1}^{\infty}$ is also a Cauchy sequence. Hence, if $f = \lim_{n \to \infty} f_n$, then $Tf = \lim_{n \to \infty} Tf_n$ is in ran T and thus ran T is a closed subspace of $\mathcal{H}$. If we assume, in addition, that ran T is dense, then ran $T = \mathcal{H}$. Since T being bounded below obviously implies that T is one-to-one, the inverse transformation T^{-1} is well defined. Moreover, if $g = Tf$, then

$$\|T^{-1}g\| = \|f\| \leqslant \frac{1}{\varepsilon} \|Tf\| = \frac{1}{\varepsilon} \|g\|$$

and hence T^{-1} is bounded.   ■

4.9 Corollary If T is an operator on the Hilbert space $\mathcal{H}$ such that both T and T^* are bounded below, then T is invertible.

Proof If T^* is bounded below, then $\ker T^* = \{0\}$. Since $(\operatorname{ran} T)^\perp = \ker T^* = \{0\}$ by Proposition 4.6, we have $(\operatorname{ran} T)^\perp = \{0\}$, which implies $\operatorname{clos}[\operatorname{ran} T] = (\operatorname{ran} T)^{\perp\perp} = \{0\}^\perp = \mathcal{H}$ by Corollary 3.22. Therefore, $\operatorname{ran} T$ is dense in $\mathcal{H}$ and the result follows from the theorem. ∎

4.10 If T is an operator on the finite dimensional Hilbert space $\mathbb{C}^n$ and $\{e_i\}_{i=1}^n$ is an orthonormal basis for $\mathbb{C}^n$, then the action of T is given by the matrix $\{a_{ij}\}_{i,j=1}^n$, where $a_{ij} = (Te_i, e_j)$. The adjoint operator T^* has the matrix $\{b_{ij}\}_{i,j=1}^n$, where $b_{ij} = \overline{a_{ji}}$ for $i, j = 1, 2, \ldots, n$.

The simplest operators on $\mathbb{C}^n$ are those for which it is possible to choose an orthonormal basis such that the corresponding matrix is diagonal, that is, such that $a_{ij} = 0$ for $i \neq j$. An operator can be shown to belong to this class if and only if it commutes with its adjoint. In one direction, this result is obvious and the other is the content of the so called "spectral theorem" for matrices.

For operators on infinite dimensional Hilbert spaces such a theorem is no longer valid. Hilbert showed, however, that a reformulation of this result holds for operators on arbitrary Hilbert spaces. This "spectral theorem" is the main theorem of this chapter.

We begin by defining the relevant classes of operators.

4.11 Definition If T is an operator on the Hilbert space $\mathcal{H}$, then:

(1) T is normal if $TT^* = T^*T$;
(2) T is self-adjoint or hermitian if $T = T^*$;
(3) T is positive if $(Tf, f) \geqslant 0$ for f in $\mathcal{H}$; and
(4) T is unitary if $T^*T = TT^* = I$.

The following is a characterization of self-adjoint operators.

4.12 Proposition An operator T on the Hilbert space $\mathcal{H}$ is self-adjoint if and only if (Tf, f) is real for f in $\mathcal{H}$.

Proof If T is self-adjoint and f is in $\mathcal{H}$, then

$$(Tf, f) = (f, T^*f) = (f, Tf) = \overline{(Tf, f)}$$

and hence (Tf, f) is real. If (Tf, f) is real for f in $\mathcal{H}$, then using Lemma 3.3, we obtain for f and g in $\mathcal{H}$ that $(Tf, g) = \overline{(Tg, f)}$ and hence $T = T^*$. ∎

4.13 Corollary If P is a positive operator on the Hilbert space $\mathscr{H}$, then P is self-adjoint.

Proof Obvious. ■

4.14 Proposition If T is an operator on the Hilbert space $\mathscr{H}$, then T^*T is a positive operator.

Proof For f in $\mathscr{H}$ we have $(T^*Tf,f) = \|Tf\|^2 \geq 0$, from which the result follows. ■

When we speak of the spectrum of an operator T defined on the Hilbert space $\mathscr{H}$, we mean its spectrum when T is considered as an element of the Banach algebra $\mathfrak{L}(\mathscr{H})$ and we use $\sigma(T)$ to denote it. On a finite-dimensional space λ is in the spectrum of T if and only if λ is an eigenvalue for T. This is no longer the case for operators on infinite-dimensional Hilbert spaces.

In linear algebra one shows that the eigenvalues of a hermitian matrix are real. The generalization to hermitian operators takes the following form.

4.15 Proposition If T is a self-adjoint operator on the Hilbert space $\mathscr{H}$, then the spectrum of T is real. Furthermore, if T is a positive operator, then the spectrum of T is nonnegative.

Proof If $\lambda = \alpha + i\beta$ with α, β real and $\beta \neq 0$, then we must show that $T - \lambda$ is invertible. The operator $K = (T - \alpha)/\beta$ is self-adjoint and $T - \lambda$ is invertible if and only if $K - i$ is invertible, since $K - i = (T - \lambda)/\beta$. Therefore, in view of Proposition 4.9, the result will follow once we show that the operators $K - i$ and $(K - i)^* = K + i$ are bounded below. However, for f in $\mathscr{H}$, we have

$$\|(K \pm i)f\|^2 = ((K \pm i)f,(K \pm i)f) = \|Kf\|^2 \mp i(Kf,f) \pm i(f,Kf) + \|f\|^2$$
$$= \|Kf\|^2 + \|f\|^2 \geq \|f\|^2$$

and hence the spectrum of a self-adjoint operator is real.

If we assume, in addition, that T is positive and $\lambda < 0$, then

$$\|(T - \lambda)f\|^2 = \|Tf\|^2 - 2\lambda(Tf,f) + \lambda^2\|f\|^2 \geq \lambda^2\|f\|^2.$$

Since $(T - \lambda)^* = (T - \lambda)$, then $T - \lambda$ is invertible by Proposition 4.9 and the proof is complete. ■

We consider now a special class of positive operators which form the building blocks for the self-adjoint operators in a sense which will be made clear in the spectral theorem.

4.16 Definition An operator P on the Hilbert space $\mathscr{H}$ is a projection if P is idempotent $(P^2 = P)$ and self-adjoint.

The following construction gives a projection and, in fact, all projections arise in this manner.

4.17 Definition Let $\mathscr{M}$ be a closed subspace of the Hilbert space $\mathscr{H}$. Define $P_{\mathscr{M}}$ to be the mapping $P_{\mathscr{M}}f = g$, where $f = g + h$ with g in $\mathscr{M}$ and h in $\mathscr{M}^{\perp}$.

4.18 Theorem If $\mathscr{M}$ is a closed subspace of $\mathscr{H}$, then $P_{\mathscr{M}}$ is a projection having range $\mathscr{M}$. Moreover, if P is a projection on $\mathscr{H}$, then there exists a closed subspace $\mathscr{M}(= \operatorname{ran} P)$ such that $P = P_{\mathscr{M}}$.

Proof First we prove that $P_{\mathscr{M}}$ is an operator on $\mathscr{H}$. If f_1, f_2 are vectors in $\mathscr{H}$ and λ_1, λ_2 complex numbers, then $f_1 = g_1 + h_1$ and $f_2 = g_2 + h_2$, where g_1, g_2 are in $\mathscr{M}$ and h_1, h_2 are in $\mathscr{M}^{\perp}$. Moreover

$$\lambda_1 f_1 + \lambda_2 f_2 = (\lambda_1 g_1 + \lambda_2 g_2) + (\lambda_1 h_1 + \lambda_2 h_2),$$

where $\lambda_1 g_2 + \lambda_2 g_2$ is in $\mathscr{M}$ and $(\lambda_1 h_1 + \lambda_2 h_2)$ is in $\mathscr{M}^{\perp}$. By the uniqueness of such a decomposition, we have

$$P_{\mathscr{M}}(\lambda_1 f_1 + \lambda_2 f_2) = \lambda_1 g_1 + \lambda_2 g_2 = \lambda_1 P_{\mathscr{M}}f_1 + \lambda_2 P_{\mathscr{M}}f_2$$

and hence $P_{\mathscr{M}}$ is a linear transformation on $\mathscr{H}$. Moreover, the inequality

$$\|P_{\mathscr{M}}f_1\|^2 = \|g_1\|^2 \leqslant \|g_1\|^2 + \|h_1\|^2 = \|f_1\|^2$$

shows that $P_{\mathscr{M}}$ is bounded and has norm at most one. Therefore, P is an operator on $\mathscr{H}$. Moreover, since

$$(P_{\mathscr{M}}f_1, f_2) = (g_1, g_2 + h_2) = (g_1, g_2) = (g_1 + h_1, g_2) = (f_1, P_{\mathscr{M}}f_2),$$

we see that $P_{\mathscr{M}}$ is self-adjoint. Lastly, if f is in $\mathscr{M}$, then $f = f + 0$ is the required decomposition of f and hence $P_{\mathscr{M}}f = f$. Since $\operatorname{ran} P_{\mathscr{M}} = \mathscr{M}$, it follows that $P_{\mathscr{M}}^2 = P_{\mathscr{M}}$ and hence $P_{\mathscr{M}}$ is idempotent. Therefore $P_{\mathscr{M}}$ is a projection with range $\mathscr{M}$.

Now suppose P is a projection on $\mathscr{H}$ and set $\mathscr{M} = \operatorname{ran} P$. If $\{Pf_n\}_{n=1}^{\infty}$ is a Cauchy sequence in $\mathscr{H}$ converging to g, then

$$g = \lim_{n \to \infty} Pf_n = \lim_{n \to \infty} P^2 f_n = P[\lim_{n \to \infty} Pf_n] = Pg.$$

Thus g is in $\mathscr{M}$ and hence $\mathscr{M}$ is a closed subspace of $\mathscr{H}$. If g is in $\mathscr{M}^{\perp}$, then $\|Pg\|^2 = (Pg, Pg) = (g, P^2 g) = 0$, since $P^2 g$ is in $\mathscr{M}$, and hence $Pg = 0$. If f is in $\mathscr{H}$, then $f = Pg + h$, where h is in $\mathscr{M}^{\perp}$ and hence $P_{\mathscr{M}}f = Pg = P^2 g + Ph = Pf$. Therefore, $P = P_{\mathscr{M}}$. ∎

Many geometrical properties of closed subspaces can be expressed in terms of the projections onto them.

4.19 Proposition If $\mathscr{H}$ is a Hilbert space, $\{\mathscr{M}_i\}_{i=1}^n$ are closed subspaces of $\mathscr{H}$, and $\{P_i\}_{i=1}^n$ are the projections onto them, then $P_1 + P_2 + \cdots + P_n = I$ if and only if the subspaces $\{\mathscr{M}_i\}_{i=1}^n$ are pairwise orthogonal and span $\mathscr{H}$, that is, if and only if each f in $\mathscr{H}$ has a unique representation $f = f_1 + f_2 + \cdots + f_n$, where f_i is in $\mathscr{M}_i$.

Proof If $P_1 + P_2 + \cdots + P_n = I$, then each f in $\mathscr{H}$ has the representation $f = P_1 f + P_2 f + \cdots + P_n f$ and hence the $\mathscr{M}_i$ span $\mathscr{H}$. Conversely, if the $\{\mathscr{M}_i\}_{i=1}^n$ span $\mathscr{H}$ and the sum $P_1 + P_2 + \cdots + P_n$ is a projection, then it must be the identity operator. Thus, we are reduced to proving that $P_1 + P_2 + \cdots + P_n$ is a projection if and only if the subspaces $\{\mathscr{M}_i\}_{i=1}^n$ are pairwise orthogonal, and for this it suffices to consider the case of two subspaces.

Therefore, suppose P_1 and P_2 are projections such that $P_1 + P_2$ is a projection. For f in $\mathscr{M}_1$, we have

$$((P_1 + P_2)f, f) = ((P_1 + P_2)^2 f, f)$$
$$= (P_1 f, P_1 f) + (P_1 f, P_2 f) + (P_2 f, P_1 f) + (P_2 f, P_2 f)$$
$$= (P_1 f, f) + (f, P_2 f) + (P_2 f, f) + (P_2 f, f)$$
$$= ((P_1 + P_2)f, f) + 2(P_2 f, f),$$

since $P_1 f = f$ and thus $2\|P_2 f\|^2 = 2(P_2 f, P_2 f) = 2(P_2 f, f) = 0$. Hence, f and therefore $\mathscr{M}_1$ is orthogonal to $\mathscr{M}_2$.

Conversely, if P_1 and P_2 are projections such that the range of P_1 is orthogonal to the range of P_2, then for f in $\mathscr{H}$, we have

$$(P_1 + P_2)^2 f = (P_1 + P_2) P_1 f + (P_1 + P_2) P_2 f = P_1^2 f + P_1^2 f$$
$$= (P_1 + P_2)f,$$

since $P_2 P_1 f = P_1 P_2 f = 0$. ∎

The proof shows that the sum of a finite number of projections onto pairwise orthogonal subspaces is itself a projection.

We next consider some examples of normal operators other than those defined by the diagonalizable matrices on a finite-dimensional Hilbert space.

4.20 Example Let $(X, \mathscr{S}, \mu)$ be a probability space. For φ in $L^\infty(\mu)$ define the mapping M_φ on $L^2(\mu)$ such that $M_\varphi f = \varphi f$ for f in $L^2(\mu)$, where φf denotes the pointwise product, and let $\mathfrak{M} = \{M_\varphi : \varphi \in L^\infty(\mu)\}$. Obviously M_φ is linear and the inequality

$$\|M_\varphi f\|_2 = \left(\int_X |\varphi f|^2 \, d\mu\right)^{1/2} \leqslant \left(\int_X (\|\varphi\|_\infty |f|)^2 \, d\mu\right)^{1/2} \leqslant \|\varphi\|_\infty \|f\|_2$$

shows that M_φ is bounded. Moreover, if E_n is the set

$$\{x \in X : |\varphi(x)| \geqslant \|\varphi\|_\infty - 1/n\},$$

then

$$\|M_\varphi I_{E_n}\| = \left(\int_X |\varphi I_{E_n}|^2 \, d\mu\right)^{1/2} \geqslant \left[\int_X \left(\|\varphi\|_\infty - \frac{1}{n}\right)^2 |I_{E_n}|^2 \, d\mu\right]^{1/2}$$

$$\geqslant \left(\|\varphi\|_\infty - \frac{1}{n}\right)\|I_{E_n}\|_2$$

and hence $\|M_\varphi\| = \|\varphi\|_\infty$. For f and g in $L^2(\mu)$ and φ in $L^\infty(\mu)$ we have

$$(M_\varphi f, g) = \int_X (\varphi f)\bar{g} \, d\mu = \int_X f(\overline{\bar{\varphi}g}) \, d\mu = (f, M_{\bar{\varphi}}g),$$

which implies $M_\varphi{}^* = M_{\bar{\varphi}}$. Lastly, the mapping defined by $\Psi(\varphi) = M_\varphi$ from $L^\infty(\mu)$ to $\mathfrak{M}$ is obviously linear and multiplicative. Therefore, Ψ is a *-isometrical isomorphism of $L^\infty(\mu)$ onto $\mathfrak{M}$. (The terminology *- is used to denote the fact that conjugation in $L^\infty(\mu)$ is transformed by Ψ to adjunction in $\mathfrak{L}(L^2(\mu))$.)

Since $L^\infty(\mu)$ is commutative, it follows that M_φ commutes with $M_\varphi{}^*$ and hence is a normal operator. For φ in $L^\infty(\mu)$ the operator M_φ is self-adjoint if and only if $M_\varphi = M_\varphi{}^*$ and hence if and only if $\varphi = \bar{\varphi}$ or φ is real. Since $M_\varphi{}^2 = M_{\varphi^2}$, the operator M_φ is idempotent if and only if $\varphi^2 = \varphi$ or φ is a characteristic function. Therefore, the self-adjoint operators in $\mathfrak{M}$ are the M_φ for which φ is real, and the projections are the M_φ for which φ is a characteristic function.

Let us now consider the spectrum of the operator M_φ. If $\varphi - \lambda$ is invertible in $L^\infty(\mu)$, then $M_\varphi - \lambda = M_{(\varphi-\lambda)}$ is invertible in $\mathfrak{L}(L^2(\mu))$ with inverse $M_{(\varphi-\lambda)^{-1}}$, and hence $\sigma(M_\varphi) \subset \mathcal{R}(\varphi)$. To assert the converse inclusion we need to know that if $M_\varphi - \lambda$ is invertible, then its inverse is in $\mathfrak{M}$. There are at least two different ways of showing this which reflect two important properties possessed by $\mathfrak{M}$.

4.21 Definition If $\mathcal{H}$ is a Hilbert space, then a subalgebra $\mathfrak{M}$ of $\mathfrak{L}(\mathcal{H})$ is said to be maximal abelian if it is commutative and is not properly contained in any commutative subalgebra of $\mathfrak{L}(\mathcal{H})$.

4.22 Proposition The algebra $\mathfrak{M} = \{M_\varphi : \varphi \in L^\infty(\mu)\}$ is maximal abelian.

Proof Let T be an operator on $L^2(\mu)$ that commutes with $\mathfrak{M}$. If we set $\psi = T1$, then ψ is in $L^2(\mu)$ and $T\varphi = TM_\varphi 1 = M_\varphi T1 = \psi\varphi$ for φ in $L^\infty(\mu)$. Moreover, if E_n is the set $\{x \in X : |\psi(x)| \geqslant \|T\| + 1/n\}$, then

$$\|T\| \|I_{E_n}\|_2 \geqslant \|TI_{E_n}\|_2 = \|\psi I_{E_n}\|_2 = \left(\int_X |\psi I_{E_n}|^2 \, d\mu\right)^{\!\!\frac{1}{2}}$$

$$\geqslant \left(\|T\| + \frac{1}{n}\right)\!\left(\int_X |I_{E_n}|^2 \, d\mu\right)^{\!\!\frac{1}{2}} \geqslant \left(\|T\| + \frac{1}{n}\right)\|I_{E_n}\|_2.$$

Therefore, $\|I_{E_n}\|_2 = 0$ and hence the set $\{x \in X : |\psi(x)| > \|T\|\}$ has measure zero. Thus ψ is in $L^\infty(\mu)$ and $T\varphi = M_\psi \varphi$ for φ in $L^\infty(\mu)$. Since $C(X)$ is dense in $L^2(\mu)$ as proved in Section 3.33 and $C(X) \subset L^\infty(\mu)$, it follows that $T = M_\psi$ is in $\mathfrak{M}$. Therefore, $\mathfrak{M}$ is maximal abelian. ∎

4.23 Proposition If $\mathscr{H}$ is a Hilbert space and $\mathfrak{A}$ is a maximal abelian subalgebra of $\mathfrak{L}(\mathscr{H})$, then $\sigma(T) = \sigma_{\mathfrak{A}}(T)$ for T in $\mathfrak{A}$.

Proof Clearly $\sigma(T) \subset \sigma_{\mathfrak{A}}(T)$ for T in $\mathfrak{A}$. If λ is not in $\sigma(T)$, then $(T-\lambda)^{-1}$ exists. Since $(T-\lambda)^{-1}$ commutes with $\mathfrak{A}$ and $\mathfrak{A}$ is maximal abelian, we have $(T-\lambda)^{-1}$ is in $\mathfrak{A}$. Therefore λ is not in $\sigma_{\mathfrak{A}}(T)$ and hence $\sigma_{\mathfrak{A}}(T) = \sigma(T)$. ∎

4.24 Corollary If φ is in $L^\infty(\mu)$, then $\sigma(M_\varphi) = \mathscr{R}(\varphi)$.

Proof Since $\mathfrak{M} = \{M_\psi : \psi \in L^\infty(\mu)\}$ is maximal abelian, it follows from the previous result that $\sigma_{\mathfrak{M}}(T) = \sigma(T)$. Since $\mathfrak{M}$ and $L^\infty(\mu)$ are isomorphic we have $\sigma_{\mathfrak{M}}(T) = \sigma_{L^\infty(\mu)}(T)$ and Lemma 2.63 completes the proof. ∎

We give the alternate approach after giving an example of a subalgebra $\mathfrak{A}$ of $\mathfrak{L}(\mathscr{H})$ and an operator T in $\mathfrak{A}$ for which $\sigma_{\mathfrak{A}}(T) \neq \sigma(T)$.

4.25 Example Let $l^2(\mathbb{Z})$ denote the Hilbert space consisting of the complex functions f on $\mathbb{Z}$ such that $\sum_{n=-\infty}^{\infty} |f(n)|^2 < \infty$. Define U on $l^2(\mathbb{Z})$ such that $(Uf)(n) = f(n-1)$ for f in $l^2(\mathbb{Z})$. The operator U is called the bilateral shift. It is clearly linear and the identity

$$\|Uf\|_2^2 = \sum_{n=-\infty}^{\infty} |(Uf)(n)|^2 = \sum_{n=-\infty}^{\infty} |f(n-1)|^2 = \|f\|^2$$

for f in $l^2(\mathbb{Z})$ shows that U preserves the norm and, in particular, is bounded.

If we define A on $l^2(\mathbb{Z})$ such that $(Af)(n) = f(n+1)$ for f in $l^2(\mathbb{Z})$, we have

$$(Uf, g) = \sum_{n=-\infty}^{\infty} (Uf)(n)\overline{g(n)} = \sum_{n=-\infty}^{\infty} f(n-1)\overline{g(n)}$$

$$= \sum_{n=-\infty}^{\infty} f(n)\overline{g(n+1)} = (f, Ag).$$

Therefore, $A = U^*$ and a computation yields $UU^* = U^*U = I$ or $U^{-1} = U^*$. Thus U is a unitary operator.

From Proposition 2.28 we have $\sigma(U) \subset \overline{\mathbb{D}}$ and $\sigma(U^{-1}) = \sigma(U^*) \subset \overline{\mathbb{D}}$. If λ is in $\overline{\mathbb{D}}, 0 < |\lambda| < 1$, then $(U-\lambda)U^{-1} = \lambda((1/\lambda) - U^{-1})$. Since $1/\lambda$ is not in $\overline{\mathbb{D}}$, the operator $\lambda((1/\lambda) - U^{-1})$ is invertible and hence so is $U - \lambda$. Therefore $\sigma(U)$ is contained in $\mathbb{T}$. For fixed θ in $[0, 2\pi]$ and n in $\mathbb{Z}^+$, let f_n be the vector in $l^2(\mathbb{Z})$ defined by $f_n(k) = (2n+1)^{-1/2} e^{-ik\theta}$ for $|k| \leq n$ and 0 otherwise. A straightforward calculation shows that f_n is a unit vector and that $\lim_{n\to\infty}(U - e^{i\theta})f_n = 0$. Therefore $U - e^{i\theta}$ is not bounded below, and hence $e^{i\theta}$ is in $\sigma(U)$. Thus $\sigma(U) = \mathbb{T}$.

Let $\mathfrak{A}_+$ denote the smallest closed subalgebra of $\mathfrak{L}(l^2(\mathbb{Z}))$ containing I and U. Then $\mathfrak{A}_+$ is the closure in the uniform norm of the collection of polynomials in U, that is,

$$\mathfrak{A}_+ = \text{clos}\left\{ \sum_{n=0}^{N} \alpha_n U^n : \alpha_n \in \mathbb{C} \right\}.$$

If $\mathcal{M}$ denotes the closed subspace

$$\{ f \in l^2(\mathbb{Z}) : f(k) = 0 \text{ for } k < 0 \}$$

of $l^2(\mathbb{Z})$, then $U\mathcal{M} \subset \mathcal{M}$, which implies $p(U)\mathcal{M} \subset \mathcal{M}$ for each polynomial p. If for A in $\mathfrak{A}_+$ we choose a sequence of polynomials $\{p_n\}_{n=1}^{\infty}$ such that $\lim_{n\to\infty} p_n(U) = A$, then $Af = \lim_{n\to\infty} p_n(U)f$ implies Af in $\mathcal{M}$, if f is, since each $p_n(U)f$ is in $\mathcal{M}$. Therefore $A\mathcal{M} \subset \mathcal{M}$ for A in $\mathfrak{A}_+$. We claim that U^{-1} is not in $\mathfrak{A}_+$. If e_0 is the vector in $l^2(\mathbb{Z})$ defined to be 1 at 0 and 0 otherwise, then

$$(U^{-1}e_0)(-1) = (U^*e_0)(-1) = 1 \neq 0.$$

Hence $U^{-1}\mathcal{M} \not\subset \mathcal{M}$ which implies that U^{-1} is not in $\mathfrak{A}_+$. Therefore, 0 is in $\sigma_{\mathfrak{A}_+}(U)$, implying $\sigma_{\mathfrak{A}_+}(U) \neq \sigma(U)$.

The algebra $\mathfrak{A}_+$ can be shown to be isometrically isomorphic to the disk algebra defined in Section 2.50 with U corresponding to χ, and hence $\sigma_{\mathfrak{A}_+}(U) = \overline{\mathbb{D}}$. We return to this later.

The algebra $\mathfrak{A}_+$ is not maximal abelian, since it is contained in the obviously larger commutative algebra $\{M_\varphi : \varphi \in L^\infty(\mathbb{T})\}$. Equally important is that $\mathfrak{A}_+$ is not a self-adjoint subalgebra of $\mathfrak{L}(l^2(\mathbb{Z}))$, since, as we will soon establish, such algebras also have the property of preserving the spectrum. We consider the abstract analog of a self-adjoint subalgebra.

4.26 Definition If $\mathfrak{A}$ is a Banach algebra, then an involution on $\mathfrak{A}$ is a mapping $T \to T^*$ which satisfies:

(i) $T^{**} = T$ for T in $\mathfrak{A}$;
(ii) $(\alpha S + \beta T)^* = \bar{\alpha} S^* + \bar{\beta} T^*$ for S, T in $\mathfrak{A}$ and α, β in $\mathbb{C}$; and
(iii) $(ST)^* = T^* S^*$ for S and T in $\mathfrak{A}$.

If, in addition, $\|T^*T\| = \|T\|^2$ for T in $\mathfrak{A}$, then $\mathfrak{A}$ is a C^*-algebra.

A closed self-adjoint subalgebra of $\mathfrak{L}(\mathscr{H})$ for $\mathscr{H}$ a Hilbert space is a C^*-algebra in view of Proposition 4.4. Every C^*-algebra can, in fact, be shown to be isometrically isomorphic to such an algebra (see Exercise 5.26).

All of the various classes of operators whose definitions are based on the adjoint can be extended to a C^*-algebra; for example, an element T in a C^*-algebra is said to be self-adjoint if $T = T^*$, normal if $TT^* = T^*T$, and unitary if $T^*T = TT^* = I$.

We now give a proof of Proposition 4.15 which is valid for C^*-algebras. Our previous proof made essential use of the fact that we were dealing with operators defined on a Hilbert space.

4.27 Theorem In a C^*-algebra a self-adjoint element has real spectrum.

Proof Observe first that if T is an element of the C^*-algebra $\mathfrak{A}$, then the inequality

$$\|T\|^2 = \|T^*T\| \leqslant \|T^*\| \|T\|$$

implies that $\|T\| \leqslant \|T^*\|$ and hence $\|T\| = \|T^*\|$, since $T^{**} = T$. Thus the involution on a C^*-algebra is an isometry.

Now let H be a self-adjoint element of $\mathfrak{A}$ and set $U = \exp iH$. Then from the fact that H is self-adjoint and the definition of the exponential function, it follows that $U^* = \exp(iH)^* = \exp(-iH)$. Moreover, from Lemma 2.12 we have

$$UU^* = \exp(iH)\exp(-iH) = \exp(iH - iH) = I = \exp(-iH)\exp(iH) = U^*U$$

and hence U is unitary. Moreover, since $1 = \|I\| = \|U^*U\| = \|U\|^2$ we see that $\|U\| = \|U^*\| = \|U^{-1}\| = 1$, and therefore $\sigma(U)$ is contained in $\mathbb{T}$. Since $\sigma(U) = \exp(i\sigma(H))$ by Corollary 2.37, we see that the spectrum of H must be real. ∎

4.28   Theorem   If $\mathfrak{B}$ is a C^*-algebra, $\mathfrak{A}$ is a closed self-adjoint subalgebra of $\mathfrak{B}$, and T is an element of $\mathfrak{A}$, then $\sigma_{\mathfrak{A}}(T) = \sigma_{\mathfrak{B}}(T)$.

Proof   Since $\sigma_{\mathfrak{A}}(T)$ contains $\sigma_{\mathfrak{B}}(T)$, it is sufficient to show that if $T - \lambda$ is invertible in $\mathfrak{B}$, then the inverse $(T-\lambda)^{-1}$ is in $\mathfrak{A}$. We can assume $\lambda = 0$ without loss of generality. Thus, T is invertible in $\mathfrak{B}$, and therefore T^*T is a self-adjoint element of $\mathfrak{A}$ which is invertible in $\mathfrak{B}$. Since $\sigma_{\mathfrak{A}}(T^*T)$ is real by the previous theorem, we see that $\sigma_{\mathfrak{A}}(T^*T) = \sigma_{\mathfrak{B}}(T^*T)$ by Corollary 2.55. Thus, T^*T is invertible in $\mathfrak{A}$ and therefore

$$T^{-1} = (T^{-1}(T^*)^{-1})T^* = (T^*T)^{-1}T^*$$

is in $\mathfrak{A}$ and the proof is complete.   ■

We are now in a position to obtain a form of the spectral theorem for normal operators. We use it to obtain a "functional calculus" for continuous functions as well as to prove many elementary results about normal operators.

Our approach is based on the following characterization of commutative C^*-algebras.

4.29   Theorem (Gelfand–Naimark)   If $\mathfrak{A}$ is a commutative C^*-algebra and M is the maximal ideal space of $\mathfrak{A}$, then the Gelfand map is a *-isometrical isomorphism of $\mathfrak{A}$ onto $C(M)$.

Proof   If Γ denotes the Gelfand map, then we must show that $\overline{\Gamma(T)} = \Gamma(T^*)$ and that $\|\Gamma(T)\|_\infty = \|T\|$. The fact that Γ is onto will then follow from the Stone–Weierstrass theorem.

If T is in $\mathfrak{A}$, then $H = \frac{1}{2}(T + T^*)$ and $K = (T - T^*)/2i$ are self-adjoint operators in $\mathfrak{A}$ such that $T = H + iK$ and $T^* = H - iK$. Since the sets $\sigma(H)$ and $\sigma(K)$ are contained in $\mathbb{R}$, by Theorem 4.27, the functions $\Gamma(H)$ and $\Gamma(K)$ are real valued by Corollary 2.36. Therefore,

$$\overline{\Gamma(T)} = \overline{\Gamma(H) + i\Gamma(K)} = \Gamma(H) - i\Gamma(K) = \Gamma(H - iK) = \Gamma(T^*),$$

and hence Γ is a *-map.

To show that Γ is an isometry, let T be an operator in $\mathfrak{A}$. Using Definition 4.26, Corollary 2.36, Theorem 2.38, and the fact that T^*T is self-adjoint, we have

$$\|T\|^2 = \|T^*T\| = \|(T^*T)^{2^k}\|^{1/2^k} = \lim_{k \to \infty} \|(T^*T)^{2^k}\|^{1/2^k}$$

$$= \|\Gamma(T^*T)\|_\infty = \|\Gamma(T^*)\Gamma(T)\|_\infty = \||\Gamma(T)|^2\|_\infty = \|\Gamma(T)\|_\infty^2.$$

Therefore Γ is an isometry and hence a *-isometrical isomorphism onto $C(M)$. ∎

If $\mathfrak{A}$ is a commutative C^* algebra and T is in $\mathfrak{A}$, then T is normal, since T^* is also in $\mathfrak{A}$ and the operators in $\mathfrak{A}$ commute. On the other hand, a normal operator generates a commutative C^*-algebra.

4.30 Theorem (Spectral Theorem) If $\mathcal{H}$ is a Hilbert space and T is a normal operator on $\mathcal{H}$, then the C^*-algebra $\mathfrak{C}_T$ generated by T is commutative. Moreover, the maximal ideal space of $\mathfrak{C}_T$ is homeomorphic to $\sigma(T)$, and hence the Gelfand map is a *-isometrical isomorphism of $\mathfrak{C}_T$ onto $C(\sigma(T))$.

Proof Since T and T^* commute, the collection of all polynomials in T and T^* form a commutative self-adjoint subalgebra of $\mathfrak{L}(\mathcal{H})$ which must be contained in the C^*-algebra generated by T. It is easily verified that the closure of this collection is a commutative C^*-algebra and hence must be $\mathfrak{C}_T$. Therefore $\mathfrak{C}_T$ is commutative.

To show that the maximal ideal space M of $\mathfrak{C}_T$ is homeomorphic to $\sigma(T)$, define ψ from M to $\sigma(T)$ by $\psi(\varphi) = \Gamma(T)(\varphi)$. Since the range of $\Gamma(T)$ is $\sigma(T)$ by Corollary 2.36, ψ is well defined and onto. If φ_1 and φ_2 are elements of M such that $\psi(\varphi_1) = \psi(\varphi_2)$, then

$$\Gamma(T)(\varphi_1) = \Gamma(T)(\varphi_2), \qquad \varphi_1(T) = \varphi_2(T),$$

and

$$\varphi_1(T^*) = \Gamma(T^*)(\varphi_1) = \overline{\Gamma(T)(\varphi_1)} = \overline{\Gamma(T)(\varphi_2)} = \Gamma(T^*)(\varphi_2) = \varphi_2(T^*),$$

and hence φ_1 and φ_2 agree on all polynomials in T and T^*. Since this collection of operators is dense in $\mathfrak{C}_T$, it follows that $\varphi_1 = \varphi_2$ and therefore ψ is one-to-one. Lastly, if $\{\varphi_\alpha\}_{\alpha \in A}$ is a net in M such that $\lim_{\alpha \in A} \varphi_\alpha = \varphi$, then

$$\lim_{\alpha \in A} \psi(\varphi_\alpha) = \lim_{\alpha \in A} \Gamma(T)(\varphi_\alpha) = \Gamma(T)(\varphi) = \psi(\varphi)$$

and hence ψ is continuous. Since M and $\sigma(T)$ are compact Hausdorff spaces, then ψ is a homeomorphism and the proof is complete. ∎

4.31 Functional Calculus If T is an operator, then a rudimentary functional calculus for T can be defined as follows: for the polynomial $p(z) = \sum_{n=0}^{N} \alpha_n z^n$, define $p(T) = \sum_{n=0}^{N} \alpha_n T^n$. The mapping $p \to p(T)$ is a homomorphism from the algebra of polynomials to the algebra of operators. If $\mathcal{H}$ is finite dimensional, then one can base the analysis of T on this functional calculus. In particular, the kernel of this mapping, that is, $\{p(z) : p(T) = 0\}$, is a nonzero principal ideal in the algebra of all polynomials and the generator of this ideal is the minimum polynomial for T. If $\mathscr{H}$ is not finite dimensional, then this functional calculus may yield little information.

The extension of this map to larger algebras of functions (see Exercise 2.18) is a problem of considerable importance in operator theory.

If T is a normal operator on the Hilbert space $\mathscr{H}$, then the Gelfand map establishes a *-isometrical isomorphism between $C(\sigma(T))$ and $\mathfrak{C}_T$. For φ in $C(\sigma(T))$, we define $\varphi(T) = \Gamma^{-1}\varphi$. It is clear that if φ is a polynomial in z, then this definition agrees with the preceding one. Moreover, if A is an operator on $\mathscr{H}$ that commutes with T and T^*, then A must commute with every operator in $\mathfrak{C}_T$ and hence, in particular, with $\varphi(T)$ for each φ in $C(\sigma(T))$.

In the remainder of this chapter we shall obtain certain results about operators using this calculus and then extend the functional calculus to a larger class of functions.

4.32 Corollary If T is a normal operator on the Hilbert space $\mathscr{H}$, then T is positive if and only if $\sigma(T)$ is nonnegative and self-adjoint if and only if $\sigma(T)$ is real.

Proof By Proposition 4.15, the spectrum of T is nonnegative if T is positive. Conversely, if T is normal, $\sigma(T)$ is nonnegative, and Γ is the Gelfand transform from $\mathfrak{C}_T$ to $C(\sigma(T))$, then $\Gamma(T) \geqslant 0$. Thus there exists a continuous function φ on $\sigma(T)$ such that $\Gamma(T) = |\varphi|^2$. Then

$$T = [\bar{\varphi}(T)][\varphi(T)] = \varphi(T)^*\varphi(T)$$

and hence T is positive by Proposition 4.14.

If T is self-adjoint, then the spectrum of T is real by Proposition 4.15. If T is normal and has real spectrum, then $\psi = \Gamma(T)$ is a real-valued function by Corollary 2.36, and hence $T = \psi(T) = \bar{\psi}(T) = \psi(T)^* = T^*$. Therefore, T is self-adjoint. ■

The preceding proposition is false without the assumption that T is normal, that is, there exist operators with spectrum consisting of just zero which are not self-adjoint.

The second half of the preceding proposition is valid in a C^*-algebra, while the first half allows us to define a positive element of a C^*-algebra to be a normal element with nonnegative spectrum.

We now show the existence and uniqueness of the square root of a positive operator.

4.33 Proposition If P is a positive operator on the Hilbert space $\mathcal{H}$, then there exists a unique positive operator Q such that $Q^2 = P$. Moreover, Q commutes with every operator which commutes with P.

Proof Since the spectrum $\sigma(P)$ is positive, the square root function $\sqrt{}$ is continuous on $\sigma(P)$. Therefore $\sqrt{P}$ is a well-defined operator on $\mathcal{H}$ that is positive by Corollary 4.32, since $\sigma(\sqrt{P}) = \sqrt{\sigma(P)}$. Moreover, $(\sqrt{P})^2 = P$ by the definition of the functional calculus, and $\sqrt{P}$ commutes with every operator which commutes with P by Section 4.31. It remains only to show that $\sqrt{P}$ is unique.

Suppose Q is a positive operator on $\mathcal{H}$ satisfying $Q^2 = P$. Since $QP = QQ^2 = Q^2Q = PQ$, it follows from the remarks in Section 4.31 that the C^*-algebra $\mathfrak{A}$ generated by $P, \sqrt{P}$, and Q is commutative. If Γ denotes the Gelfand transform of $\mathfrak{A}$ onto $C(M_{\mathfrak{A}})$, then $\Gamma(\sqrt{P})$ and $\Gamma(Q)$ are nonnegative functions by Proposition 2.36, while $\Gamma(\sqrt{P})^2 = \Gamma(P) = \Gamma(Q)^2$, since Γ is a homomorphism. Thus $\Gamma(\sqrt{P}) = \Gamma(Q)$ implying $Q = \sqrt{P}$ and hence the uniqueness of the positive square root. ∎

4.34 Corollary If T is an operator on $\mathcal{H}$, then T is positive if and only if there exists an operator S on $\mathcal{H}$ such that $T = S^*S$.

Proof If T is positive, take $S = \sqrt{T}$. If $T = S^*S$, then Proposition 4.14 yields that T is positive. ∎

Every complex number can be written as the product of a nonnegative number and a number of modulus one. A polar form for linear transformations on $\mathbb{C}^n$ persists in which a positive operator is one factor and a unitary operator the other. For operators on an infinite-dimensional Hilbert space, a similar result is valid and the representation obtained is, under suitable hypotheses, unique. Before proving this result we need to introduce the notion of a partial isometry.

4.35 Definition An operator V on a Hilbert space $\mathcal{H}$ is a partial isometry if $\|Vf\| = \|f\|$ for f orthogonal to the kernel of V; if, in addition, the kernel of V is $\{0\}$, then V is an isometry. The initial space of a partial isometry is the closed subspace orthogonal to the kernel.

On a finite-dimensional space every isometry is, in fact a unitary operator. However, on an infinite-dimensional Hilbert space this is no longer the case.

Let us consider an important example of this which is related to the bilateral shift.

4.36 Example Define the operator U_+ on $l^2(\mathbb{Z}^+)$ by $(U_+f)(n) = f(n-1)$ for $n > 0$ and 0 otherwise. The operator is called the unilateral shift and an easy calculation shows that U_+ is an isometry. Moreover, since the function e_0 defined to be 1 at 0 and 0 otherwise is orthogonal to the range of U_+, we see that U_+ is not unitary. A straightforward verification shows that the adjoint U_+^* is defined by $(U_+^*f)(n) = f(n+1)$.

Let us next consider the spectrum of U_+. Since U_+ is a contraction, that is, $\|U_+\| = 1$, we have $\sigma(U_+) \subset \overline{\mathbb{D}}$. Moreover, for z in $\mathbb{D}$ the function f_z defined by $f_z(n) = z^n$ is in $l^2(\mathbb{Z}^+)$ and $U_+^*f_z = \bar{z}f_z$. Thus z is in $\sigma(U_+)$ and hence $\sigma(U_+) = \overline{\mathbb{D}}$.

The question of whether a partial isometry exists with given subspaces for initial space and range depends only on the dimension, as the following result shows.

4.37 Proposition If $\mathcal{M}$ and $\mathcal{N}$ are closed subspaces of the Hilbert space $\mathcal{H}$ such that $\dim \mathcal{M} = \dim \mathcal{N}$, then there exists a partial isometry V with initial space $\mathcal{M}$ and range $\mathcal{N}$.

Proof If $\mathcal{M}$ and $\mathcal{N}$ have the same dimension, then there exist orthonormal bases $\{e_\alpha\}_{\alpha \in A}$ and $\{f_\alpha\}_{\alpha \in A}$ for $\mathcal{M}$ and $\mathcal{N}$ with the same index set. Define an operator V on $\mathcal{H}$ as follows: for g in $\mathcal{H}$ write $g = h + \sum_{\alpha \in A} \lambda_\alpha e_\alpha$ with $h \perp \mathcal{M}$ and set $Vg = \sum_{\alpha \in A} \lambda_\alpha f_\alpha$. Then the kernel of V is $\mathcal{M}^\perp$ and $\|Vg\| = \|g\|$ for g in $\mathcal{M}$. Thus, V is a partial isometry with initial space $\mathcal{M}$, and range $\mathcal{N}$. ∎

We next consider a useful characterization of partial isometries which allows us to define partial isometries in a C^*-algebra.

4.38 Proposition Let V be an operator on the Hilbert space $\mathcal{H}$. The following are equivalent:

(1) V is a partial isometry;
(2) V^* is a partial isometry;
(3) VV^* is a projection; and
(4) V^*V is a projection.

Moreover, if V is a partial isometry, then VV^* is the projection onto the range of V, while V^*V is the projection onto the initial space.

Proof Since a partial isometry V is a contraction, we have for f in $\mathscr{H}$ that

$$((I-V^*V)f,f) = (f,f) - (V^*Vf,f) = \|f\|^2 - \|Vf\|^2 \geqslant 0.$$

Thus $I-V^*V$ is a positive operator. Now if f is orthogonal to $\ker V$, then $\|Vf\| = \|f\|$ which implies that $((I-V^*V)f,f) = 0$. Since $\|(I-V^*V)^{1/2}f\|^2 = ((I-V^*V)f,f) = 0$, we have $(I-V^*V)f = 0$ or $V^*Vf = f$. Therefore, V^*V is the projection onto the initial space of V.

Conversely, if V^*V is a projection and f is orthogonal to $\ker(V^*V)$, then $V^*Vf = f$. Therefore,

$$\|Vf\|^2 = (V^*Vf,f) = (f,f) = \|f\|^2,$$

and hence V preserves the norm on $\ker(V^*V)^\perp$. Moreover, if $V^*Vf = 0$, then $0 = (V^*Vf,f) = \|Vf\|^2$ and consequently $\ker(V^*V) = \ker V$. Therefore, V is a partial isometry, and thus (1) and (4) are equivalent. Reversing the roles of V and V^*, we see that (2) and (3) are equivalent. Moreover, if V^*V is a projection, then $(VV^*)^2 = V(V^*V)V^* = VV^*$, since $V(V^*V) = V$. Therefore, VV^* is a projection, which completes the proof. ∎

We now obtain the polar decomposition for an operator.

4.39 Theorem If T is an operator on the Hilbert space $\mathscr{H}$, then there exists a positive operator P and a partial isometry V such that $T = VP$. Moreover, V and P are unique if $\ker P = \ker V$.

Proof If we set $P = (T^*T)^{1/2}$, then

$$\|Pf\|^2 = (Pf, Pf) = (P^2 f,f) = (T^*Tf,f) = \|Tf\|^2 \qquad \text{for } f \text{ in } \mathscr{H}.$$

Thus, if we define $\tilde{V}$ on $\operatorname{ran} P$ such that $\tilde{V}Pf = Tf$, then $\tilde{V}$ is well defined and is, in fact, isometric. Hence, $\tilde{V}$ can be extended uniquely to an isometry from $\operatorname{clos}[\operatorname{ran} P]$ to $\mathscr{H}$. If we further extend $\tilde{V}$ to $\mathscr{H}$ by defining it to be the zero operator on $[\operatorname{ran} P]^\perp$, then the extended operator V is a partial isometry satisfying $T = VP$ and $\ker V = [\operatorname{ran} P]^\perp = \ker P$ by Proposition 4.6.

We next consider uniqueness. Suppose $T = WQ$, where W is a partial isometry, Q is a positive operator, and $\ker W = \ker Q$, Then $P^2 = T^*T = QW^*WQ = Q^2$, since W^*W is the projection onto

$$[\ker W]^\perp = [\ker Q]^\perp = \operatorname{clos}[\operatorname{ran} Q]\cdot$$

by Propositions 4.38 and 4.6. Thus, by the uniqueness of the square root, Proposition 4.33, we have $P = Q$ and hence $WP = VP$. Therefore, $W = V$

on ran P. But

$$[\operatorname{ran} P]^{\perp} = \ker P = \ker W = \ker V,$$

and hence $W = V$ on $[\operatorname{ran} P]^{\perp}$. Therefore, $V = W$ and the proof is complete. ∎

Although the positive operator will be in every closed self-adjoint subalgebra of $\mathfrak{L}(\mathscr{H})$ which contains T, the same is not true of the partial isometry. Consider, for example, the operator $T = M_{\varphi} M_{\psi}$ in $\mathfrak{L}(L^2(\mathbb{T}))$, where φ is a continuous nonnegative function on $\mathbb{T}$ while ψ has modulus one, is not continuous but the product $\varphi\psi$ is.

In many instances a polar form in which the order of the factors is reversed is useful.

4.40   Corollary   If T is an operator on the Hilbert space $\mathscr{H}$, then there exists a positive operator Q and a partial isometry W such that $T = QW$. Moreover, W and Q are unique if $\operatorname{ran} W = [\ker Q]^{\perp}$.

Proof   From the theorem we obtain a partial isometry V and a positive operator P such that $T^* = VP$. Taking adjoints we have $T = PV^*$, which is the form we desire with $W = V^*$ and $Q = P$. Moreover, the uniqueness also follows from the theorem since $\operatorname{ran} W = [\ker Q]^{\perp}$ if and only if

$$\ker V = \ker W^* = [\operatorname{ran} W]^{\perp} = [\ker Q]^{\perp\perp} = \ker P. \quad \blacksquare$$

It T is a normal operator on a finite-dimensional Hilbert space, then the subspace spanned by the eigenvectors belonging to a certain eigenvalue reduces the operator, and these subspaces can be used to put the operator in diagonal form. If T is a not necessarily normal operator still on a finite-dimensional space, then the appropriate subspaces to consider are those spanned by the generalized eigenvectors belonging to an eigenvalue. These subspaces do not, in general, reduce the operator but are only invariant for it.

Although no analogous structure theory exists for operators on an infinite-dimensional Hilbert space, the notions of invariant and reducing subspace remain important.

4.41   Definition   If T is an operator on the Hilbert space $\mathscr{H}$ and $\mathscr{M}$ is a closed subspace of $\mathscr{H}$, then $\mathscr{M}$ is an *invariant subspace* for T if $T\mathscr{M} \subset \mathscr{M}$ and a *reducing subspace* if, in addition, $T(\mathscr{M}^{\perp}) \subset \mathscr{M}^{\perp}$.

We begin with the following elementary facts.

4.42   Proposition   If T is an operator on $\mathscr{H}$, $\mathscr{M}$ is a closed subspace of $\mathscr{H}$, and $P_{\mathscr{M}}$ is the projection onto $\mathscr{M}$, then $\mathscr{M}$ is an invariant subspace for T if and only if $P_{\mathscr{M}} T P_{\mathscr{M}} = T P_{\mathscr{M}}$ if and only if $\mathscr{M}^{\perp}$ is an invariant subspace for T^*; further, $\mathscr{M}$ is a reducing subspace for T if and only if $P_{\mathscr{M}} T = T P_{\mathscr{M}}$ if and only if $\mathscr{M}$ is an invariant subspace for both T and T^*.

Proof   If $\mathscr{M}$ is invariant for T, then for f in $\mathscr{H}$, we have $T P_{\mathscr{M}} f$ in $\mathscr{M}$ and hence $P_{\mathscr{M}} T P_{\mathscr{M}} f = T P_{\mathscr{M}} f$; thus $P_{\mathscr{M}} T P_{\mathscr{M}} = T P_{\mathscr{M}}$. Conversely, if $P_{\mathscr{M}} T P_{\mathscr{M}} = T P_{\mathscr{M}}$, then for f in $\mathscr{M}$, we have $Tf = T P_{\mathscr{M}} f = P_{\mathscr{M}} T P_{\mathscr{M}} f = P_{\mathscr{M}} Tf$, and hence Tf is in $\mathscr{M}$. Therefore, $T\mathscr{M} \subset \mathscr{M}$ and $\mathscr{M}$ is invariant for T. Further, since $I - P_{\mathscr{M}}$ is the projection onto $\mathscr{M}^{\perp}$ and the identity

$$T^*(I - P_{\mathscr{M}}) = (I - P_{\mathscr{M}}) T^*(I - P_{\mathscr{M}})$$

is equivalent to $P_{\mathscr{M}} T^* = P_{\mathscr{M}} T^* P_{\mathscr{M}}$, we see that $\mathscr{M}^{\perp}$ is invariant for T^* if and only if $\mathscr{M}$ is invariant for T. Finally, if $\mathscr{M}$ reduces T, then $P_{\mathscr{M}} T = P_{\mathscr{M}} T P_{\mathscr{M}} = T P_{\mathscr{M}}$ by the preceding result, which completes the proof.   ∎

4.43   In the remainder of this chapter we want to extend the functional calculus obtained in Section 4.31 for continuous functions on the spectrum to a larger algebra of functions. This larger algebra of functions is related to the algebra of bounded Borel functions on the spectrum.

Before beginning let us give some consideration to the uses of a functional calculus and why we might be interested in extending it to a larger algebra of functions. Some of the details in this discussion will be omitted.

Suppose T is a normal operator on the Hilbert space $\mathscr{H}$ with finite spectrum $\sigma(T) = \{\lambda_1, \lambda_2, \cdots, \lambda_N\}$ and let $\varphi \to \varphi(T)$ be the functional calculus defined for φ in $C(\sigma(T))$. If λ_i is a point in the spectrum, then the characteristic function $I_{\{\lambda_i\}}$ is continuous on $\sigma(T)$ and hence in $C(\sigma(T))$. If we let E_i denote the operator $I_{\{\lambda_i\}}(T)$, then it follows from the fact that mapping $\varphi \to \varphi(T)$ is a *-isomorphism from $C(\sigma(T))$ onto $\mathfrak{C}_T$, that each E_i is a projection and that $E_1 + E_2 + \cdots + E_N = I$. If $\mathscr{M}_i$ denotes the range of E_i, then the $\{\mathscr{M}_i\}_{i=1}^n$ are pairwise orthogonal, their linear span is $\mathscr{H}$ and $\mathscr{M}_i$ reduces T by the preceding proposition. Moreover, since

$$T = z(T) = \left[\sum_{i=1}^N \lambda_i I_{\{\lambda_i\}} \right](T) = \sum_{i=1}^N \lambda_i E_i,$$

we see that T acts on each $\mathscr{M}_i$ as multiplication by λ_i. Thus the space $\mathscr{H}$ decomposes into a finite orthogonal direct sum such that T is multiplication by a scalar on each direct summand. Thus the functional calculus has enabled us to diagonalize T in the case where the spectrum of T is finite.

If the spectrum of T is not discrete, but is totally disconnected, then a slight modification of the preceding argument shows that $\mathscr{H}$ can be decomposed for such a T into a finite orthogonal direct sum of reducing subspaces for T such that the action of T on each direct summand is approximately (in the sense of the norm) multiplication by a scalar. Thus in this case T can be approximated by diagonal operators.

If the spectrum of T is connected, then this approach fails, since $C(\sigma(T))$ contains no nontrivial characteristic functions. Hence we seek to enlarge the functional calculus to an algebra of functions generated by its characteristic functions. We do this by considering the Gelfand transform on a larger commutative self-adjoint subalgebra of $\mathfrak{L}(\mathscr{H})$. This algebra will be obtained as the closure of $\mathfrak{C}_T$ in a weaker topology. Hence we begin by considering certain weaker topologies on $\mathfrak{L}(\mathscr{H})$.

4.44   Definition   Let $\mathscr{H}$ be a Hilbert space and $\mathfrak{L}(\mathscr{H})$ be the algebra of operators on $\mathscr{H}$. The weak operator topology is the weak topology defined by the collection of functions $T \to (Tf, g)$ from $\mathfrak{L}(\mathscr{H})$ to $\mathbb{C}$ for f and g in $\mathscr{H}$. The strong operator topology on $\mathfrak{L}(\mathscr{H})$ is the weak topology defined by the collection of functions $T \to Tf$ from $\mathfrak{L}(\mathscr{H})$ to $\mathscr{H}$ for f in $\mathscr{H}$.

Thus a net of operators $\{T_\alpha\}_{\alpha \in A}$ converges to T in the weak [strong] operator topology if

$$\lim_{\alpha \in A}(T_\alpha f, g) = (Tf, g) \quad [\lim_{\alpha \in A} T_\alpha f = Tf] \quad \text{for every } f \text{ and } g \text{ in } \mathscr{H}.$$

Clearly, the weak operator topology is weaker than the strong operator topology which is weaker than the uniform topology. We shall indicate examples in the problems to show that these topologies are all distinct.

The continuity of addition, multiplication, and adjunction in the weak operator topology is considered in the following lemma. We leave the corresponding questions for the strong operator topology to the exercises.

4.45   Lemma   If $\mathscr{H}$ is a Hilbert space and A and B are in $\mathfrak{L}(\mathscr{H})$, then the functions:

(1) $\alpha(S, T) = S + T$ from $\mathfrak{L}(\mathscr{H}) \times \mathfrak{L}(\mathscr{H})$ to $\mathfrak{L}(\mathscr{H})$,
(2) $\beta(T) = AT$ from $\mathfrak{L}(\mathscr{H})$ to $\mathfrak{L}(\mathscr{H})$,
(3) $\gamma(T) = TB$ from $\mathfrak{L}(\mathscr{H})$ to $\mathfrak{L}(\mathscr{H})$, and
(4) $\delta(T) = T^*$ from $\mathfrak{L}(\mathscr{H})$ to $\mathfrak{L}(\mathscr{H})$,

are continuous in the weak operator topology.

Proof   Compute.   ∎

The enlarged functional calculus will be based on the closure of the C^*-algebra $\mathfrak{C}_T$ in the weak operator topology.

4.46   Definition   If $\mathcal{H}$ is a Hilbert space, then a subset $\mathfrak{A}$ of $\mathfrak{L}(\mathcal{H})$ is said to be a W^*-algebra on $\mathcal{H}$ if $\mathfrak{A}$ is a C^*-algebra which is closed in the weak operator topology.

The reader should note that a W^*-algebra is a C^*-subalgebra of $\mathfrak{L}(\mathcal{H})$ which is weakly closed. In particular, a W^*-algebra is an algebra of operators. Moreover, if Φ is a *-isometrical isomorphism from the W^*-algebra $\mathfrak{A}$ contained in $\mathfrak{L}(\mathcal{H})$ to the C^*-algebra $\mathfrak{B}$ contained in $\mathfrak{L}(\mathcal{K})$, then it does not follow that $\mathfrak{B}$ is weakly closed in $\mathfrak{L}(\mathcal{K})$. We shall not consider such questions further and refer the reader to [27] or [28].

The following proposition shows one method of obtaining W^*-algebras.

4.47   Proposition   If $\mathcal{H}$ is a Hilbert space and $\mathfrak{M}$ is a self-adjoint sub-algebra of $\mathfrak{L}(\mathcal{H})$, then the closure $\mathfrak{A}$ of $\mathfrak{M}$ in the weak operator topology is a W^*-algebra. Moreover, $\mathfrak{A}$ is commutative if $\mathfrak{M}$ is.

Proof   That the closure of $\mathfrak{M}$ is a W^*-algebra follows immediately from Lemma 4.45. Moreover, assume that $\mathfrak{M}$ is commutative and let $\{S_\alpha\}_{\alpha \in A}$ and $\{T_\beta\}_{\beta \in B}$ be nets of operators in $\mathfrak{M}$ which converge in the weak operator topology to S and T, respectively. Then for f and g in $\mathcal{H}$ and β in B, we have

$$(ST_\beta f, g) = \lim_{\alpha \in A}(S_\alpha T_\beta f, g) = \lim_{\alpha \in A}(T_\beta S_\alpha f, g) = (T_\beta Sf, g).$$

Therefore, $ST_\beta = T_\beta S$ for each β in B and a similar argument establishes $ST = TS$. Hence, $\mathfrak{A}$ is commutative if $\mathfrak{M}$ is.   ∎

4.48   Corollary   If $\mathcal{H}$ is a Hilbert space and T is a normal operator on $\mathcal{H}$, then the W^*-algebra $\mathfrak{W}_T$ generated by T is commutative. Moreover, if Λ_T is the maximal idea space of $\mathfrak{W}_T$, then the Gelfand map is a *-isometrical isomorphism of $\mathfrak{W}_T$ onto $C(\Lambda_T)$.

Proof   This follows immediately from the preceding proposition along with Theorem 4.29.   ∎

4.49   If T is a normal operator on the *separable* Hilbert space $\mathcal{H}$ with spectrum Λ contained in $\mathbb{C}$, then we want to show that there is a unique L^∞

space on Λ and a unique *-isometrical isomorphism $\Gamma^* : \mathfrak{W}_T \to L^\infty$ which extends the functional calculus Γ of Section 4.31, that is, such that the accompanying diagram commutes, where the vertical arrows denote inclusion

$$
\begin{array}{ccc}
\mathfrak{C}_T & \xrightarrow{\Gamma} & C(\Lambda) \\
\downarrow & & \downarrow \\
\mathfrak{W}_T & \xrightarrow{\Gamma^*} & L^\infty
\end{array}
$$

maps. Thus, the functional calculus for T can be extended to $\mathfrak{W}_T$ and $\mathfrak{W}_T = \{\varphi(T) : \varphi \in L^\infty\}$.

We begin with some measure theoretic preliminaries concerning the following illustrative example. Let Λ be a compact subset of the complex plane and v be a finite positive regular Borel measure on Λ with support Λ. (The latter condition is equivalent to the assumption that the inclusion mapping of $C(\Lambda)$ into $L^\infty(v)$ is an isometrical isomorphism.) Recall that for each φ in $L^\infty(v)$ we define M_φ to be the multiplication operator defined on $L^2(v)$ by $M_\varphi f = \varphi f$ and that the mapping $\varphi \to M_\varphi$ is a *-isometrical isomorphism from $L^\infty(v)$ into $\mathfrak{L}(L^2(v))$. Thus we can identify the elements of $L^\infty(v)$ with operators in $\mathfrak{L}(L^2(v))$.

The following propositions give several important relationships between v, $C(\Lambda)$, $L^\infty(v)$, and $\mathfrak{L}(L^2(v))$. The first completes the presentation in Section 4.20.

4.50   Proposition   If $(X, \mathscr{S}, \mu)$ is a probability space, then $L^\infty(\mu)$ is a maximal abelian W^*-algebra in $\mathfrak{L}(L^2(\mu))$.

Proof   In view of Section 4.19, only the fact that $L^\infty(\mu)$ is weakly closed remains to be proved and that follows from Proposition 4.47, since the weak closure is commutative and hence must coincide with $L^\infty(\mu)$. ∎

The next result identifies the weak operator topology on $L^\infty(\mu)$ as a familiar one.

4.51   Proposition   If $(X, \mathscr{S}, \mu)$ is a probability space, then the weak operator topology and the w^*-topology coincide on $L^\infty(\mu)$.

Proof   We first recall that a function f on X is in $L^1(\mu)$ if and only if it can be written in the form $f = g\bar{h}$, where g and h are in $L^2(\mu)$.

Therefore a net $\{\varphi_\alpha\}_{\alpha \in A}$ in $L^\infty(\mu)$ converges in the w^*-topology to φ if and only if

$$\lim_{\alpha \in A} \int_X \varphi_\alpha f \, d\mu = \int_X \varphi f \, d\mu$$

if and only if

$$\lim_{\alpha \in A}(M_{\varphi_\alpha} g, h) = \lim_{\alpha \in A} \int_X \varphi_\alpha g\bar{h} \, d\mu = \int_X \varphi g\bar{h} \, d\mu = (M_\varphi g, h)$$

and therefore if and only if the net $\{M_{\varphi_\alpha}\}_{\alpha \in A}$ converges to M_φ in the weak operator topology. ■

The next proposition shows that the $W*$-algebra generated by multiplication by z on $L^2(v)$ is $L^\infty(v)$.

4.52   Proposition   If X is a compact Hausdorff space and μ is a finite positive regular Borel measure on X, then the unit ball of $C(X)$ is $w*$-dense in the unit ball of $L^\infty(\mu)$.

Proof   A simple step function ψ in the unit ball of $L^\infty(\mu)$ has the form $\psi = \sum_{i=1}^n \alpha_i I_{E_i}$, where $|\alpha_i| \leqslant 1$ for $i = 1, 2, \ldots, n$, the $\{E_i\}_{i=1}^n$ are pairwise disjoint, and $\bigcup_{i=1}^n E_i = X$. For $i = 1, 2, \ldots, n$, let K_i be a compact subset of E_i. By the Tietze extension theorem there exists φ in $C(X)$ such that $\|\varphi\|_\infty \leqslant 1$ and $\varphi(x) = \alpha_i$ for x in K_i. Then for f in $L^1(\mu)$, we have

$$\left| \int_X f(\varphi - \psi) \, d\mu \right| \leqslant \int_X |f| \, |\varphi - \psi| \, d\mu$$

$$= \sum_{i=1}^n \int_{E_i \backslash K_i} |f| \, |\varphi - \psi| \, d\mu \leqslant 2 \sum_{i=1}^n \int_{E_i \backslash K_i} |f| \, d\mu.$$

Because μ is regular, for $f_1, \ldots, f_m$ in $L^1(\mu)$ and $\varepsilon > 0$, there exist compact sets $K_i \subset E_i$ such that

$$\int_{E_i \backslash K_i} |f_j| \, d\mu < \frac{\varepsilon}{2n} \qquad \text{for} \quad j = 1, 2, \ldots, m.$$

This completes the proof. ■

4.53   Corollary   If X is a compact Hausdorff space and μ is a finite positive regular Borel measure on X, then $C(X)$ is $w*$-dense in $L^\infty(\mu)$.

Proof   Immediate. ■

We now consider the measure theoretic aspects of the uniqueness problem. We begin by recalling a definition.

4.54   **Definition**   Two positive measures v_1 and v_2 defined on a sigma algebra $(X, \mathcal{S})$ are mutually absolutely continuous, denoted $v_1 \sim v_2$, if v_1 is absolutely continuous with respect to v_2 and v_2 is absolutely continuous with respect to v_1.

4.55   **Theorem**   If v_1 and v_2 are finite positive regular Borel measures on the compact metric space X and there exists a *-isometrical isomorphism $\Phi : L^\infty(v_1) \to L^\infty(v_2)$ which is the identity on $C(X)$, then $v_1 \sim v_2$, $L^\infty(v_1) = L^\infty(v_2)$, and Φ is the identity.

Proof   If E is a Borel set in X, then $\Phi(I_E)$ is an idempotent and therefore a characteristic function. If we set $\Phi(I_E) = I_F$, then it suffices to show that $E = F$ v_2—a.e. (almost everywhere), since this would imply $v_1(E) = 0$ if and only if $I_E = 0$ in $L^\infty(v_1)$ if and only if $I_F = 0$ in $L^\infty(v_2)$ if and only if $v_2(F) = 0$, and therefore if and only if $v_2(E) = 0$. Thus we would have $v_1 \sim v_2$ and $L^\infty(v_1) = L^\infty(v_2)$, since the L^∞ space is determined by the sets of measure zero. Finally, Φ would be the identity since it is the identity on characteristic functions and the linear span of the characteristic functions is dense in L^∞.

To show that $E = F$ v_2—a.e. it suffices to prove that $F \subset E$ v_2—a.e., since we would also have $(X \backslash F) \subset (X \backslash E)$ v_2—a.e. Further, we may assume that E is compact. For suppose it is known for compact sets. Since v_1 and v_2 are regular, there exists a sequence of compact sets $\{K_n\}_{n=1}^\infty$ contained in E such that $E = \bigcup_{n=1}^\infty K_n$ v_1—a.e. and $E = \bigcup_{n=1}^\infty K_n$ v_2—a.e. Thus, since Φ and Φ^{-1} are *-linear and multiplicative and hence order preserving, Φ preserves suprema and we have

$$I_F = \sup_n \Phi(I_{K_n}) \leqslant \sup_n I_{K_n} = I_{\bigcup_{n=1}^\infty K_n} = I_E \ v_2\text{—a.e.}$$

and therefore F is contained in E v_2—a.e. Therefore, suppose E is closed and for n in $\mathbb{Z}^+$ let φ_n be the function in $C(X)$ defined by

$$\varphi_n(x) = \begin{cases} 1 - n \cdot d(x, E) & \text{if} \quad d(x, E) \leqslant \dfrac{1}{n}, \\ \\ 0 & \text{if} \quad d(x, E) \geqslant \dfrac{1}{n}, \end{cases}$$

where $d(x, E) = \inf\{\rho(x,y) : y \in E\}$ and ρ is the metric on X. Then $I_E \leqslant \varphi_n$ for each n and the sequence $\{\varphi_n\}_{n=1}^{\infty}$ converges pointwise to I_E. Since Φ is order preserving and the identity on continuous functions, we have $I_F \leqslant \varphi_n \ v_2$—a.e. and thus F is contained in $E \ v_2$—a.e. ∎

After giving the following definition and proving an elementary lemma we obtain the functional calculus we want under the assumption the operator has a cyclic vector.

4.56   Definition   If $\mathscr{H}$ is a Hilbert space and $\mathfrak{A}$ is a subalgebra of $\mathfrak{L}(\mathscr{H})$, then a vector f in $\mathscr{H}$ is cyclic for $\mathfrak{A}$ if $\mathrm{clos}\,[\mathfrak{A}f] = \mathscr{H}$ and separating for $\mathfrak{A}$ if $Af = 0$ for A in $\mathfrak{A}$ implies $A = 0$.

4.57   Lemma   If $\mathscr{H}$ is a Hilbert space, $\mathfrak{A}$ is a commutative subalgebra of $\mathfrak{L}(\mathscr{H})$, and f is a cyclic vector for $\mathfrak{A}$, then f is a separating vector for $\mathfrak{A}$.

Proof   If B is an operator in $\mathfrak{A}$ and $Bf = 0$, then $BAf = ABf = 0$ for every A in $\mathfrak{A}$. Therefore, we have $\mathfrak{A}f \subset \ker B$, which implies $B = 0$. ∎

The following theorem gives a spatial isomorphism between $\mathfrak{W}_T$ and L^{∞}, if T is a normal operator on $\mathscr{H}$ having a cyclic vector. (Note that such an $\mathscr{H}$ is necessarily separable.)

4.58   Theorem   If T is a normal operator on the Hilbert space $\mathscr{H}$ such that $\mathfrak{C}_T$ has a cyclic vector, then there is a positive regular Borel measure v on $\mathbb{C}$ having support $\Lambda = \sigma(T)$ and an isometrical isomorphism γ from $\mathscr{H}$ onto $L^2(v)$ such that the map Γ^* defined from $\mathfrak{W}_T$ to $\mathfrak{L}(L^2(v))$ by $\Gamma^* A = \gamma A \gamma^{-1}$ is a *-isometrical isomorphism from $\mathfrak{W}_T$ onto $L^{\infty}(v)$. Moreover, Γ^* is an extension of the Gelfand transform Γ from $\mathfrak{C}_T$ onto $C(\Lambda)$. Lastly, if v_1 is a positive regular Borel measure on $\mathbb{C}$ and Γ_1^* is a *-isometrical isomorphism from $\mathfrak{W}_T$ onto $L^{\infty}(v_1)$ which extends the Gelfand transform, then $v_1 \sim v$, $L^{\infty}(v_1) = L^{\infty}(v)$, and $\Gamma_1^* = \Gamma^*$.

Proof   Let f be a cyclic vector for $\mathfrak{C}_T$ of norm one and consider the functional defined on $C(\Lambda)$ by $\psi(\varphi) = (\varphi(T)f, f)$. Since ψ is obviously linear and positive and

$$|\psi(\varphi)| = |(\varphi(T)f, f)| \leqslant \|\varphi(T)\| \ \|f\|^2 = \|\varphi\|_{\infty},$$

there exists by the Riesz representation theorem (see Section 1.38) a positive regular Borel measure v on Λ such that

$$\int_\Lambda \varphi \, dv = (\varphi(T)f,f) \qquad \text{for} \quad \varphi \text{ in } C(\Lambda).$$

Now suppose that the support of v were not all of Λ, that is, suppose there exists an open subset V of Λ such that $v(V) = 0$. By Urysohn's lemma there is a nonzero function φ in $C(\Lambda)$ which vanishes outside of V. Since, however, we have

$$\|\varphi(T)f\|^2 = (\varphi(T)f, \varphi(T)f) = (|\varphi|^2(T)f,f) = \int_\Lambda |\varphi|^2 \, dv = 0,$$

and f is a separating vector for $\mathfrak{C}_T$ by the previous lemma we arrive at a contradiction. Thus the support of v is Λ.

If we define γ_0 from $\mathfrak{C}_T f$ to $L^2(v)$ such that $\gamma_0(\varphi(T)f) = \varphi$, then the computation

$$\|\varphi\|_2^2 = \int_\Lambda |\varphi|^2 \, dv = (|\varphi|^2(T)f,f) = \|\varphi(T)f\|^2$$

shows that γ_0 is a well-defined isometry. Since $\mathfrak{C}_T f$ is dense in $\mathscr{H}$ by assumption and $C(\Lambda)$ is dense in $L^2(v)$ by Section 3.33, the mapping γ_0 can be extended to a unique isometrical isomorphism γ from $\mathscr{H}$ onto $L^2(v)$. Moreover, if we define Γ^* from $\mathfrak{W}_T$ into $\mathfrak{L}(L^2(v))$ by $\Gamma^*(A) = \gamma A \gamma^{-1}$, then Γ^* is a *-isometrical isomorphism of $\mathfrak{W}_T$ into $\mathfrak{L}(L^2(v))$. We want to first show that Γ^* extends the Gelfand transform Γ on $\mathfrak{C}_T$. If ψ is in $C(\Lambda)$, then for all φ in $C(\Lambda)$, we have

$$[\Gamma^*(\psi(T))]\varphi = \gamma \psi(T) \gamma^{-1} \varphi = \gamma \psi(T) \varphi(T)f = \gamma[(\psi\varphi)(T)f] = \psi\varphi = M_\psi \varphi$$

and since $C(\Lambda)$ is dense in $L^2(v)$, it follows that $\Gamma^*(\psi(T)) = M_\psi = \Gamma(\psi(T))$. Thus Γ^* extends the Gelfand transform.

To show that $\Gamma^*(\mathfrak{W}_T) = L^\infty(v)$, we note that since Γ^* is defined spatially by γ, it follows from Proposition 4.51 that Γ^* is a continuous map from $\mathfrak{W}_T$ with the weak operator topology to $L^\infty(v)$ with the w^*-topology. Therefore, by Corollary 4.53, we have

$$\Gamma^*(\mathfrak{W}_T) = \Gamma^*(\text{weak oper clos}[\mathfrak{C}_T])$$

$$= w^*\text{-clos}[C(\Lambda)] = L^\infty(v),$$

and thus Γ^* is a *-isometrical isomorphism mapping $\mathfrak{W}_T$ onto $L^\infty(v)$.

Finally, if v_1 is a positive regular Borel measure on Λ and Γ_1^* is a *-isometrical isomorphism from $\mathfrak{W}_T$ onto $L^\infty(v_1)$ which extends the Gelfand transform, then $\Gamma^*\Gamma_1^{*-1}$ is a *-isometrical isomorphism from $L^\infty(v_1)$ onto

$L^\infty(v)$ which is the identity on $C(\Lambda)$. Hence, by Theorem 4.55, we have $v_1 \sim v$, $L^\infty(v_1) = L^\infty(v)$, and $\Gamma^* \Gamma_1^{*-1}$ is the identity. Therefore the proof is complete. ∎

4.59 The preceding result gives very precise information concerning normal operators possessing a cyclic vector. Unfortunately, most normal operators do not have a cyclic vector. Consider the example:

$$\mathscr{H} = L^2([0,1]) \oplus L^2([0,1])$$

and $T = M_g \oplus M_g$, where $g(t) = t$. It is easy to verify that $\mathfrak{C}_T$ has no cyclic vector and this is left as an exercise. Thus the preceding result does not apply to T. Notice, however, that

$$\mathfrak{C}_T = \{M_\varphi \oplus M_\varphi : \varphi \in C([0,1])\} \quad \text{and} \quad \mathfrak{W}_T = \{M_\varphi \oplus M_\varphi : \varphi \in L^\infty([0,1])\},$$

and thus there still exists a *-isometrical isomorphism Γ^* from $\mathfrak{W}_T$ onto $L^\infty([0,1])$ which extends the Gelfand transform on $\mathfrak{C}_T$. The difference is that Γ^* is no longer spatially implemented.

To see how to obtain Γ^* in the general case, observe that $f = 1 \oplus 0$ is a separating vector for $\mathfrak{W}_T$ and that the space $\mathscr{M} = \text{clos}[\mathfrak{W}_T f]$ is just $L^2([0,1]) \oplus \{0\}$; moreover, $\mathscr{M}$ is a reducing subspace for T and the mapping $A \to A \,|\, \mathscr{M}$ defines a *-isometrical isomorphism from $\mathfrak{W}_T$ onto the W^*-algebra generated by $T \,|\, \mathscr{M}$. Lastly, the operator $T \,|\, \mathscr{M}$ is normal and $\mathfrak{C}_{T|\mathscr{M}}$ has a cyclic vector; hence the preceding theorem applies to it.

Our program is as follows: For a normal T we show $\mathfrak{W}_T$ has a separating vector, that $\mathfrak{W}_T$ and $\mathfrak{W}_{T|\mathscr{M}}$ are naturally isomorphic, and that the theorem applies to $T|\mathscr{M}$. Thus we obtain the desired result for arbitrary normal operators. To show that $\mathfrak{W}_T$ has a separating vector requires some preliminary results on W^*-algebras.

4.60 Proposition If $\mathfrak{A}$ is an abelian C^*-algebra contained in $\mathfrak{L}(\mathscr{H})$, then there exists a maximal abelian W^*-algebra in $\mathfrak{L}(\mathscr{H})$ containing $\mathfrak{A}$.

Proof The commutative C^*-subalgebras of $\mathfrak{L}(\mathscr{H})$ which contain $\mathfrak{A}$ are partially ordered by inclusion. Since the norm closure of the union of any chain is a commutative C^*-algebra of $\mathfrak{L}(\mathscr{H})$ containing $\mathfrak{A}$, then there is a maximal element by Zorn's lemma. Since the closure of such an element in the weak operator topology is commutative by Proposition 4.47, it follows that such an element is a W^*-algebra. ∎

The following notion is of considerable importance in any serious study of W^*-algebras.

4.61 Definition If $\mathfrak{A}$ is a subset of $\mathfrak{L}(\mathcal{H})$, then the commutant of $\mathfrak{A}$, denoted $\mathfrak{A}'$, is the set of operators in $\mathfrak{L}(\mathcal{H})$ which commute with every operator in $\mathfrak{A}$.

It is easy to show that if $\mathfrak{A}$ is a self-adjoint subset of $\mathfrak{L}(\mathcal{H})$, then $\mathfrak{A}'$ is a W^*-algebra.

The following proposition gives an algebraic characterization of maximal abelian W^*-algebras.

4.62 Proposition A C^*-subalgebra $\mathfrak{A}$ of $\mathfrak{L}(\mathcal{H})$ is a maximal abelian W^*-algebra if and only if $\mathfrak{A} = \mathfrak{A}'$.

Proof If $\mathfrak{A} = \mathfrak{A}'$, then $\mathfrak{A}$ is a W^*-algebra by our previous remarks. Moreover, by definition each operator in $\mathfrak{A}'$ commutes with every operator in $\mathfrak{A}$ and therefore $\mathfrak{A}$ is abelian. Moreover, if A is an operator commuting with $\mathfrak{A}$, then A is in $\mathfrak{A}'$ and hence already in $\mathfrak{A}$. Thus $\mathfrak{A}$ is a maximal abelian W^*-algebra.

Conversely, if $\mathfrak{A}$ is an abelian W^*-algebra, then $\mathfrak{A} \subset \mathfrak{A}'$. Moreover, if T is in $\mathfrak{A}'$, then $T = H + iK$, where H and K are self-adjoint and in $\mathfrak{A}'$. Since the W^*-algebra generated by $\mathfrak{A}$ and either H or K is an abelian W^*-algebra, then for $\mathfrak{A}$ to be maximal abelian, it is necessary for H and K to be in $\mathfrak{A}$ and hence $\mathfrak{A} = \mathfrak{A}'$. ■

4.63 Lemma If $\mathfrak{A}$ is a C^*-algebra contained in $\mathfrak{L}(\mathcal{H})$ and f is a vector in $\mathcal{H}$, then the projection onto the closure of $\mathfrak{A}f$ is in $\mathfrak{A}'$.

Proof In view of Proposition 4.42 it suffices to show that $\operatorname{clos}[\mathfrak{A}f]$ is invariant for both A and A^* for each A in $\mathfrak{A}$. That is obvious from the definition of the subspace and the fact that $\mathfrak{A}$ is self-adjoint. ■

The following proposition shows one of the reasons for the importance of the strong operator topology.

4.64 Proposition If $\mathcal{H}$ is a Hilbert space and $\{P_\alpha\}_{\alpha \in A}$ is a net of positive operators on $\mathcal{H}$ such that $0 \leqslant P_\alpha \leqslant P_\beta \leqslant I$ for α and β in A with $\alpha \leqslant \beta$, then there exists P in $\mathfrak{L}(\mathcal{H})$ such that $0 \leqslant P_\alpha \leqslant P \leqslant I$ for α in A and the net $\{P_\alpha\}_{\alpha \in A}$ converges to P in the strong operator topology.

Proof If Q is in $\mathfrak{L}(\mathcal{H})$ and $0 \leqslant Q \leqslant I$, then $0 \leqslant Q^2 \leqslant Q \leqslant I$ since Q commutes with $(I-Q)^{1/2}$ by Proposition 4.33 and since

$$((Q-Q^2)f,f) = (Q(I-Q)^{1/2}f, (I-Q)^{1/2}f) \geqslant 0 \qquad \text{for } f \text{ in } \mathcal{H}.$$

Further, for each f in $\mathcal{H}$ the net $\{(P_\alpha f, f)\}_{\alpha \in A}$ is increasing and hence a Cauchy net. Since for $\beta \geqslant \alpha$, we have

$$\|(P_\beta - P_\alpha)f\|^2 = ((P_\beta - P_\alpha)^2 f, f) \leqslant ((P_\beta - P_\alpha)f, f) = (P_\beta f, f) - (P_\alpha f, f),$$

it therefore follows that $\{P_\alpha f\}_{\alpha \in A}$ is a Cauchy net in the norm of $\mathcal{H}$. If we define $Pf = \lim_{\alpha \in A} P_\alpha f$, then P is linear,

$$\|Pf\| \leqslant \lim_{\alpha \in A} \|P_\alpha f\| \leqslant \|f\|, \qquad \text{and} \qquad 0 \leqslant \lim_{\alpha \in A}(P_\alpha f, f) = (Pf, f).$$

Therefore, P is a bounded positive operator, $0 \leqslant P_\alpha \leqslant P \leqslant I$, for α in A, and $\{P_\alpha\}_{\alpha \in A}$ converges strongly to P, and the proof is complete. ∎

The converse of the following theorem is also valid but is left as an exercise.

4.65 Theorem If $\mathfrak{A}$ is a maximal abelian W^*-algebra on the separable Hilbert space $\mathcal{H}$, then $\mathfrak{A}$ has a cyclic vector.

Proof Let $\mathcal{E}$ denote the set of all collections of projections $\{E_\alpha\}_{\alpha \in A}$ in $\mathfrak{A}$ such that:

(1) For each α in A there exists a nonzero vector f_α in $\mathcal{H}$ such that E_α is the projection onto $\mathrm{clos}[\mathfrak{A}f_\alpha]$; and

(2) the subspaces $\{\mathrm{clos}[\mathfrak{A}f_\alpha]\}_{\alpha \in A}$ are pairwise orthogonal.

We want to show there is an element $\{E_\alpha\}_{\alpha \in A}$ in $\mathcal{E}$ such that the span of the ranges of the E_α is all of $\mathcal{H}$.

Order $\mathcal{E}$ by inclusion, that is, $\{E_\alpha\}_{\alpha \in A}$ is greater than or equal to $\{F_\beta\}_{\beta \in B}$ if for each β_0 in B there exists α_0 in A such that $F_{\beta_0} = E_{\alpha_0}$. To show that $\mathcal{E}$ is nonempty observe that the one element set $\{P_{\mathrm{clos}[\mathfrak{A}f]}\}$ is in $\mathcal{E}$ for each nonzero vector f in $\mathcal{H}$. Moreover, since the union of a chain of elements in $\mathcal{E}$ is in $\mathcal{E}$, then $\mathcal{E}$ has a maximal element $\{E_\alpha\}_{\alpha \in A}$ by Zorn's lemma.

Let $\mathcal{F}$ denote the collection of all finite subsets of the index set A partially ordered by inclusion and let $\{P_F\}_{F \in \mathcal{F}}$ denote the net of operators defined by $P_F = \sum_{\alpha \in F} E_\alpha$. By the remark after Proposition 4.19, each P_F is a projection and hence the net is increasing. Therefore, by the previous proposition there exists a positive operator P such that $0 \leqslant P_F \leqslant P \leqslant I$ for every F in $\mathcal{F}$ and $\{P_F\}_{F \in \mathcal{F}}$ converges to P in the strong operator topology. Therefore, $\{P_F^2\}_{F \in \mathcal{F}}$ converges strongly to P^2 and P is seen to be a projection.

The projection P is in $\mathfrak{A}$ since $\mathfrak{A}$ is weakly closed and the range $\mathcal{M}$ of P reduces $\mathfrak{A}$ since $\mathfrak{A}$ is abelian. Thus, if f is a nonzero vector in $\mathcal{M}^\perp$, then clos$[\mathfrak{A}f]$ is orthogonal to the range of each E_α. If we let E_β denote the projection onto clos$[\mathfrak{A}f]$, then $\{E_\alpha\}_{\alpha \in A \cup \{\beta\}}$ is an element in $\mathscr{E}$ larger than $\{E_\alpha\}_{\alpha \in A}$. This contradiction shows that $\mathcal{M}^\perp = \{0\}$ and hence that $P = I$.

Since $\mathscr{H}$ is separable, $\dim \operatorname{ran} E_\alpha > 0$, and $\dim \mathscr{H} = \sum_{\alpha \in A} \dim \operatorname{ran} E_\alpha$, we see that A is countable. Enumerate A such that $A = \{\alpha_1, \alpha_2, \ldots\}$ and set $f = \sum_{i \geqslant 1} 2^{-i} f_{\alpha_i}/\|f_{\alpha_i}\|$. Since $E_{\alpha_i} f = 2^{-i} f_{\alpha_i}/\|f_{\alpha_i}\|$, we see that the range of E_{α_i} is contained in clos$[\mathfrak{A}f]$. Therefore, since the ranges of the E_{α_i} span $\mathscr{H}$, we see that f is a cyclic vector for $\mathfrak{A}$ and the proof is complete. ■

The assumption that $\mathscr{H}$ is separable is needed only to conclude that A is countable.

The following result is what we need in our study of normal operators.

4.66   Corollary   If $\mathfrak{A}$ is an abelian C^*-algebra defined on the separable Hilbert space $\mathscr{H}$, then $\mathfrak{A}$ has a separating vector.

Proof   By Proposition 4.60 $\mathfrak{A}$ is contained in a maximal abelian W^*-algebra $\mathfrak{B}$ which has a cyclic vector f by the previous theorem. Finally, by Lemma 4.57 the vector f is a separating vector for $\mathfrak{B}$ and hence also for the subalgebra $\mathfrak{A}$. ■

The appropriate setting for the last two results is in W^*-algebras having the property that every collection of pairwise disjoint projections is countable. While for algebras defined on a separable Hilbert space this is always the case, it is not necessarily true for W^*-algebras defined on a nonseparable Hilbert space $\mathscr{H}$; for example, consider $\mathfrak{L}(\mathscr{H})$.

Before we can proceed we need one more technical result concerning *-homomorphisms between C^*-algebras.

4.67   Proposition   If $\mathfrak{A}$ and $\mathfrak{B}$ are C^*-algebras and Φ is a *-homomorphism from $\mathfrak{A}$ to $\mathfrak{B}$, then $\|\Phi\| \leqslant 1$ and Φ is an isometry if and only if Φ is one-to-one.

Proof   If H is a self-adjoint element of $\mathfrak{A}$, then $\mathfrak{C}_H$ is an abelian C^*-algebra contained in $\mathfrak{A}$ and $\Phi(\mathfrak{C}_H)$ is an abelian *-algebra contained in $\mathfrak{B}$. If ψ is a multiplicative linear functional on the closure of $\Phi(\mathfrak{C}_H)$ then $\psi \circ \Phi$ defines a multiplicative linear functional on $\mathfrak{C}_H$. If ψ is chosen such that $|\psi(\Phi(H))| = \|\Phi(H)\|$, which we can do by Theorem 4.29, then we have

$\|H\| \geqslant |\psi(\Phi(H))| = \|\Phi(H)\|$ and hence Φ is a contraction on the self-adjoint elements of $\mathfrak{A}$. Thus we have for T in $\mathfrak{A}$ that

$$\|T\|^2 = \|T^*T\| \geqslant \|\Phi(T^*T)\| = \|\Phi(T)^*\Phi(T)\| = \|\Phi(T)\|^2$$

and hence $\|\Phi\| \leqslant 1$.

Now for the second statement. Clearly, if Φ is an isometry, then it is one-to-one. Therefore, assume that Φ is not an isometry and that T is an element of $\mathfrak{A}$ for which $\|T\| = 1$ and $\|\Phi(T)\| < 1$. If we set $A = T^*T$, then $\|A\| = 1$ and $\|\Phi(A)\| = 1 - \varepsilon$ with $\varepsilon > 0$. Let f be a function in $C([0,1])$ chosen such that $f(1) = 1$ and $f(x) = 0$ for $0 \leqslant x \leqslant 1 - \varepsilon$. Then using the functional calculus on $\mathfrak{C}_A$ we can define $f(A)$ and since by Corollary 2.37 we have $\sigma(f(A)) = \text{range}\,\Gamma(f(A)) = f(\sigma(A))$, we conclude that 1 is in $\sigma(f(A))$ and thus $f(A) \neq 0$. Since Φ is a contractive *-homomorphism, it is clear that $\Phi(p(A)) = p(\Phi(A))$ for each polynomial and hence $\Phi(f(A)) = f(\Phi(A))$. Since, however, we have $\|\Phi(A)\| = 1 - \varepsilon$, it follows that $\sigma(\Phi(A)) \subset [0, 1-\varepsilon]$ and therefore

$$\sigma(\Phi(f(A))) = f(\sigma(\Phi(A))) \subset f([0, 1-\varepsilon]) = \{0\}.$$

Since $\Phi(A)$ is self-adjoint, we have $\Phi(A) = 0$, which shows that Φ is not one-to-one. ■

Now we are prepared to extend the functional calculus for an arbitrary normal operator on a separable Hilbert space.

4.68 Let $\mathscr{H}$ be a separable Hilbert space and T be a normal operator on $\mathscr{H}$. By Corollary 4.66, the abelian W^*-algebra $\mathfrak{W}_T$ has a separating vector f. If we set $\mathscr{M} = \text{clos}\lceil\mathfrak{W}_T f\rceil$, then $\mathscr{M}$ is a reducing subspace for each A in $\mathfrak{W}_T$, and hence we can define the mapping Φ from $\mathfrak{W}_T$ to $\mathfrak{L}(\mathscr{M})$ by $\Phi(A) = A \,|\, \mathscr{M}$ for A in $\mathfrak{W}_T$. It is clear that Φ is a *-homomorphism. We use the previous result to show that Φ is a *-isometrical isomorphism.

4.69 Lemma If $\mathfrak{A}$ is a C^*-algebra contained in $\mathfrak{L}(\mathscr{H})$, f is a separating vector for $\mathfrak{A}$ and $\mathscr{M}$ is the closure of $\mathfrak{A}f$, then the mapping Φ defined $\Phi(A) = A \,|\, \mathscr{M}$ for A in $\mathfrak{A}$ is a *-isometrical isomorphism from $\mathfrak{A}$ into $\mathfrak{L}(\mathscr{M})$. Moreover, $\sigma(A) = \sigma(A \,|\, \mathscr{M})$ for A in $\mathfrak{A}$.

Proof Since Φ is obviously a *-homomorphism, it is sufficient to show that Φ is one-to-one. If A is in $\mathfrak{A}$ and $\Phi(A) = 0$, then $Af = 0$, which implies $A = 0$ since f is a separating vector for $\mathfrak{A}$. The last remark follows from

Theorem 4.28, since we have

$$\sigma_{\mathfrak{L}(\mathscr{H})}(A) = \sigma_{\mathfrak{A}}(A) = \sigma_{\Phi(\mathfrak{A})}(A|\mathscr{M}) = \sigma_{\mathfrak{L}(\mathscr{M})}(A|\mathscr{M}). \quad \blacksquare$$

Finally, we shall need the following result whose proof is similar to that of Theorem 1.23 and hence is left as an exercise.

4.70   Proposition   If $\mathfrak{A}$ is a W*-algebra contained in $\mathfrak{L}(\mathscr{H})$, then the unit ball of $\mathfrak{A}$ is compact in the weak operator topology.

Our principal result on normal operators can now be given.

4.71   Theorem (Extended Functional Calculus)   If T is a normal operator on the separable Hilbert space $\mathscr{H}$ with spectrum Λ, and Γ is the Gelfand transform from $\mathfrak{C}_T$ onto $C(\Lambda)$, then there exists a positive regular Borel measure ν having support Λ and a *-isometrical isomorphism Γ^* from $\mathfrak{W}_T$ onto $L^\infty(\nu)$ which extends Γ. Moreover, the measure ν is unique up to mutual absolute continuity, while the space $L^\infty(\nu)$ and Γ^* are unique.

Proof   Let f be a separating vector for $\mathfrak{W}_T$, $\mathscr{M}$ be the closure of $\mathfrak{W}_T f$ and Φ be the *-isometrical isomorphism defined from $\mathfrak{W}_T$ into $\mathfrak{L}(\mathscr{M})$ by $\Phi(A) = A|\mathscr{M}$ as in Section 4.68. Further, let $\mathfrak{W}_\mathscr{M}$ be the W*-algebra generated by $T|\mathscr{M}$. Since Φ is defined by restricting the domain of the operators, it follows that Φ is continuous from the weak operator topology on $\mathfrak{W}_T$ to the weak operator topology on $\mathfrak{L}(\mathscr{M})$, and hence $\Phi(\mathfrak{W}_T) \subset \mathfrak{W}_\mathscr{M}$. Moreover, it is obvious that if Γ_0 is the Gelfand transform from $\mathfrak{C}_{T|\mathscr{M}}$ onto $C(\Lambda)$, then $\Gamma = \Gamma_0 \circ \Phi$.

Since $T|\mathscr{M}$ is normal and has the cyclic vector f, there exist by Theorem 4.58 a positive regular measure ν with support equal to $\sigma(T|\mathscr{M}) = \Lambda$ by the previous lemma and a *-isometrical isomorphism Γ_0^* from $\mathfrak{W}_\mathscr{M}$ onto $L^\infty(\nu)$ which extends the Gelfand transform Γ_0 on $\mathfrak{C}_{T|\mathscr{M}}$. Moreover, Γ_0^* is continuous from the weak operator topology on $\mathfrak{W}_\mathscr{M}$ to the w*-topology on $L^\infty(\nu)$. Therefore, the composition $\Gamma^* = \Gamma_0^* \circ \Phi$ is a *-isometrical isomorphism from $\mathfrak{W}_T$ into $L^\infty(\nu)$ which is continuous from the weak operator topology on $\mathfrak{W}_T$ to the w*-topology on $L^\infty(\nu)$ and extends the Gelfand transform Γ on $\mathfrak{C}_T$.

The only thing remaining is to show that Γ^* takes $\mathfrak{W}_T$ onto $L^\infty(\nu)$. To do this we argue as follows: Since the unit ball of $\mathfrak{W}_T$ is compact in the weak operator topology, it follows that its image is w*-compact in $L^\infty(\nu)$ and hence w*-closed. Since this image contains the unit ball of $C(\Lambda)$, it follows from Proposition 4.52 that Γ^* takes the unit ball of $\mathfrak{W}_T$ onto the unit ball of $L^\infty(v)$. Thus, Γ^* is onto.

The uniqueness assertion follows as in Theorem 4.58 from Theorem 4.55. ∎

4.72 Definition If T is a normal operator on the separable Hilbert space $\mathscr{H}$, then there exists a unique equivalence class of measures v on Λ such that there is a *-isometrical isomorphism Γ^* from $\mathfrak{W}_T$ onto $L^\infty(v)$ such that $\Gamma^*(\varphi(T)) = \varphi$ for φ in $C(\Lambda)$. Any such measure is called a scalar spectral measure for T. The extended functional calculus for T is defined for φ in $L^\infty(v)$ such that $\Gamma^*(\varphi(T)) = \varphi$. If I_Λ is a characteristic function in $L^\infty(v)$, then $I_\Lambda(T)$ is a projection in $\mathfrak{W}_T$ called a spectral projection for T, and its range is called a spectral subspace for T.

We conclude this chapter with some remarks concerning normal operators on nonseparable Hilbert spaces and a proposition which will enable us to make use of certain aspects of the extended functional calculus for such operators. We begin with the proposition which we have essentially proved.

4.73 Proposition If $\mathfrak{A}$ is a norm separable C^*-subalgebra of $\mathfrak{L}(\mathscr{H})$, then $\mathscr{H} = \sum_{\alpha \in A} \oplus \mathscr{H}_\alpha$ such that each $\mathscr{H}_\alpha$ is separable and is a reducing subspace for $\mathfrak{A}$.

Proof It follows from the first three paragraphs of the proof of Theorem 4.65 that $\mathscr{H} = \sum_{\alpha \in A} \oplus \mathscr{H}_\alpha$, where each $\mathscr{H}_\alpha$ is the closure of $\mathfrak{A}f_\alpha$ for some f_α in $\mathscr{H}$. Since $\mathfrak{A}$ is norm separable, each $\mathscr{H}_\alpha$ is separable and the result follows. ∎

4.74 From this result it follows that if T is normal on $\mathscr{H}$, then each $T_\alpha = T | \mathscr{H}_\alpha$ is normal on a separable Hilbert space and hence has a scalar spectral measure v_α. If there exists a measure v such that each v_α is absolutely continuous with respect to v, then Theorem 4.71 can be shown to hold for T. If no such v exists, then the functional calculus for T is usually based on the algebra of bounded Borel functions on Λ. In particular, one defines $\varphi(T) = \sum_{\alpha \in A} \oplus \varphi(T_\alpha)$ for each Borel function φ on Λ. The primary deficiencies in this approach is that the range of this functional calculus is no longer $\mathfrak{W}_T$ and the norm of $\varphi(T)$ is less accessible.

Sometimes the spectral measure $E(\cdot)$ for T is made the principal object of study. If Δ is a Borel subset of Λ, then the spectral measure is defined by

$$E(\Delta) = I_\Delta(T) = \sum_{a \in A} \oplus I_\Delta(T_a)$$

and is a projection-valued measure such that $(E(\Delta)f, f)$ is countably additive for each f in $\mathcal{H}$. Moreover, the Stieltjes integral can be defined and $T = \int_\Lambda z \, dE$. We shall not develop these ideas further except to show that the range of each spectral measure for T lies in $\mathfrak{W}_T$.

4.75   Proposition   If T is a normal operator on the Hilbert space $\mathcal{H}$ and $E(\cdot)$ is a spectral measure for T, then $E(\Delta)$ is in $\mathfrak{W}_T$ for each Borel set Δ in Λ.

Proof   Let $T = \sum_{\alpha \in A} \oplus T_\alpha$ be the decomposition of T relative to which the spectral measure $E(\cdot)$ is defined, where each T_α acts on the separable space $\mathcal{H}_\alpha$ with scalar spectral measure μ_α. If $\mathcal{F}$ denotes the collection of finite subsets of Λ, then $\bigcup_{F \in \mathcal{F}} \sum_{\alpha \in F} \oplus \mathcal{H}_\alpha$ is a dense linear manifold in $\mathcal{H}$. Thus if Δ is a Borel subset of Λ, it is sufficient to show that for $f_1, f_2, ..., f_n$ lying in $\sum_{\alpha \in F_0} \oplus \mathcal{H}_\alpha$ for some F_0 in $\mathcal{F}$ and $\varepsilon > 0$, there exists a continuous function φ such that $0 \leqslant \varphi \leqslant 1$, and $\|(\varphi(T) - E(\Delta))f_i\| < \varepsilon$ for $i = 1, 2, ..., n$. Since

$$\|(\varphi(T) - E(\Delta))f_i\|^2 = \sum_{\alpha \in F_0} \|(\varphi(T_\alpha) - I_\Delta(T_\alpha)) P_{\mathcal{H}_\alpha} f_i\|^2$$

$$= \sum_{\alpha \in F_0} \int_\Lambda |\varphi - I_\Delta|^2 |P_{\mathcal{H}_\alpha} f_i|^2 \, d\mu_\alpha,$$

this is possible using Proposition 4.52.   ∎

We conclude this chapter with an important complimentary result. The following ingenious proof is due to Rosenblum [93].

4.76   Theorem (Fuglede)   If T is a normal operator on the Hilbert space $\mathcal{H}$ and X is an operator in $\mathfrak{L}(\mathcal{H})$ for which $TX = XT$, then X lies in $\mathfrak{W}_T'$.

Proof   Since $\mathfrak{W}_T$ is generated by T and T^*, the result will follow once it is established that $T^*X = XT^*$.

Since $T^k X = XT^k$ for each $k \geqslant 0$, it follows that $\exp(i\lambda T) X = X \exp(i\lambda T)$ for each λ in $\mathbb{C}$. Therefore, we have $X = \exp(i\lambda T) X \exp(-i\lambda T)$, and hence

$$F(\lambda) = \exp(i\lambda T^*) X \exp(-i\lambda T^*)$$

$$= \exp[i(\bar{\lambda}T + \lambda T^*)] X \exp[-i(\bar{\lambda}T + \lambda T^*)]$$

by Lemma 2.12, since $TT^* = T^*T$. Since $\bar{\lambda}T + \lambda T^*$ is self-adjoint, it follows that $\exp[i(\lambda T+\lambda T^*)]$ and $\exp[-i(\lambda T+\lambda T^*)]$ are unitary operators for λ in $\mathbb{C}$. Thus the operator-valued entire function $F(\lambda)$ is bounded and hence by Liouville's theorem must be constant (see the proof of Theorem 2.29). Lastly, differentiating with respect to λ yields

$$F'(\lambda) = i\lambda T^* \exp(i\lambda T^*) X \exp(-i\lambda T^*)$$

$$+ \exp(i\lambda T^*) X \exp(-i\lambda T^*)(-i\lambda T^*) = 0.$$

Canceling λ and then setting $\lambda = 0$ yields $T^*X = XT^*$, which completes the proof. ∎

Notes

The spectral theorem for self-adjoint operators is due to Hilbert, but elementary operator theory is the joint work of a number of authors including Hilbert, Riesz, Weyl, von Neumann, Stone, and others. Among the early works which are still of interest are von Neumann's early papers [81], [82], and the book of Stone [104]. More recent books include Akhieser and Glazman [2], Halmos [55], [58], Kato [70], Maurin [79], Naimark [80], Riesz and Sz.-Nagy [92], and Yoshida [117].

We have only introduced the most elementary results from the theory of C^*- and W^*-algebras. The interested reader should consult the two books of Dixmier [27], [28] for further information on the subject as well as a guide to the vast literature on this subject.

Exercises

4.1   If T is a linear transformation defined on the Hilbert space $\mathscr{H}$, then T is bounded if and only if

$$\sup\{|(Tf,f)| : f \in \mathscr{H}, \|f\| = 1\} < \infty.$$

Definition   If T is an operator defined on the Hilbert space $\mathscr{H}$, then the numerical range $W(T)$ of T is the set $\{(Tf,f) : f \in \mathscr{H}, \|f\| = 1\}$ and the numerical radius $w(T)$ is $\sup\{|\lambda| : \lambda \in W(T)\}$.

4.2   (*Hausdorff–Toeplitz*) If T is an operator on $\mathscr{H}$, then $W(T)$ is a convex set. Moreover, if $\mathscr{H}$ is finite dimensional, then $W(T)$ is compact. (Hint: Consider $W(T)$ for T compressed to the two-dimensional subspaces of $\mathscr{H}$.)

4.3   If T is a normal operator on $\mathcal{H}$, then the closure of $W(T)$ is the closed convex hull of $\sigma(T)$. Further, an extreme point of the closure of $W(T)$ belongs to $W(T)$ if and only if it is an eigenvalue for T.

4.4   If T is an operator on $\mathcal{H}$, then $\sigma(T)$ is contained in the closure of $W(T)$. (Hint: Show that if the closure of $W(T)$ lies in the open right-half plane, then T is invertible.)

4.5   If T is an operator on the Hilbert space $\mathcal{H}$, then $r(T) \leqslant w(T) \leqslant \|T\|$ and both inequalities can be strict.

4.6   (*Hellinger–Toeplitz*)   If S and T are linear transformations defined on the Hilbert space $\mathcal{H}$ such that $(Sf, g) = (f, Tg)$ for f and g in $\mathcal{H}$, then S and T are bounded and $T = S^*$.

4.7   If T is an operator on $\mathcal{H}$, then the graph $\{\langle f, Tf \rangle : f \in \mathcal{H}\}$ of T is a closed subspace of $\mathcal{H} \oplus \mathcal{H}$ with orthogonal complement $\{\langle -T^*g, g \rangle : g \in \mathcal{H}\}$.

4.8   If T is an operator on $\mathcal{H}$, then T is normal if and only if $\|Tf\| = \|T^*f\|$ for f in $\mathcal{H}$. Moreover, a complex number λ is an eigenvalue for a normal operator T if and only if $\bar{\lambda}$ is an eigenvalue for T^*. The latter statement is not valid for general operators on infinite-dimensional Hilbert spaces.

4.9   If S and T are self-adjoint operators on $\mathcal{H}$, then ST is self-adjoint if and only if S and T commute. If P and Q are projections on $\mathcal{H}$, then PQ is a projection if and only if P and Q commute. Determine the range of PQ in this case.

4.10   If $\mathcal{H}$ and $\mathcal{K}$ are Hilbert spaces and A is an operator on $\mathcal{H} \oplus \mathcal{K}$, then there exist unique operators A_{11}, A_{12}, A_{21}, and A_{22} in $\mathfrak{L}(\mathcal{H})$, $\mathfrak{L}(\mathcal{K}, \mathcal{H})$, $\mathfrak{L}(\mathcal{H}, \mathcal{K})$, and $\mathfrak{L}(\mathcal{K})$, respectively, such that

$$A\langle h, k \rangle = \langle A_{11}h + A_{12}k, A_{21}h + A_{22}k \rangle.$$

In other words A is given by the matrix

$$\begin{bmatrix} A_{11} & A_{12} \\ A_{21} & A_{22} \end{bmatrix}.$$

Moreover show that such a matrix defines an operator $\mathcal{H} \oplus \mathcal{K}$.

4.11   If $\mathcal{H}$ and $\mathcal{K}$ are Hilbert spaces, A is an operator on $\mathfrak{L}(\mathcal{K}, \mathcal{H})$, and J is the operator on $\mathcal{H} \oplus \mathcal{K}$ defined by the matrix

$$\begin{bmatrix} I & A \\ 0 & 0 \end{bmatrix},$$

then J is an idempotent. Moreover, J is a projection if and only if $A = 0$. Further, every idempotent on a Hilbert space $\mathscr{L}$ can be written in this form for some decomposition $\mathscr{L} = \mathscr{H} \oplus \mathscr{K}$.

Definition   If T_1 and T_2 are operators on the Hilbert space $\mathscr{H}_1$ and $\mathscr{H}_2$, respectively, then T_1 is similar (unitarily equivalent) to T_2 if and only if there exists an invertible operator (isometrical isomorphism) S from $\mathscr{H}_1$ onto $\mathscr{H}_2$ such that $T_2 S = S T_1$.

4.12   If $\mathscr{H}$ is a Hilbert space and J is an idempotent on $\mathscr{H}$ with range $\mathscr{M}$, then J and $P_{\mathscr{M}}$ are similar operators.

4.13   If $(X, \mathscr{S}, \mu)$ is a probability space and φ a function in $L^\infty(\mu)$, then λ is an eigenvalue for M_φ if and only if the set $\{x \in X : \varphi(x) = \lambda\}$ has positive measure.

4.14   Show that the unitary operator U defined in Section 4.25 has no eigenvalues, while the eigenvalues of the coisometry $U_+{}^*$ defined in Section 4.36 have multiplicity one.

4.15   If $\mathfrak{A}$ is a C^*-algebra, then the set $\mathscr{P}$ of positive elements in $\mathfrak{A}$ forms a closed convex cone such that $\mathscr{P} \cap -\mathscr{P} = \{0\}$. (Hint: Show that a self-adjoint contraction H in $\mathfrak{A}$ is positive if and only if $\|I - H\| \leqslant 1$.)

4.16   If $\mathfrak{A}$ is a C^*-algebra, then an element H in $\mathfrak{A}$ is positive if and only if there exists T in $\mathfrak{A}$ such that $H = T^*T$.* (Hint: Express T^*T as the difference of two positive operators and show that the second is zero.)

4.17   If P and Q are projections on $\mathscr{H}$ such that $\|P - Q\| < 1$, then $\dim \operatorname{ran} P = \dim \operatorname{ran} Q$. (Hint: Show that $(I - P) + Q$ is invertible, that $\ker P \cap \operatorname{ran} Q = \{0\}$, and that $P[\operatorname{ran} Q] = \operatorname{ran} P$.)

4.18   An operator V on $\mathscr{H}$ is an isometry if and only if $V^*V = I$. If V is an isometry on $\mathscr{H}$, then V is a unitary operator if and only if V^* is an isometry if and only if $\ker V^* = \{0\}$.

4.19   If $\mathscr{H}$ and $\mathscr{K}$ are Hilbert spaces and A is an operator on $\mathscr{H} \oplus \mathscr{K}$ given by the matrix

$$
\begin{bmatrix}
A_{11} & A_{12} \\
A_{21} & A_{22}
\end{bmatrix},
$$

then $\mathscr{H} \oplus \{0\}$ is an invariant subspace for A if and only if $A_{21} = 0$, and $\mathscr{H} \oplus \{0\}$ reduces A if and only if $A_{21} = A_{12} = 0$.

4.20   If U_+ is the unilateral shift, then the sequence $\{U_+^n\}_{n=1}^{\infty}$ converges to 0 in the weak operator topology but not in the strong. Moreover, the sequence $\{U_+^{*n}\}_{n=1}^{\infty}$ converges to 0 in the strong operator topology but not in the norm.

4.21   Show that multiplication is not continuous in both variables in either the weak or strong operator topologies. Show that it is in the relative strong operator topology on the unit ball of $\mathfrak{L}(\mathscr{H})$.

4.22   If $\mathscr{H}$ is a Hilbert space, then the unit ball of $\mathfrak{L}(\mathscr{H})$ is compact in the weak operator topology but not in the strong. (Hint: Compare the proof of Theorem 1.23.)

4.23   If $\mathfrak{A}$ is a W^*-algebra contained in $\mathfrak{L}(\mathscr{H})$, then the unit ball of $\mathfrak{A}$ is compact in the weak operator topology.

4.24   If $\mathfrak{A}$ is a *-subalgebra of $\mathfrak{L}(\mathscr{H})$, then $\mathfrak{A}_{(2)} = \{[\begin{smallmatrix} A & 0 \\ 0 & A \end{smallmatrix}] : A \in \mathfrak{A}\}$ is a *-subalgebra of $\mathfrak{L}(\mathscr{H} \oplus \mathscr{H})$. Similarly, $\mathfrak{A}_{(N)}$ is a *-subalgebra of $\mathfrak{L}(\mathscr{H} \oplus \cdots \oplus \mathscr{H})$ for any integer N. Moreover, $\mathfrak{A}_{(N)}$ is closed in the norm, strong, or weak operator topologies if and only if $\mathfrak{A}$ is. Further, we have the identity $\mathfrak{A}_{(N)}{}'' = \mathfrak{A}_{(N)}''$.

4.25   If $\mathfrak{A}$ is a *-subalgebra of $\mathscr{H}$, A is in $\mathfrak{A}''$, $x_1, x_2, \ldots, x_N$ are vectors in $\mathscr{H}$, and $\varepsilon > 0$, then there exists B in $\mathfrak{A}$ such that $\|Ax_i - Bx_i\| < \varepsilon$ for $i = 1, 2, \ldots, N$. (Hint: Show first that for $\mathscr{M}$ a subspace of $\mathscr{H} \oplus \cdots \oplus \mathscr{H}$, we have $\operatorname{clos}[\mathfrak{A}_{(N)}' \mathscr{M}] = \operatorname{clos}[\mathfrak{A}_{(N)} \mathscr{M}]$.)

4.26   (*von Neumann Double Commutant Theorem*)   If $\mathfrak{A}$ is a *-subalgebra of $\mathfrak{L}(\mathscr{H})$, then $\mathfrak{A}$ is a W^*-algebra if and only if $\mathfrak{A} = \mathfrak{A}''$.*

4.27   If $\mathfrak{A}$ is a C^*-subalgebra of $\mathfrak{L}(\mathscr{H})$, then $\mathfrak{A}$ is a W^*-algebra if and only if it is closed in the strong operator topology.

4.28   If S and T are operators in the Hilbert spaces $\mathscr{H}$ and $\mathscr{K}$, respectively, then an operator $S \otimes T$ can be defined on $\mathscr{H} \otimes \mathscr{K}$ in a natural way such that $\|S \otimes T\| = \|S\| \|T\|$ and $(S \otimes T)^* = S^* \otimes T^*$.

In Exercises (4.29–4.34) we are considering linear transformations defined on only a linear subspace of the Hilbert space.

Definition   A linear transformation L defined on the linear subspace $\mathscr{D}_L$ of the Hilbert space $\mathscr{H}$ is closable if the closure in $\mathscr{H} \oplus \mathscr{H}$ of the graph $\{\langle f, Lf \rangle : f \in \mathscr{D}_L\}$ of L is the graph of a linear transformation $\bar{L}$ called the closure of L. If L has a dense domain, is closable, and $L = \bar{L}$, then L is said to be closed.

4.29 Give an example of a densely defined linear transformation which is not closable. If T is a closable linear transformation with $\mathcal{D}_T = \mathcal{H}$, then T is bounded.

4.30 If L is a closable, densely defined linear transformation on $\mathcal{H}$, then these exists a closed, densely defined linear transformation M on $\mathcal{H}$ such that $(Lf, g) = (f, Mg)$ for f in $\mathcal{D}_L$ and g in $\mathcal{D}_M$. Moreover, if N is any linear transformation on $\mathcal{H}$ for which $(Lf, g) = (f, Ng)$ for f in $\mathcal{D}_L$ and g in $\mathcal{D}_N$, then $\mathcal{D}_N \subset \mathcal{D}_M$ and $Ng = Mg$ for g in $\mathcal{D}_N$. (Hint: Show that the graph of M can be obtained as in Exercise 4.7 as the orthogonal complement of the graph of L.)

Definition If L is a closable, densely defined linear transformation on $\mathcal{H}$, then the operator given in the preceding problem is called the adjoint of L and is denoted L^*. A densely defined linear transformation H is symmetric if (Hf, f) is real for f in $\mathcal{D}_H$ and self-adjoint if $H = H^*$.

4.31 If T is a closed, densely defined linear transformation on $\mathcal{H}$, then $T = T^{**}$. * (This includes the fact that $\mathcal{D}_T = \mathcal{D}_{T^{**}}$). If H is a densely defined symmetric transformation on $\mathcal{H}$, then H is closable and H^* extends H.

4.32 If H is a densely defined symmetric linear transformation on $\mathcal{H}$ with range equal to $\mathcal{H}$, then H is self-adjoint.

4.33 If T is a closed, densely defined linear transformation on $\mathcal{H}$, then T^*T is a densely defined, symmetric operator. (Note: T^*Tf is defined for those f for which Tf is in $\mathcal{D}_{T^*}$.)

4.34 If T is a closed, densely defined linear transformation on $\mathcal{H}$, then T^*T is self-adjoint. (Hint: Show that the range of $I + T^*T$ is dense in $\mathcal{H}$ and closed.)

4.35 If $\mathfrak{A}$ is a commutative W^*-algebra on $\mathcal{H}$ with $\mathcal{H}$ separable, then $\mathfrak{A}$ is a *-isometrically isomorphic to $L^\infty(\mu)$ for some probability space $(X, \mathcal{S}, \mu)$.

4.36 Show that there exist W^*-algebras $\mathfrak{A}$ and $\mathfrak{B}$ such that $\mathfrak{A}$ and $\mathfrak{B}$ are *-isomorphic but $\mathfrak{A}'$ and $\mathfrak{B}'$ are not.

4.37 If $\mathfrak{A}$ is an abelian W^*-algebra on $\mathcal{H}$ for $\mathcal{H}$ separable, then $\mathfrak{A}$ is maximal abelian if and only if $\mathfrak{A}$ has a cyclic vector.

4.38   Give an alternate proof of Fuglede's theorem for normal operators on a separable Hilbert space as follows: Show that it is enough to prove that $E(\Delta_1) X E(\Delta_2) = 0$ for Δ_1 and Δ_2 disjoint Borel sets; show this first for Borel sets at positive distance from each other and then approximate from within by compact sets in the general case.

4.39   (*Putnam*) If T_1 and T_2 are normal operators on the Hilbert spaces $\mathcal{H}_1$ and $\mathcal{H}_2$, respectively, and X in $\mathfrak{L}(\mathcal{H}_1, \mathcal{H}_2)$ satisfies $T_2 X = X T_1$, then $T_2^* X = X T_1^*$. (Hint: Consider the normal operator $\begin{bmatrix} T_1 & 0 \\ 0 & T_2 \end{bmatrix}$ on $\mathcal{H}_1 \oplus \mathcal{H}_2$ together with the operator $\begin{bmatrix} 0 & 0 \\ X & 0 \end{bmatrix}$).

4.40   If T_1 and T_2 are normal operators on the Hilbert spaces $\mathcal{H}_1$ and $\mathcal{H}_2$, respectively, then T_1 is similar to T_2 if and only if T_1 is unitarily equivalent to T_2.

4.41   Let X be a compact Hausdorff space, $\mathcal{H}$ be a Hilbert space, and Φ be a *-isomorphism from $C(X)$ into $\mathfrak{L}(\mathcal{H})$. Show that if there exists a vector f in $\mathcal{H}$ for which $\Phi(C(X))$ is dense in $\mathcal{H}$, then there exists a probability measure μ on X and an isometrical isomorphism Ψ from $L^2(\mu)$ onto $\mathcal{H}$ such that $\Psi M_\varphi \Psi^* = \Phi(\varphi)$ for φ in $C(X)$. (Hint: repeat the argument for Theorem 4.58.)

4.42   Let X be a compact Hausdorff space, $\mathcal{H}$ be a Hilbert space, and Φ be a *-isomorphism from $C(X)$ into $\mathfrak{L}(\mathcal{H})$, then there exists a *-homomorphism Φ^* from the algebra $\mathcal{B}(X)$ of bounded Borel functions on X which extends Φ. Moreover, the range of Φ^* is contained in the von Neumann algebra generated by the range of Φ. (Hint: use the arguments of 4.74 and 4.75 together with the preceding exercise.)

5 *Compact Operators, Fredholm Operators, and Index Theory*

5.1 In the preceding chapter we studied operators on Hilbert space and obtained, in particular, the spectral theorem for normal operators. As we indicated this result can be viewed as the appropriate generalization to infinite-dimensional spaces of the diagonalizability of matrices on finite-dimensional spaces. There is another class of operators which are a generalization in a topological sense of operators on a finite-dimensional space. In this chapter we study these operators and a certain related class. The organization of our study is somewhat unorthodox and is arranged so that the main results are obtained as quickly as possible. We first introduce the class of compact operators and show that this class coincides with the norm closure of the finite rank operators. After that we give some concrete examples of compact operators and then proceed to introduce the notion of a Fredholm operator. We begin with a definition.

5.2 Definition If $\mathcal{H}$ is a Hilbert space, then an operator T in $\mathfrak{L}(\mathcal{H})$ is a finite rank operator if the dimension of the range of T is finite and a compact operator if the image of the unit ball of $\mathcal{H}$ under T is a compact subset of $\mathcal{H}$. Let $\mathfrak{L}\mathfrak{F}(\mathcal{H})$, respectively, $\mathfrak{L}\mathfrak{C}(\mathcal{H})$ denote the set of finite rank, respectively, compact operators.

In the definition of compact operator it is often assumed only that

$T[(\mathcal{H})_1]$ has a compact closure. The equivalence of these two notions follows from the corollary to the next lemma.

5.3 Lemma If $\mathcal{H}$ is a Hilbert space and T is in $\mathfrak{L}(\mathcal{H})$, then T is a continuous function from $\mathcal{H}$ with the weak topology to $\mathcal{H}$ with the weak topology.

Proof If $\{f_\alpha\}_{\alpha \in A}$ is a net in $\mathcal{H}$ which converges weakly to f and g is a vector in $\mathcal{H}$, then

$$\lim_{\alpha \in A}(Tf_\alpha, g) = \lim_{\alpha \in A}(f_\alpha, T^*g) = (f, T^*g) = (Tf, g),$$

and hence the net $\{Tf_\alpha\}_{\alpha \in A}$ converges weakly to Tf. Thus T is weakly continuous. ∎

5.4 Corollary If $\mathcal{H}$ is a Hilbert space and T is in $\mathfrak{L}(\mathcal{H})$, then $T[(\mathcal{H})_1]$ is a closed subset of $\mathcal{H}$.

Proof Since $(\mathcal{H})_1$ is weakly compact and T is weakly continuous, it follows that $T[(\mathcal{H})_1]$ is weakly compact. Hence, $T[(\mathcal{H})_1]$ is a weakly closed subset of $\mathcal{H}$ and therefore is also norm closed. ∎

The following proposition summarizes most of the elementary facts about finite rank operators.

5.5 Proposition If $\mathcal{H}$ is a Hilbert space, then $\mathfrak{LF}(\mathcal{H})$ is the minimal two-sided *-ideal in $\mathfrak{L}(\mathcal{H})$.

Proof If S and T are finite rank operators, then the inclusion

$$\operatorname{ran}(S+T) \subset \operatorname{ran} S + \operatorname{ran} T$$

implies that $S+T$ is finite rank. Thus, $\mathfrak{LF}(\mathcal{H})$ is a linear subspace. If S is a finite rank operator and T is an operator in $\mathfrak{L}(\mathcal{H})$, then the inclusion $\operatorname{ran} ST \subset \operatorname{ran} S$ shows that $\mathfrak{LF}(\mathcal{H})$ is a left ideal in $\mathfrak{L}(\mathcal{H})$. Further, if T is a finite rank operator, then the identity

$$\operatorname{ran} T^* = T^*[(\ker T^*)^\perp] = T^*[\operatorname{clos} \operatorname{ran} T]$$

which follows from Corollary 3.22 and Proposition 4.6, shows that T^* is also a finite rank operator. Lastly, if S is in $\mathfrak{L}(\mathcal{H})$ and T is in $\mathfrak{LF}(\mathcal{H})$, then T^* is in $\mathfrak{LF}(\mathcal{H})$ which implies that T^*S^* is in $\mathfrak{LF}(\mathcal{H})$ and hence that $ST = (T^*S^*)^*$ is in $\mathfrak{LF}(\mathcal{H})$. Therefore, $\mathfrak{LF}(\mathcal{H})$ is a two-sided *-ideal in $\mathfrak{L}(\mathcal{H})$.

To show that $\mathfrak{LF}(\mathcal{H})$ is minimal, assume that $\mathfrak{I}$ is an ideal in $\mathfrak{L}(\mathcal{H})$

not (0). Thus there exists an operator $T \neq 0$ in $\mathfrak{I}$, and hence there is a non-zero vector f and a unit vector g in $\mathcal{H}$ such that $Tf = g$. Now let h and k be arbitrary nonzero vectors in $\mathcal{H}$ and A and B be the operators defined on $\mathcal{H}$ by $Al = (l,g)k$ and $Bl = (l,h)f$ for l in $\mathcal{H}$. Then, $S = ATB$ is the rank one operator in $\mathfrak{I}$ which takes h to k. It is now clear that $\mathfrak{I}$ contains all finite rank operators and hence $\mathfrak{L}\mathfrak{F}(\mathcal{H})$ is the minimal two-sided ideal in $\mathfrak{L}(\mathcal{H})$. ∎

The following proposition provides an alternative characterization of compact operators.

5.6   Proposition   If $\mathcal{H}$ is a Hilbert space and T is in $\mathfrak{L}(\mathcal{H})$, then T is compact if and only if for every bounded net $\{f_\alpha\}_{\alpha \in A}$ in $\mathcal{H}$ which converges weakly to f it is true that $\{Tf_\alpha\}_{\alpha \in A}$ converges in norm to Tf.

Proof   If T is compact and $\{f_\alpha\}_{\alpha \in A}$ is a bounded net in $\mathcal{H}$ which converges weakly to f, then $\{Tf_\alpha\}_{\alpha \in A}$ converges weakly to Tf by Lemma 5.3 and lies in a norm compact subset by the definition of compactness. Since any norm Cauchy subnet of $\{Tf_\alpha\}_{\alpha \in A}$ must converge to Tf, it follows that $\lim_{\alpha \in A} Tf_\alpha = Tf$ in the norm topology.

Conversely, suppose T is an operator in $\mathfrak{L}(\mathcal{H})$ for which the conclusion of the statement is valid. If $\{Tf_\alpha\}_{\alpha \in A}$ is a net of vectors in $T[(\mathcal{H})_1]$, then there exists a subnet $\{f_{\alpha_\beta}\}_{\beta \in B}$ which converges weakly to an f. Moreover, since each f_{α_β} is in the unit ball of $\mathcal{H}$, it follows that $\{Tf_{\alpha_\beta}\}_{\beta \in B}$ converges in norm to Tf. Therefore, $T[(\mathcal{H})_1]$ is a compact subset of $\mathcal{H}$ and hence T is compact. ∎

5.7   Lemma   The unit ball of a Hilbert space $\mathcal{H}$ is compact in the norm topology if and only if $\mathcal{H}$ is finite dimensional.

Proof   If $\mathcal{H}$ is finite dimensional, then $\mathcal{H}$ is isometrically isomorphic to $\mathbb{C}^n$ and the compactness of $(\mathcal{H})_1$ follows. On the other hand if $\mathcal{H}$ is infinite dimensional, then there exists an orthonormal subset $\{e_n\}_{n=1}^\infty$ contained in $(\mathcal{H})_1$ and the fact that $\|e_n - e_m\| = \sqrt{2}$ for $n \neq m$ shows that $(\mathcal{H})_1$ is not compact in the norm topology. ∎

The following property actually characterizes compact operators on a Hilbert space, but the proof of the converse is postponed until after the next theorem. Whether this property characterizes compact operators on a Banach space is unknown.

5.8   Lemma   If $\mathcal{H}$ is an infinite-dimensional Hilbert space and T is a compact operator, then the range of T contains no closed infinite-dimensional subspace.

Proof   Let $\mathcal{M}$ be a closed subspace contained in the range of T and let $P_{\mathcal{M}}$ be the projection onto $\mathcal{M}$. It follows easily from Proposition 5.6 that the operator $P_{\mathcal{M}}T$ is also compact. Let A be the operator defined from $\mathcal{H}$ to $\mathcal{M}$ by $Af = P_{\mathcal{M}}Tf$ for f in $\mathcal{H}$. Then A is bounded and onto and hence by the open mapping theorem is also open. Therefore, $A[(\mathcal{H})_1]$ contains the open ball in $\mathcal{M}$ of radius δ centered at 0 for some $\delta > 0$. Since the closed ball of radius δ is contained in the compact set $P_{\mathcal{M}}T[(\mathcal{H})_1]$, it follows from the preceding lemma that $\mathcal{M}$ is finite dimensional. ∎

We are now in a position to show that $\mathfrak{LC}(\mathcal{H})$ is the norm closure of $\mathfrak{LF}(\mathcal{H})$. The corresponding result for Banach spaces is unknown.

5.9   Theorem   If $\mathcal{H}$ is an infinite-dimensional Hilbert space, then $\mathfrak{LC}(\mathcal{H})$ is the norm closure of $\mathfrak{LF}(\mathcal{H})$.

Proof   We first show that the closure of $\mathfrak{LF}(\mathcal{H})$ is contained in $\mathfrak{LC}(\mathcal{H})$. Firstly, it is obvious that $\mathfrak{LF}(\mathcal{H})$ is contained in $\mathfrak{LC}(\mathcal{H})$. Secondly, to prove that $\mathfrak{LC}(\mathcal{H})$ is closed, assume that $\{K_n\}_{n=1}^{\infty}$ is a sequence of compact operators which converges in norm to an operator K. If $\{f_\alpha\}_{\alpha \in A}$ is a bounded net in $\mathcal{H}$ that converges weakly to f, and

$$M = \sup\{1, \|f_\alpha\| : \alpha \in A\},$$

then choose N such that $\|K - K_N\| < \varepsilon/3M$. Since K_N is a compact operator, we have by Proposition 5.6 that there exists α_0 in A such that $\|K_N f_\alpha - K_N f\| < \varepsilon/3$ for $\alpha \geq \alpha_0$. Thus we have

$$\|Kf_\alpha - Kf\| \leq \|(K - K_N)f_\alpha\| + \|K_N f_\alpha - K_N f\| + \|(K_N - K)f\|$$

$$\leq \frac{\varepsilon}{3} + \frac{\varepsilon}{3} + \frac{\varepsilon}{3} = \varepsilon \qquad \text{for} \quad \alpha \geq \alpha_0,$$

and hence K is compact by Proposition 5.6. Therefore, the closure of $\mathfrak{LF}(\mathcal{H})$ is contained in $\mathfrak{LC}(\mathcal{H})$.

To show that $\mathfrak{LF}(\mathcal{H})$ is dense in $\mathfrak{LC}(\mathcal{H})$, let K be a compact operator on $\mathcal{H}$ and let $K = PV$ be the polar decomposition for K. Let $\mathcal{H} = \sum_{\alpha \in A} \oplus \mathcal{H}_\alpha$ be a decomposition of $\mathcal{H}$ into separable reducing subspaces for P given by Proposition 4.73, and set $P_\alpha = P|\mathcal{H}_\alpha$. Further, let $\mathfrak{W}_\alpha$ be the abelian W^*-

algebra generated by P_α on $\mathscr{H}_\alpha$ and consider the extended functional calculus obtained from Theorem 4.71 and defined for functions in $L^\infty(v_\alpha)$ for some positive regular Borel measure v_α with support contained in $[0, \|P\|]$. If χ_ε denotes the characteristic function of the set $(\varepsilon, \|P\|]$, then χ_ε is in $L^\infty(v_\alpha)$ and $E_\alpha^\varepsilon = \chi_\varepsilon(P_\alpha)$ is a projection on $\mathscr{H}_\alpha$. If we define ψ_ε on $[0, \|P\|]$ such that $\psi_\varepsilon(x) = 1/x$ for $\varepsilon < x \leqslant \|P\|$ and 0 otherwise, then the operator $Q_\alpha^\varepsilon = \psi_\varepsilon(P_\alpha)$ satisfies $Q_\alpha^\varepsilon P_\alpha = P_\alpha Q_\alpha^\varepsilon = E_\alpha^\varepsilon$. Thus we have

$$\operatorname{ran}\Big(\sum_{\alpha \in A} \oplus E_\alpha^\varepsilon\Big) = \operatorname{ran} P\Big(\sum_{\alpha \in A} \oplus Q_\alpha^\varepsilon\Big) \subset \operatorname{ran} P = \operatorname{ran} K$$

and therefore the range of the projection $\sum_{\alpha \in A} \oplus E_\alpha^\varepsilon$ is finite dimensional by Lemma 5.7. Hence, $P_\varepsilon = P(\sum_{\alpha \in A} \oplus E_\alpha^\varepsilon)$ is in $\mathfrak{LF}(\mathscr{H})$ and thus so is $P_\varepsilon V$. Finally, we have

$$\|K - P_\varepsilon V\| = \|PV - P_\varepsilon V\| \leqslant \|P - P_\varepsilon\| = \sup_{\alpha \in A} \|P_\alpha - P_\alpha E_\alpha^\varepsilon\|$$

$$= \sup_{\alpha \in A} \sup_{0 \leqslant x \leqslant \|P\|} \|x - x\chi_\varepsilon(x)\|_\infty \leqslant \varepsilon,$$

and therefore the theorem is proved. ∎

Notice that in the last paragraph of the preceding proof we used only the fact that the range of K contained no closed infinite-dimensional subspace. Thus we have proved the remaining half of the following result.

5.10 Corollary If $\mathscr{H}$ is an infinite-dimensional Hilbert space and T is an operator on $\mathscr{H}$, then T is compact if and only if the range of T contains no closed infinite-dimensional subspaces.

5.11 Corollary If $\mathscr{H}$ is an infinite-dimensional Hilbert space, then $\mathfrak{LC}(\mathscr{H})$ is a minimal closed two-sided *-ideal in $\mathfrak{L}(\mathscr{H})$. Moreover, if $\mathscr{H}$ is separable, then $\mathfrak{LC}(\mathscr{H})$ is the only proper closed two-sided ideal in $\mathfrak{L}(\mathscr{H})$.

Proof For the first statement combine the theorem with Proposition 5.5. For the second, note by the previous corollary that if T is not compact, then the range of T contains a closed infinite-dimensional subspace $\mathscr{M}$. A simple application of the open mapping theorem yields the existence of an operator S on $\mathscr{H}$ such that $TS = P_\mathscr{M}$, and hence any two-sided ideal containing T must also contain I. This completes the proof. ∎

5.12 Example Let K be a complex function on the unit square $[0,1] \times [0,1]$ which is measurable and square-integrable with respect to planar

Lebesgue measure. We define a transformation T_K on $L^2([0,1])$ such that

$$(T_K f)(x) = \int_0^1 K(x,y) f(y) \, dy \qquad \text{for } f \text{ in } L^2([0,1]).$$

The computation

$$\begin{aligned}
\int_0^1 |(T_K f)(x)|^2 \, dx &= \int_0^1 \left| \int_0^1 K(x,y) f(y) \, dy \right|^2 dx \\
&\leqslant \int_0^1 \left\{ \int_0^1 |K(x,y)|^2 \, dy \right\} \left\{ \int_0^1 |f(y)|^2 \, dy \right\} dx \\
&= \|f\|_2^2 \int_0^1 \int_0^1 |K(x,y)|^2 \, dy \, dx
\end{aligned}$$

which uses the Cauchy–Schwarz inequality, shows that T_K is a bounded operator on $L^2([0,1])$ with

$$\|T_K\| \leqslant \|K\|_2 = \left\{ \int_0^1 \int_0^1 |K(x,y)|^2 \, dx \, dy \right\}^{1/2}.$$

The operator T_K is called an integral operator with kernel K. We want to show next that T_K is a compact operator.

If we let Φ denote the mapping from $L^2([0,1]) \times [0,1]$ to $\mathfrak{L}(L^2([0,1]))$ defined by $\Phi(K) = T_K$, then Φ is a contractive linear transformation. Let $\mathscr{D}$ be the subspace of $L^2([0,1] \times [0,1])$ consisting of the functions of the form

$$K(x,y) = \sum_{i=1}^N f_i(x) g_i(y),$$

where each f_i and g_i is continuous on $[0,1]$. Since $\mathscr{D}$ is obviously a self-adjoint subalgebra of the algebra of continuous functions on $[0,1] \times [0,1]$ which contains the identity and separates points, it follows from the Stone–Weierstrass theorem that $C([0,1] \times [0,1])$ is the uniform closure of $\mathscr{D}$. Moreover, an obvious modification of the argument given in Section 3.33 shows that $\mathscr{D}$ is dense in $L^2([0,1] \times [0,1])$ in the L^2-norm. Thus the range of Φ is contained in the norm closure of $\Phi(\mathscr{D})$ in $\mathfrak{L}(L^2([0,1]))$.

Let f and g be continuous functions on $[0,1]$ and let T be the integral operator with kernel $f(x)g(y)$. For h in $L^2([0,1])$, we have

$$(Th)(x) = \int_0^1 f(x) g(y) h(y) \, dy = f(x) \left(\int_0^1 g(y) h(y) \, dy \right),$$

and hence the range of T consists of multiples of f. Therefore, T is a rank one operator and all the operators in $\Phi(\mathscr{D})$ are seen to have finite rank. Thus we have by Theorem 5.9 that

$$\Phi(L^2([0,1] \times [0,1])) \subset \text{clos } \Phi(\mathcal{D}) \subset \text{clos } \mathfrak{L}\mathfrak{F}(L^2([0,1])) = \mathfrak{L}\mathfrak{C}(L^2([0,1])),$$

and hence each integral operator T_K is compact.

We will obtain results on the nature of the spectrum of a compact operator after proving some elementary facts about Fredholm operators.

If $\mathcal{H}$ is finite dimensional, then $\mathfrak{L}\mathfrak{C}(\mathcal{H}) = \mathfrak{L}(\mathcal{H})$. *Hence, in the remainder of this chapter we assume that $\mathcal{H}$ is infinite dimensional.*

5.13 Definition If $\mathcal{H}$ is a Hilbert space, then the quotient algebra $\mathfrak{L}(\mathcal{H})/\mathfrak{L}\mathfrak{C}(\mathcal{H})$ is a Banach algebra called the Calkin algebra. The natural homomorphism from $\mathfrak{L}(\mathcal{H})$ onto $\mathfrak{L}(\mathcal{H})/\mathfrak{L}\mathfrak{C}(\mathcal{H})$ is denoted by π.

That the Calkin algebra is actually a C^*-algebra will be established later in this chapter.

The Calkin algebra is of considerable interest in several phases of analysis. Our interest is in its connection with the collection of Fredholm operators. The following definition of Fredholm operator is convenient for our purposes and will be shown to be equivalent to the classical definition directly.

5.14 Definition If $\mathcal{H}$ is a Hilbert space, then T in $\mathfrak{L}(\mathcal{H})$ is a Fredholm operator if $\pi(T)$ is an invertible element of $\mathfrak{L}(\mathcal{H})/\mathfrak{L}\mathfrak{C}(\mathcal{H})$. The collection of Fredholm operators on $\mathcal{H}$ is denoted by $\mathscr{F}(\mathcal{H})$.

The following properties are immediate from the definition.

5.15 Proposition If $\mathcal{H}$ is a Hilbert space, then $\mathscr{F}(\mathcal{H})$ is an open subset of $\mathfrak{L}(\mathcal{H})$, which is self-adjoint, closed under multiplication, and invariant under compact perturbations.

Proof If Δ denotes a group of invertible elements in $\mathfrak{L}(\mathcal{H})/\mathfrak{L}\mathfrak{C}(\mathcal{H})$, then Δ is open by Proposition 2.7 and hence so is $\mathscr{F}(\mathcal{H}) = \pi^{-1}(\Delta)$, since π is continuous. That $\mathscr{F}(\mathcal{H})$ is closed under multiplication follows from the fact that π is multiplicative and Δ is a group. Further, if T is in $\mathscr{F}(\mathcal{H})$ and K is compact, then $T+K$ is in $\mathscr{F}(\mathcal{H})$ since $\pi(T) = \pi(T+K)$. Lastly, if T is in $\mathscr{F}(\mathcal{H})$, then there exist S in $\mathfrak{L}(\mathcal{H})$ and compact operators K_1 and K_2 such that $ST = I + K_1$ and $TS = I + K_2$. Taking adjoints we see that $\pi(T^*)$ is invertible in the Calkin algebra and hence $\mathscr{F}(\mathcal{H})$ is self-adjoint. ∎

The usual characterization of Fredholm operators is obtained after we prove the following lemma about the linear span of subspaces. If $\mathcal{M}$ and $\mathcal{N}$

are closed subspaces of a Hilbert space, then the linear span $\mathscr{M} + \mathscr{N}$ is, in general, not a *closed* subspace (see Exercise 3.18) unless one of the spaces is finite dimensional.

5.16 Lemma   If $\mathscr{H}$ is a Hilbert space, $\mathscr{M}$ is a closed subspace of $\mathscr{H}$, and $\mathscr{N}$ is a finite-dimensional subspace of $\mathscr{H}$, then the linear span $\mathscr{M} + \mathscr{N}$ is a closed subspace of $\mathscr{H}$.

Proof   Replacing $\mathscr{N}$, if necessary, by the orthogonal complement of $\mathscr{M} \cap \mathscr{N}$ in $\mathscr{N}$, we may assume that $\mathscr{M} \cap \mathscr{N} = \{0\}$. To show that

$$\mathscr{M} + \mathscr{N} = \{f + g : f \in \mathscr{M}, \, g \in \mathscr{N}\}$$

is closed, assume that $\{f_n\}_{n=1}^{\infty}$ and $\{g_n\}_{n=1}^{\infty}$ are sequences of vectors in $\mathscr{M}$ and $\mathscr{N}$, respectively, such that $\{f_n + g_n\}_{n=1}^{\infty}$ is a Cauchy sequence in $\mathscr{H}$. We want to prove first that the sequence $\{g_n\}_{n=1}^{\infty}$ is bounded. If it were not, there would exist a subsequence $\{g_{n_k}\}_{k=1}^{\infty}$ and a unit vector h in $\mathscr{N}$ such that

$$\lim_{k \to \infty} \|g_{n_k}\| = \infty \qquad \text{and} \qquad \lim_{k \to \infty} \frac{g_{n_k}}{\|g_{n_k}\|} = h.$$

(This depends, of course, on the compactness of the unit ball of $\mathscr{N}$.) However, since the sequence $\{(1/\|g_{n_k}\|)(f_{n_k} + g_{n_k})\}_{k=1}^{\infty}$ would converge to 0, we would have

$$\lim_{k \to \infty} \frac{f_{n_k}}{\|g_{n_k}\|} = -h.$$

This would imply that h is in both $\mathscr{M}$ and $\mathscr{N}$ and hence a contradiction.

Since the sequence $\{g_n\}_{n=1}^{\infty}$ is bounded, we may extract a subsequence $\{g_{n_k}\}_{k=1}^{\infty}$ such that $\lim_{k \to \infty} g_{n_k} = g$ for some g in $\mathscr{N}$. Therefore, since

$$\{(f_{n_k} + g_{n_k})\}_{k=1}^{\infty}$$

is a Cauchy sequence, we see that $\{f_{n_k}\}_{k=1}^{\infty}$ is a Cauchy sequence and hence converges to a vector f in $\mathscr{M}$. Therefore, $\lim_{n \to \infty} (f_n + g_n) = f + g$ and thus $\mathscr{M} + \mathscr{N}$ is a closed subspace of $\mathscr{H}$.   ∎

The following theorem contains the usual definition of Fredholm operators.

5.17 Theorem (Atkinson)   If $\mathscr{H}$ is a Hilbert space, then T in $\mathfrak{L}(\mathscr{H})$ is a Fredholm operator if and only if the range of T is closed, dim ker T is finite, and dim ker T^* is finite.

Proof If T is a Fredholm operator, then there exists an operator A in $\mathfrak{L}(\mathscr{H})$ and a compact operator K such that $AT = I + K$. If f is a vector in the kernel of $I + K$, then $(I + K)f = 0$ implies that $Kf = -f$, and hence f is in the range of K. Thus,

$$\ker T \subset \ker AT = \ker(I + K) \subset \operatorname{ran} K$$

and therefore by Lemma 5.8, the dimension of $\ker T$ is finite. By symmetry, the dimension of $\ker T^*$ is also finite. Moreover, by Theorem 5.9 there exists a finite rank operator F such that $\|K - F\| < \frac{1}{2}$. Hence for f in $\ker F$, we have

$$\|A\| \, \|Tf\| \geqslant \|ATf\| = \|f + Kf\| = \|f + Ff + Kf - Ff\|$$
$$\geqslant \|f\| - \|Kf - Ff\| \geqslant \|f\|/2.$$

Therefore, T is bounded below on $\ker F$, which implies that $T(\ker F)$ is a closed subspace of $\mathscr{H}$ (see the proof of Proposition 4.8). Since $(\ker F)^\perp$ is finite dimensional, it follows from the preceding lemma that $\operatorname{ran} T = T(\ker F) + T[(\ker F)^\perp]$ is a closed subspace of $\mathscr{H}$.

Conversely, assume that the range of T is closed, $\dim \ker T$ is finite, and $\dim \ker T^*$ is finite. The operator T_0 defined $T_0 f = Tf$ from $(\ker T)^\perp$ to $\operatorname{ran} T$ is one-to-one and onto and hence by Theorem 1.42, is invertible. If we define the operator S on $\mathscr{H}$ such that $Sf = T_0^{-1} f$ for f in $\operatorname{ran} T$ and $Sf = 0$ for f orthogonal to $\operatorname{ran} T$, then S is bounded, $ST = I - P_1$, and $TS = I - P_2$, where P_1 is the projection onto $\ker T$ and P_2 is the projection onto $(\operatorname{ran} T)^\perp = \ker T^*$. Therefore, $\pi(S)$ is the inverse of $\pi(T)$ in $\mathfrak{L}(\mathscr{H})/\mathfrak{LC}(\mathscr{H})$, and hence T is a Fredholm operator which completes the proof. ∎

5.18 As we mentioned previously, the conclusion of the preceding theorem is the classical definition of a Fredholm operator. Early in this century several important classes of operators were shown to consist of Fredholm operators. Moreover, if T is a Fredholm operator, then the solvability of the equation $Tf = g$ for a given g is equivalent to determining whether g is orthogonal to the finite-dimensional subspace $\ker T^*$. Lastly, the space of solutions of the equation $Tf = g$ for a given g is finite dimensional.

At first thought the numbers $\dim \ker T$ and $\dim \ker T^*$ would seem to describe important properties of T, and indeed they do. It turns out, however, that the difference of these two integers is of even greater importance, since it is invariant under small perturbations of T. We shall refer to this difference as the classical index, since we shall also introduce an abstract index. We will eventually show that the two indices coincide.

5.19 Definition If $\mathscr{H}$ is a Hilbert space, then the classical index j is the function defined from $\mathscr{F}(\mathscr{H})$ to $\mathbb{Z}$ such that $j(T) = \dim \ker T - \dim \ker T^*$. For n in $\mathbb{Z}$, set $\mathscr{F}_n = \{T \in \mathscr{F}(\mathscr{H}) : j(T) = n\}$.

We first show that $\mathscr{F}_0$ is invariant under compact perturbations, after which we obtain the classical Fredholm alternative for compact operators.

5.20 Lemma If $\mathscr{H}$ is a Hilbert space, T is in $\mathscr{F}_0$, and K is in $\mathfrak{LC}(\mathscr{H})$, then $T + K$ is in $\mathscr{F}_0$.

Proof Since T is in $\mathscr{F}_0$, there exists a partial isometry V by Proposition 4.37 with initial space equal to $\ker T$ and final space equal to $\ker T^*$. For f in $\ker T$ and g orthogonal to $\ker T$, we have $(T+V)(f+g) = Tg + Vf$, and since Tg is in $\operatorname{ran} T$ and Vf is orthogonal to $\operatorname{ran} T$ then $(T+V)(f+g) = 0$ implies $f + g = 0$. Thus $T + V$ is one-to-one. Moreover, since $T + V$ is onto, it follows that $T + V$ is invertible by the open mapping theorem.

Let F be a finite rank operator chosen such that $\|K - F\| < 1/\|(T+V)^{-1}\|$. Then $T + V + K - F$ is invertible by Proposition 2.7, and hence $T + K$ is the perturbation of the invertible operator $S = T + V + K - F$ by the finite rank operator $G = F - V$. Now $T + K$ is a Fredholm operator by Proposition 5.15, and $j(T+K) = j(S+G) = j(I + S^{-1}G)$, since

$$\ker(S+G) = \ker(S(I + S^{-1}G))$$

and

$$\ker((S+G)^*) = \ker((I + S^{-1}G)^* S^*) = S^{*-1} \ker((I + S^{-1}G)^*).$$

Thus it is sufficient to show that $j(I + S^{-1}G) = 0$.

Since $S^{-1}G$ is a finite rank operator the subspaces $\operatorname{ran}(S^{-1}G)$ and $\operatorname{ran}(S^{-1}G)^*$ are finite dimensional, and hence the subspace $\mathscr{M}$ spanned by them is finite dimensional. Clearly, $(I + S^{-1}G)\mathscr{M} \subset \mathscr{M}$, $(I + S^{-1}G)^*\mathscr{M} \subset \mathscr{M}$, and $(I + S^{-1}G)f = f$ for f orthogonal to $\mathscr{M}$. If A is the operator on $\mathscr{M}$ defined by $Ag = (I + S^{-1}G)g$ for g in $\mathscr{M}$, then $\ker A = \ker(I + S^{-1}G)$ and $\ker A^* = \ker((I + S^{-1}G)^*)$. Since A is an operator on a finite-dimensional space, we have $\dim \ker A = \dim \ker A^*$, and therefore

$$\dim(I + S^{-1}G) = \dim((I + S^{-1}G)^*).$$

Thus, $j(I + S^{-1}G) = 0$ and the proof is complete. ∎

After recalling the definition of generalized eigenspace we will prove a theorem describing properties of the spectrum of a compact operator.

5.21 Definition If $\mathscr{H}$ is a Hilbert space and T is an operator in $\mathfrak{L}(\mathscr{H})$, then the generalized eigenspace $\mathscr{E}_\lambda$ for the complex number λ is the collection of vectors f such that $(T-\lambda)^n f = 0$ for some integer n.

5.22 Theorem (Fredholm Alternative) If K is a compact operator on the Hilbert space $\mathscr{H}$, then $\sigma(K)$ is countable with 0 the only possible limit point, and if λ is a nonzero element of $\sigma(K)$, then λ is an eigenvalue of K with finite multiplicity and $\bar{\lambda}$ is an eigenvalue of K^* with the same multiplicity. Moreover, the generalized eigenspace $\mathscr{E}_\lambda$ for λ is finite dimensional and has the same dimension as the generalized eigenspace for K^* for $\bar{\lambda}$.

Proof If λ is a nonzero complex number, then $-\lambda I$ is invertible, and hence $K - \lambda$ is in $\mathscr{F}$ by Proposition 5.15. Therefore, if λ is in $\sigma(K)$, then $\ker(K-\lambda) \neq \{0\}$, and hence λ is an eigenvalue of K of finite multiplicity. Moreover, since $j(K-\lambda) = 0$, we see that $\bar{\lambda}$ is an eigenvalue of K^* of the same multiplicity.

If $\mathscr{E}_\lambda \neq \ker(K-\lambda)^N$ for any integer N, then there exists an infinite orthonormal sequence $\{k_{n_j}\}_{j=1}^\infty$ such that k_{n_j} is in $\mathscr{E}_{n+1} \ominus \mathscr{E}_n$. Since $\|K k_{n_j}\|^2 = |\lambda|^2 + \|(K-\lambda)k_{n_j}\|^2$ and the sequence $\{k_{n_j}\}_{j=1}^\infty$ converges weakly to 0, it follows from Proposition 5.6 that $0 = \lim_{j=\infty} \|K k_{n_j}\| \geq |\lambda|$ which is a contradiction. Thus $\mathscr{E}_\lambda = \ker(K-\lambda)^N$ for some integer N. Moreover, since $(K-\lambda)^N$ is a compact perturbation of $(-\lambda)^N I$, $(K-\lambda)^N$ is a Fredholm operator with index 0. Therefore,

$$\dim \ker[(K-\lambda)^N] = \dim \ker[(K^* - \bar{\lambda})^N],$$

and since $\mathscr{E}_{\bar{\lambda}} = \ker(K-\lambda)^N$ for some N by the finite dimensionality of $\mathscr{E}_\lambda$, it follows that the dimension of the generalized eigenspace for K for λ is the same as that of the generalized eigenspace for K^* for $\bar{\lambda}$.

Finally, to show that $\sigma(K)$ is countable with 0 the only possible limit point, it is sufficient to show that any sequence of distinct eigenvalues converges to 0. Thus let $\{\lambda_n\}_{n=1}^\infty$ be a sequence of distinct eigenvalues and let f_n be an eigenvector of λ_n. If we let $\mathscr{M}_N$ denote the subspace spanned by $\{f_1, f_2, \cdots, f_N\}$, then $\mathscr{M}_1 \subsetneqq \mathscr{M}_2 \subsetneqq \mathscr{M}_3 \subsetneqq \cdots$, since the eigenvectors for distinct eigenvalues are linearly independent. Let $\{g_n\}_{n=1}^\infty$ be a sequence of unit vectors chosen such that g_n is in $\mathscr{M}_n$ and g_n is orthogonal to $\mathscr{M}_{n-1}$. If h is a vector in $\mathscr{H}$, then $h = \sum_{n=1}^\infty (h, g_n) g_n + g_0$, where g_0 is orthogonal to all the $\{g_n\}_{n=1}^\infty$. Since $\|h\|^2 = \sum_{n=1}^\infty |(h, g_n)|^2 + \|g_0\|^2$ by Theorem 3.25, it follows that $\lim_{n\to\infty}(g_n, h) = 0$. Therefore, the sequence $\{g_n\}_{n=1}^\infty$ converges weakly to 0, and hence by Proposition 5.6 the sequence $\{Kg_n\}_{n=1}^{\infty}$ converges to 0 in norm. Since g_n is in $\mathcal{M}_n$, there exist scalars $\{\alpha_i\}_{i=1}^{n}$ such that $g_n = \sum_{i=1}^{n} \alpha_i f_i$, and hence

$$Kg_n = \sum_{i=1}^{n} \alpha_i Kf_i = \sum_{i=1}^{n} \alpha_i \lambda_i f_i = \lambda_n \sum_{i=1}^{n} \alpha_i f_i + \sum_{i=1}^{n-1} \alpha_i (\lambda_i - \lambda_n) f_i$$

$$= \lambda_n g_n + h_n,$$

where h_n is in $\mathcal{M}_{n-1}$. Therefore,

$$\lim_{n \to \infty} |\lambda_n|^2 \leqslant \lim_{n \to \infty} (|\lambda_n|^2 \|g_n\|^2 + \|h_n\|^2) = \lim_{n \to \infty} \|Kg_n\|^2$$

and the theorem follows. ∎

5.23   Example   We now return to a special integral operator, the Volterra integral operator, and compute its spectrum. Let

$$K(x,y) = \begin{cases} 1 & \text{if } x \geqslant y \\ 0 & \text{if } x < y \end{cases} \qquad \text{for } (x,y) \text{ in } [0,1] \times [0,1]$$

and let V be the corresponding integral operator defined on $L^2([0,1])$ in Section 5.12. We want to show that $\sigma(V) = \{0\}$. If λ were a nonzero number in $\sigma(V)$, then since V is compact by Section 5.12 it follows that λ is an eigenvalue for V. If f is an eigenvector of V for the eigenvalue λ, then $\int_0^x f(y)\, dy = \lambda f(x)$. Thus, we have

$$|\lambda|\,|f(x)| \leqslant \int_0^x |f(y)|\, dy \leqslant \int_0^1 |f(y)|\, dy \leqslant \|f\|_2$$

using the Cauchy–Schwarz inequality. Hence, for x_1 in $[0,1]$ and $n > 0$, we have

$$|f(x_1)| \leqslant \frac{1}{|\lambda|} \int_0^{x_1} |f(x_2)|\, dx_2 \leqslant \frac{1}{|\lambda|^2} \int_0^{x_1} \int_0^{x_2} |f(x_3)|\, dx_3\, dx_2 \leqslant \cdots$$

$$\leqslant \frac{1}{|\lambda|^{n+1}} \|f\|_2 \int_0^{x_1} \int_0^{x_2} \cdots \int_0^{x_n} dx_{n+1} \cdots dx_2$$

$$= \frac{\|f\|_2}{|\lambda|^{n+1}} \frac{x_1^{\,n}}{n!}.$$

Since

$$\lim_{n \to \infty} \frac{\|f\|_2}{|\lambda|^{n+1}} \frac{x_1^{\,n}}{n!} = 0,$$

it follows that $f = 0$. Therefore, a nonzero λ cannot be an eigenvalue, and hence $\sigma(V) = \{0\}$.

We digress to make a couple of comments.

5.24 Definition If $\mathscr{H}$ is a Hilbert space, then an operator T in $\mathfrak{L}(\mathscr{H})$ is quasinilpotent if $\sigma(T) = \{0\}$.

Thus the Volterra operator is compact and quasinilpotent.

5.25 Example Let T be a quasinilpotent operator and $\mathfrak{A}$ be the commutative Banach subalgebra of $\mathfrak{L}(\mathscr{H})$ generated by I, T, and $(T - \lambda)^{-1}$ for all nonzero λ. Since T is not invertible, there exists a maximal ideal in $\mathfrak{A}$ which contains T, and thus the corresponding multiplicative linear functional φ satisfies $\varphi(I) = 1$ and $\varphi(T) = 0$. Moreover, since the values of a multiplicative linear functional on $\mathfrak{A}$ are determined by its values on the generators and these are determined by its value at T, it follows from Corollary 2.36 that the maximal ideal space of $\mathfrak{A}$ consists of just φ. In particular, this example shows that the Gelfand representation for a commutative Banach algebra may provide little aid in studying this algebra.

We now return to the study of Fredholm operators and define the abstract index.

5.26 Definition If $\mathscr{H}$ is a Hilbert space, then the abstract index i is defined from $\mathscr{F}(\mathscr{H})$ to $\Lambda_{\mathfrak{L}(\mathscr{H})/\mathfrak{LC}(\mathscr{H})}$ such that $i = \gamma \circ \pi$, where γ is the abstract index for the Banach algebra $\mathfrak{L}(\mathscr{H})/\mathfrak{LC}(\mathscr{H})$.

The following properties of the abstract index are immediate.

5.27 Proposition If $\mathscr{H}$ is a Hilbert space and i is the abstract index, then i is continuous and multiplicative, and $i(T + K) = i(T)$ for T in $\mathscr{F}(\mathscr{H})$ and K in $\mathfrak{LC}(\mathscr{H})$.

Proof Straightforward. ∎

5.28 We have defined two notions of index on the collection of Fredholm operators: the classical index j from $\mathscr{F}(\mathscr{H})$ to $\mathbb{Z}$ and the abstract index i from $\mathscr{F}(\mathscr{H})$ to Λ. Our objective is to show that these two notions are essentially the same. That is, we want to produce an isomorphism α from the additive group $\mathbb{Z}$ onto Λ such that the following diagram commutes.

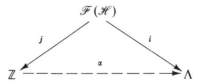

To produce α we will show that for each n, the set $\mathscr{F}_n = j^{-1}(n)$ is connected. Since i is continuous and Λ is discrete, i must be constant on $\mathscr{F}_n$. Thus the mapping defined by $\alpha(n) = i(T)$ for T in $\mathscr{F}_n$ is well defined. The mapping α is onto, since i is onto. Further, consideration of a special class of operators shows that α is a homomorphism. Finally, the fact that $\mathscr{F}_0$ is invariant under compact perturbations will be used to show that $\ker i = \mathscr{F}_0$ and hence that α is one-to-one.

Once this isomorphism is established, then the following results are immediate corollaries: j is continuous and multiplicative, and is invariant under compact perturbation.

We begin this program with the following proposition.

5.29 Proposition If $\mathfrak{A}$ is a W^*-algebra of operators in $\mathfrak{L}(\mathscr{H})$, then the set of unitary operators in $\mathfrak{A}$ is arcwise connected.

Proof Let U be a unitary operator in $\mathfrak{A}$ and let $\mathfrak{W}_U$ be the W^*-algebra generated by U. As in the proof of Theorem 4.65 there exists a decomposition of $\mathscr{H} = \sum_{\alpha \in A} \oplus \mathscr{H}_\alpha$ such that each $\mathscr{H}_\alpha$ reduces U and $U_\alpha = U | \mathscr{H}_\alpha$ has a cyclic vector. Moreover, by Theorem 4.58 there exists a positive regular Borel measure ν_α with support contained in $\mathbb{T}$ and a functional calculus defined from $L^\infty(\nu_\alpha)$ onto $\mathfrak{W}_{U_\alpha}$ by an isometrical isomorphism from $L^2(\nu_\alpha)$ onto $\mathscr{H}_\alpha$.

If we define the function ψ on $\mathbb{T}$ such that $\psi(e^{it}) = t$ for $-\pi < t \leqslant \pi$, then ψ is in each $L^\infty(\nu_\alpha)$, and there exists a sequence of polynomials $\{p_n\}_{n=1}^\infty$ such that $\|p_n\| \leqslant \pi$ and

$$\sup\left\{|p_n(e^{it}) - \psi(e^{it})| : -\pi + \frac{1}{n} < t \leqslant \pi\right\} < \frac{1}{n}.$$

(Use the Stone–Weierstrass theorem to approximate a function which agrees with ψ on $\{e^{it} : -\pi + (1/n) < t \leqslant \pi\}$ and is continuous.) Consider the sequence of operators

$$\{p_n(U)\}_{n=1}^\infty = \left\{\sum_{\alpha \in A} \oplus p_n(U_\alpha)\right\}_{n=1}^\infty$$

in $\mathfrak{W}_U$ and the operator $H = \sum_{\alpha \in A} \oplus \psi(U_\alpha)$ defined on $\mathscr{H}$. Using the identification of $\mathscr{H}_\alpha$ as $L^2(\nu_\alpha)$, it is easy to check that the sequence $\{p_n(U_\alpha)\}_{n=1}^\infty$ converges to $\psi(U_\alpha)$ in the strong operator topology. Since the operators

$$\left\{ \sum_{\alpha \in A} \oplus p_n(U_\alpha) \right\}_{n=1}^\infty$$

are uniformly bounded, it follows that the sequence converges strongly to H. Therefore, H is in $\mathfrak{W}_U$ and moreover $e^{iH} = U$, since $e^{i\psi(U_\alpha)} = U_\alpha$ for each α in A. If we define the function $U_\lambda = e^{i\lambda H}$ for λ in $[0,1]$, then for λ_1 and λ_2 in $[0,1]$, we have

$$\|U_{\lambda_1} - U_{\lambda_2}\| = \|\exp i\lambda_1 H - \exp i\lambda_2 H\| = \|\exp i\lambda_1 H(I - \exp i(\lambda_2 - \lambda_1)H)\|$$

$$= \sup_{\alpha \in A} \|I_\alpha - \exp i(\lambda_2 - \lambda_1)\psi(U_\alpha)\|$$

$$\leqslant \|1 - \exp i(\lambda_2 - \lambda_1)\psi\|_\infty = |\exp i\lambda_1 - \exp i\lambda_2|,$$

and therefore U_λ is a continuous function of λ. Thus the unitary operator U is arcwise connected to $U_0 = I$ by unitary operators in $\mathfrak{A}$. ∎

The following corollary is now easy to prove.

5.30   Corollary   If $\mathscr{H}$ is a Hilbert space, then the collection of invertible operators in $\mathfrak{L}(\mathscr{H})$ is arcwise connected.

Proof   Let T be an invertible operator in $\mathfrak{L}(\mathscr{H})$ with the polar decomposition $T = UP$. Since T is invertible, U is unitary and P is an invertible positive operator. For λ in $[0,1]$ let U_λ be an arc of unitary operators connecting the identity operator U_0 to $U - U_1$ and $P_\lambda = (1 - \lambda)I + \lambda P$. Since each P_λ is bounded below, it is invertible and hence $U_\lambda P_\lambda$ is an arc connecting the identity operator to T. ∎

Much more is true; a theorem of Kuiper [73] states that the collection of invertible operators in $\mathfrak{L}(\mathscr{H})$ for a countably infinite-dimensional $\mathscr{H}$ is a contractible topological space.

5.31   Corollary   If $\mathscr{H}$ is a Hilbert space and T is an invertible operator in $\mathfrak{L}(\mathscr{H})$, then $i(T)$ is the identity in Λ.

Proof   Since $\pi(T)$ is in the connected component Δ_0 of the identity in Δ, it is clear that $i(T)$ is the identity in Λ. ∎

We now consider the connectedness of $\mathscr{F}_n$.

5.32   Theorem   If $\mathscr{H}$ is a Hilbert space, then for each n in $\mathbb{Z}$ the set $\mathscr{F}_n(\mathscr{H})$ is arcwise connected.

Proof   Since $(\mathscr{F}_n(\mathscr{H}))^* = \mathscr{F}_{-n}(\mathscr{H})$, it is sufficient to consider $n \geqslant 0$. If $n \geqslant 0$ and T is in $\mathscr{F}_n$, then $\dim \ker T \geqslant \dim \ker T^*$; thus there exists a partial isometry V with initial space contained in $\ker T$ and range equal to $\ker T^* = (\operatorname{ran} T)^\perp$. For each $\varepsilon > 0$ it is clear that $T + \varepsilon V$ is onto and hence is right invertible and

$$\ker(T + \varepsilon V) = \ker(T) \ominus \operatorname{init}(V).$$

Hence it is sufficient to prove that if S and T are right invertible with $\dim \ker S = \dim \ker T$, then S and T can be connected by an arc of right invertible operators in $\mathscr{F}_n$.

Let S and T be right invertible operators with $\dim \ker S = \dim \ker T$. Let U be a unitary operator chosen such that $\ker SU = \ker T$ and U_λ be an arc of unitary operators such that $U_0 = I$ and $U_1 = U$. Then SU is connected to S in $\mathscr{F}_n$ and hence we can assume that kernels of our two right invertible operators are equal.

Hence, assume that S and T are right invertible operators with $\ker S = \ker T$. By Proposition 4.37, there exists an isometry W with range $W = (\ker S)^\perp = (\ker T)^\perp$. Then the operators SW and TW are invertible and hence by Corollary 5.30 there exists an arc of operators J_λ for $0 \leqslant \lambda \leqslant 1$ such that $J_0 = SW$ and $J_1 = TW$. Since WW^* is the projection onto the range of W, we see that $SWW^* = S$ and $TWW^* = T$. Hence $J_\lambda W^*$ is an arc of operators connecting S and T, and the proof will be completed once we show that each $J_\lambda W^*$ is in $\mathscr{F}_n$. Since $(J_\lambda W^*)(WJ_\lambda^{-1}) = I$, it follows that $J_\lambda W^*$ is right invertible and hence $\ker((J_\lambda W^*)^*) = \{0\}$. Further since $\ker(J_\lambda W^*) = \ker W^*$ we see that $j(J_\lambda W^*) = n$ for all λ. This completes the proof.   ∎

5.33   Recall now the unilateral shift operator U_+ on $l^2(\mathbb{Z}^+)$ introduced in Section 4.36. It is easily established that $\ker U_+ = \{0\}$, while $\ker U_+^* = \{e_0\}$. Since U_+ is an isometry, its range is closed, and thus U_+ is a Fredholm operator and $j(U_+) = -1$.

Now define

$$U_+^{(n)} = \begin{cases} U_+^{\,n} & n \geqslant 0, \\ U_+^{*-n} & n < 0. \end{cases}$$

Since for $n \geqslant 0$ we have $\ker U_+^{(n)} = \{0\}$ and

$$\ker U_+^{(n)*} = \ker U_+^{*(n)} = \bigvee \{e_0, e_1, ..., e_{n-1}\},$$

it follows that $j(U_+^{(n)}) = -n$. Similarly since $U_+^{(n)*} = U_+^{-(n)}$, we have $j(U_+^{(n)}) = -n$ for all integers n. The following extension of this formula will be used in showing that the map to be constructed is a homomorphism.

If m and n are integers, then $j(U_+^{(m)} U_+^{(n)}) = j(U_+^{(m)}) + j(U_+^{(n)})$. We prove this one case at a time. If $m \geqslant 0$ and $n \geqslant 0$ or $m < 0$ and $n < 0$, then $U_+^{(m)} U_+^{(n)} = U_+^{(m+n)}$, and hence

$$j(U_+^{(m)} U_+^{(n)}) = -m - n = j(U_+^{(m)}) + j(U_+^{(n)}).$$

If $m < 0$ and $n \geqslant 0$, then

$$U_+^{(m)} U_+^{(n)} = U_+^{*-m} U_+^{n} = \begin{cases} U_+^{n+m} = U_+^{(n+m)} & -m \leqslant n, \\ U_+^{*-(n+m)} = U_+^{(n+m)} & -m > n, \end{cases}$$

and again the formula holds. Lastly, if $m \geqslant 0$ and $n < 0$, then

$$\ker U_+^{(m)} U_+^{(n)} = \ker U_+^{*-n} = \bigvee \{e_0, e_1, ..., e_{-n-1}\}$$

and

$$\ker [U_+^{(m)} U_+^{(n)}]^* = \ker (U_+{}^m U_+^{*-n})^* = \ker U_+^{-n} U_+^{*m} = \bigvee \{e_0, ..., e_{m-1}\},$$

and hence

$$j(U_+^{(m)} U_+^{(n)}) = -n - m = j(U_+^{(m)}) + j(U_+^{(n)}).$$

The next lemma will be used in the proof of the main theorem to show that each of the $\mathscr{F}_n$ is open.

5.34 Lemma If $\mathscr{H}$ is a Hilbert space, then each of the sets $\mathscr{F}_0$ and $\bigcup_{n \neq 0} \mathscr{F}_n$ is open in $\mathfrak{L}(\mathscr{H})$.

Proof Let T be in $\mathscr{F}_0$ and let F be a finite rank operator chosen such that $T + F$ is invertible. Then if X is an operator in $\mathfrak{L}(\mathscr{H})$ which satisfies $\|T - X\| < 1/\|(T+F)^{-1}\|$, then $X + F$ is invertible by the proof of Proposition 2.7, and hence X is in $\mathscr{F}_0$ by Lemma 5.20. Therefore, $\mathscr{F}_0$ is an open set.

If T is a Fredholm operator not in $\mathscr{F}_0$, then there exists as in the proof of Lemma 5.20 a finite rank operator F such that $T + F$ is either left or right invertible. By Proposition 2.7 there exists $\varepsilon > 0$ such that if X is an operator in $\mathfrak{L}(\mathscr{H})$ such that $\|T + F - X\| < \varepsilon$, then X is either left or right invertible but not invertible. Thus X is a Fredholm operator of index not equal to 0 and therefore so is $X - F$ by Lemma 5.20. Hence $\bigcup_{n \neq 0} \mathscr{F}_n$ is also an open subset of $\mathfrak{L}(\mathscr{H})$. ∎

We now state and prove the main theorem of the chapter.

5.35   Theorem   If $\mathscr{H}$ is a Hilbert space, j is the classical index from $\mathscr{F}(\mathscr{H})$ onto $\mathbb{Z}$, and i is the abstract index from $\mathscr{F}(\mathscr{H})$ onto Λ, then there exists an isomorphism α from $\mathbb{Z}$ onto Λ such that $\alpha \circ j = i$.

Proof   Since the dimension of $\mathscr{H}$ is infinite, we have $\mathscr{H} \oplus l^2(\mathbb{Z}^+)$ isomorphic to $\mathscr{H}$, and hence there is an operator on $\mathscr{H}$ unitarily equivalent to $I \oplus U_+^{(n)}$. Therefore, each $\mathscr{F}_n$ is nonempty and we can define $\alpha(n) = i(T)$, where T is any operator in $\mathscr{F}_n$. Moreover, α is well defined by Theorem 5.32. Since i is onto, it follows that α is onto. Further, by the formula in Section 5.33, we have

$$\alpha(m+n) = i(I \oplus U_+^{(-m-n)}) = i((I \oplus U^{(-m)})(I \oplus U^{(-n)}))$$
$$= i(I \oplus U^{(-m)})\, i(I \oplus U^{(-n)}) = \alpha(m) \cdot \alpha(n),$$

and hence α is a homomorphism.

It remains only to show that α is one-to-one. Observe first that $\pi(\mathscr{F}_0)$ is disjoint from $\pi(\bigcup_{n \neq 0} \mathscr{F}_n)$, since if T is in $\mathscr{F}_0$ and S is in $\mathscr{F}_n$ with $\pi(S) = \pi(T)$, then there exists K in $\mathfrak{LC}(\mathscr{H})$ such that $S + K = T$. However, Lemma 5.20 implies that $j(T) = 0$, and hence $\pi(\mathscr{F}_0)$ is disjoint from $\pi(\bigcup_{n \neq 0} \mathscr{F}_n)$. Since $\mathscr{F}_0$ and $\bigcup_{n \neq 0} \mathscr{F}_n$ are open and π is an open map, it follows that $\pi(\mathscr{F}_0)$ and $\pi(\bigcup_{n \neq 0} \mathscr{F}_n)$ are disjoint open sets. Therefore, $\pi(\mathscr{F}_0)$ is an open and closed subset of Δ and hence is equal to the connected component Δ_0 of the identity in Δ. Therefore, π takes $\mathscr{F}_0$ onto Δ_0 and hence i takes $\mathscr{F}_0$ onto the identity of Λ. Thus α is an isomorphism.   ■

We now summarize what we have proved in the following theorem.

5.36   Theorem   If $\mathscr{H}$ is a Hilbert space, then the components of $\mathscr{F}(\mathscr{H})$ are precisely the sets $\{\mathscr{F}_n : n \in \mathbb{Z}\}$. Moreover, the classical index defined by

$$j(T) = \dim \ker T - \dim \ker T^*$$

is a continuous homomorphism from $\mathscr{F}(\mathscr{H})$ onto $\mathbb{Z}$ which is invariant under compact perturbation.

We continue now with the study of $\mathfrak{LC}(\mathscr{H})$ and $\mathfrak{L}(\mathscr{H})/\mathfrak{LC}(\mathscr{H})$ as C^*-algebras. (Strictly speaking, $\mathfrak{LC}(\mathscr{H})$ is not a C^*-algebra by our definition, since it has no identity.) This requires that we first show that the quotient of a C^*-algebra by a two-sided ideal is again a C^*-algebra. While this is indeed true, it is much less trivial to prove than our previous results on quotient objects. We begin by considering the abelian case which will be used as a lemma in the proof of the main result.

5.37 Lemma If $\mathfrak{A}$ is an abelian C^*-algebra and $\mathfrak{I}$ is a closed ideal in $\mathfrak{A}$, then $\mathfrak{I}$ is self-adjoint and the natural map π induces an involution on the quotient algebra $\mathfrak{A}/\mathfrak{I}$ with respect to which it is a C^*-algebra.

Proof Let X be the maximal ideal space of $\mathfrak{A}$ and M be the maximal ideal space of the commutative Banach algebra $\mathfrak{A}/\mathfrak{I}$. Each m in M defines a multiplicative linear function $m \circ \pi$ on $\mathfrak{A}$. The map $\psi(m) = m \circ \pi$ is a homeomorphism of M onto a closed subset of X. Further, this homeomorphism ψ defines a homomorphism Ψ from $C(X)$ to $C(M)$ by

$$\Psi(k) = k \circ \psi \qquad \text{and} \qquad \ker \Psi = \{k \in C(X): k(\psi(m)) = 0 \text{ for } m \in M\}.$$

Moreover, $\Psi \circ \Gamma_{\mathfrak{A}} = \Gamma_{\mathfrak{A}/\mathfrak{I}} \circ \pi$ and by the Gelfand–Naimark theorem (4.29), we know that $\Gamma_{\mathfrak{A}}$ is a *-isometrical isomorphism. Therefore, $\ker \pi = \Gamma_{\mathfrak{A}}^{-1}(\ker \psi)$ and thus the kernel is self-adjoint and π induces an involution on $\mathfrak{A}/\mathfrak{I}$. Moreover, it is clear that $\mathfrak{A}/\mathfrak{I}$ is *-isometrically isomorphic to $C(M)$ and hence a C^*-algebra. ∎

We now proceed to the main result about quotient algebras.

5.38 Theorem If $\mathfrak{A}$ is a C^*-algebra and $\mathfrak{I}$ is a closed two-sided ideal in $\mathfrak{A}$, then $\mathfrak{I}$ is self-adjoint and the quotient algebra $\mathfrak{A}/\mathfrak{I}$ is a C^*-algebra with respect to the involution induced by the natural map.

Proof We begin by showing that $\mathfrak{I}$ is self-adjoint. For T an element of $\mathfrak{I}$, set $H = T^*T$. For $\lambda > 0$ the element λH^2 is positive, since it is the square of a self-adjoint element, and therefore $\lambda H^2 + I$ is invertible in $\mathfrak{A}$. Moreover, rearranging the identity $(\lambda H^2 + I)(\lambda H^2 + I)^{-1} = I$ shows that the element defined by

$$U_\lambda = (\lambda H^2 + I)^{-1} - I = -(\lambda H^2 + I)^{-1} \lambda H^2$$

is in $\mathfrak{I}$. Moreover, if we set $S_\lambda = TU_\lambda + T$, then $S_\lambda^* S_\lambda = (\lambda H^2 + I)^{-2} H$ and from the functional calculus for $\mathfrak{C}_H$, we have

$$\|S_\lambda^* S_\lambda\| = \|(\lambda H^2 + I)^{-2} H\| \leqslant \sup \left\{ \frac{x}{(\lambda x^2 + 1)^2} : x \in \sigma(H) \right\}$$

$$\leqslant \sup \left\{ \frac{x}{(\lambda x^2 + 1)^2} : x \geqslant 0 \right\} \leqslant \frac{9}{16(3\lambda)^{1/2}},$$

where the last inequality is obtained by maximizing the function $\varphi(x) = x(\lambda x^2 + 1)^{-2}$ on $\mathbb{R}$. Taking adjoints of the equation $S_\lambda = TU_\lambda + T$ and rearranging yields

$$\lim_{\lambda \to \infty} \|T^* + U_\lambda T^*\| = \lim_{\lambda \to \infty} \|S_\lambda\| \leqslant \lim_{\lambda \to \infty} \frac{3}{4(3\lambda)^{1/4}} = 0.$$

Since each $-U_\lambda T^*$ is in $\mathfrak{I}$, and $\mathfrak{I}$ is closed, we have T^* in $\mathfrak{I}$ and hence $\mathfrak{I}$ is self-adjoint.

Now $\mathfrak{A}/\mathfrak{I}$ is a Banach algebra and the mapping $(A + \mathfrak{I})^* = A^* + \mathfrak{I}$ is the involution induced by the natural map. Since we have $\|(A + \mathfrak{I})^*\| = \|A + \mathfrak{I}\|$, it follows that

$$\|(A + \mathfrak{I})^*(A + \mathfrak{I})\| \leqslant \|(A + \mathfrak{I})^*\| \, \|A + \mathfrak{I}\| = \|A + \mathfrak{I}\|^2,$$

and hence only the reverse inequality remains to be proved before we can conclude that $\mathfrak{A}/\mathfrak{I}$ is a C^*-algebra.

Returning to the previous notation, if we set $\mathfrak{R} = \mathfrak{C}_H \cap \mathfrak{I}$, then $\mathfrak{R}$ is a closed two-sided self-adjoint ideal in the abelian C^*-algebra $\mathfrak{C}_H$, and hence $\mathfrak{C}_H/\mathfrak{R}$ is a C^*-algebra by the previous lemma. If we consider the subalgebra $\pi(\mathfrak{C}_H) = \mathfrak{C}_H/\mathfrak{I}$ of $\mathfrak{A}/\mathfrak{I}$, then there is a natural map π' from $\mathfrak{C}_H/\mathfrak{R}$ onto $\mathfrak{C}_H/\mathfrak{I}$. Moreover, it follows from the definition of $\mathfrak{R}$ and the quotient norm that π' is a contractive isomorphism. Therefore, for A in $\mathfrak{C}_H$, we have

$$\sigma_{\mathfrak{C}_H/\mathfrak{R}}(A + \mathfrak{R}) = \sigma_{\mathfrak{C}_H/\mathfrak{I}}(A + \mathfrak{I})$$

and hence

$$\|A + \mathfrak{R}\|_{\mathfrak{C}_H/\mathfrak{R}} \geqslant \|A + \mathfrak{I}\|_{\mathfrak{C}_H/\mathfrak{I}} \geqslant \rho_{\mathfrak{C}_H/\mathfrak{I}}(A + \mathfrak{I}) = \rho_{\mathfrak{C}_H/\mathfrak{R}}(A + \mathfrak{R}) = \|A + \mathfrak{R}\|_{\mathfrak{C}_H/\mathfrak{R}}.$$

Thus, π' is an isometry and $\mathfrak{C}_H/\mathfrak{I}$ is an abelian C^*-algebra. Lastly, it follows from the functional calculus for $\mathfrak{C}_H/\mathfrak{I}$ and the special form of the function $\psi(x) = \lambda x^2/(1 + \lambda x^2)$, that

$$\|\pi(U_\lambda)\| = \lambda \|\pi(H)\|^2 (1 + \lambda \|\pi(H)\|^2)^{-1}.$$

To complete the proof we use the identity $T = S_\lambda - TU_\lambda$ to obtain

$$\|\pi(T)\| \leqslant \|\pi(S_\lambda)\| + \|\pi(T)\| \, \|\pi(U_\lambda)\| \leqslant \frac{3}{4(3\lambda)^{1/4}} + \frac{\lambda \|\pi(T)\| \, \|\pi(H)\|^2}{1 + \lambda \|\pi(H)\|^2}.$$

Setting $\lambda = 1/(3 \|\pi(H)\|^2)$, we further obtain the inequality

$$\|\pi(T)\| \leqslant \frac{3(\|\pi(H)\|)^{1/2}}{4} + \frac{1}{4} \|\pi(T)\|$$

and finally

$$\|\pi(T)\|^2 \leqslant \|\pi(H)\| = \|\pi(T)^*\pi(T)\|.$$

Therefore, $\mathfrak{A}/\mathfrak{J}$ is a C^*-algebra. ∎

We now consider the algebra $\mathfrak{LC}(\mathscr{H})$ which plays a fundamental role in the study of C^*-algebras. The following result has several important consequences. A subset $\mathfrak{S}$ of $\mathfrak{L}(\mathscr{H})$ is said to be irreducible if no proper closed subspace is reducing for all S in $\mathfrak{S}$.

5.39 Theorem If $\mathfrak{A}$ is an irreducible C^*-algebra contained in $\mathfrak{L}(\mathscr{H})$ such that $\mathfrak{A} \cap \mathfrak{LC}(\mathscr{H}) \neq 0$, then $\mathfrak{LC}(\mathscr{H})$ is contained in $\mathfrak{A}$.

Proof If K is a nonzero compact operator in $\mathfrak{A}$, then $(K+K^*)$ and $(1/i)(K-K^*)$ are compact self-adjoint operators in $\mathfrak{A}$. Moreover, since at least one is not zero, there exists a nonzero compact self-adjoint operator H in $\mathfrak{A}$. If λ is a nonzero eigenvalue for H which it must have, then the projection onto the eigenspace for λ is in $\mathfrak{C}_H$ and hence in $\mathfrak{A}$ using the functional calculus. Moreover, since this eigenspace is finite dimensional by Theorem 5.22, we see that $\mathfrak{A}$ must contain a nonzero finite rank projection.

Let E be a nonzero finite rank projection in $\mathfrak{A}$ of minimum rank. Consider the closed subalgebra $\mathfrak{A}_E = \{EAE: A \in \mathfrak{A}\}$ of $\mathfrak{A}$ as a subalgebra of $E\mathscr{H}$. If any self-adjoint operator in $\mathfrak{A}_E$ were not a constant multiple of a scalar, then $\mathfrak{A}_E$ and hence $\mathfrak{A}$ would contain a spectral projection for this operator and hence a projection of smaller rank than E. Therefore, the algebra $\mathfrak{A}_E$ must consist of scalar multiples of E. Suppose the rank of E is greater than one and x and y are linearly independent vectors in its range. Since the closure of $\{Ax: A \in \mathscr{H}\}$ is a reducing subspace for $\mathfrak{A}$, it follows that it must be dense in $\mathscr{H}$. Therefore, there must exist a sequence $\{A_n\}_{n=1}^{\infty}$ in $\mathfrak{A}$ such that $\lim_{n \to \infty} \|A_n x - y\| = 0$, and hence $\lim_{n \to \infty} \|EA_n Ex - y\| = 0$. Since x and y are linearly independent, the sequence $\{EA_n E\}_{n=1}^{\infty}$ cannot consist of scalar multiples of E. Therefore, E must have rank one.

We next show that every rank one operator is in $\mathfrak{A}$ which will imply by Theorem 5.9 that $\mathfrak{LC}(\mathscr{H})$ is in $\mathfrak{A}$. For x and y in $\mathscr{H}$, let $T_{y,x}$ be the rank one operator defined by $T_{y,x}(z) = (z,x)y$. For a unit vector x_0 in $E\mathscr{H}$, if $\{A_n\}_{n=1}^{\infty}$ is chosen as above such that $\lim_{n \to \infty} \|A_n x_0 - y\| = 0$, then the sequence $\{A_n T_{x_0,x_0}\}_{n=1}^{\infty}$ is contained in $\mathfrak{A}$ and $\lim_{n \to \infty} A_n T_{x_0,x_0} = T_{y,x_0}$. Similarly, using adjoints, we obtain that $T_{x_0,x}$ is in $\mathfrak{A}$ and hence finally that $T_{y,x} = T_{y,x_0} T_{x_0,x}$ is in $\mathfrak{A}$. Thus $\mathfrak{LC}(\mathscr{H})$ is contained in $\mathfrak{A}$ and the proof is complete. ∎

One of the consequences of this result is that we are able to determine all representations of the algebra $\mathfrak{LC}(\mathscr{H})$.

5.40 Theorem If Φ is a *-homomorphism of $\mathfrak{LC}(\mathscr{H})$ into $\mathfrak{L}(\mathscr{K})$, then there exists a unique direct sum decomposition $\mathscr{K} = \mathscr{K}_0 \oplus \sum_{\alpha \in A} \oplus \mathscr{K}_\alpha$, such that each $\mathscr{K}_\alpha$ reduces $\Phi(\mathfrak{LC}(\mathscr{H}))$, the restriction $\Phi(T)|\mathscr{K}_0 = 0$ for T in $\mathfrak{LC}(\mathscr{H})$, and there exists an isometrical isomorphism U_α from $\mathscr{H}$ onto $\mathscr{K}_\alpha$ for α in A such that $\Phi(T)|\mathscr{K}_\alpha = U_\alpha T U_\alpha^*$ for T in $\mathfrak{LC}(\mathscr{H})$.

Proof If Φ is not an isomorphism, then $\ker \Phi$ is a closed two-sided ideal in $\mathfrak{LC}(\mathscr{H})$ and hence must equal $\mathfrak{LC}(\mathscr{H})$, in which case $\Phi(T) = 0$ for T in $\mathfrak{LC}(\mathscr{H})$. Thus, if we set $\mathscr{K}_0 = \mathscr{K}$, the theorem is proved. Hence, we may assume that Φ is an isometrical isomorphism.

Now let $\{e_i\}_{i \in I}$ be an orthonormal basis for $\mathscr{H}$ and let P_i be the projection onto the subspace spanned by e_i. Then $E_i = \Phi(P_i)$ is a projection on $\mathscr{K}$. Choose a distinguished element 0 in I and define V_i on $\mathscr{H}$ for i in I such that $V_i(\sum_{j \in I} \lambda_j e_j) = \lambda_0 e_i$. It is obvious that V_i is a partial isometry with $V_i V_i^* = P_i$ and $V_i^* V_i = P_0$. Hence W_i is a partial isometry on $\mathscr{K}$ and $W_i^* W_i = E_0$ and $W_i W_i^* = E_i$. Let $\{x_0^\alpha\}_{\alpha \in A}$ be an orthonormal basis for the range of E_0 and set $x_i^\alpha = W_i x_0^\alpha$. It is easy to see that each x_i^α is in the range of E_i and that $\{x_i^\alpha\}_{i \in I, \alpha \in A}$ is an orthonormal subset of $\mathscr{K}$. Let $\mathscr{K}_\alpha$ denote the closed subspace of $\mathscr{K}$ spanned by the $\{x_i^\alpha\}_{i \in I}$. The $\{\mathscr{K}_\alpha\}_{\alpha \in A}$ are pairwise orthogonal and hence we can consider the closed subspace $\sum_{\alpha \in A} \oplus \mathscr{K}_\alpha$ of $\mathscr{K}$. Lastly, let $\mathscr{K}_0$ denote the orthogonal complement of this subspace. We want to show that the subspaces $\{\mathscr{K}_\alpha\}_{\alpha \in A \cup \{0\}}$ have the properties ascribed to them in the statement of the theorem.

Since $V_i V_j^*$ is the rank one operator on $\mathscr{H}$ taking e_j onto e_i, it is clear that the norm-closed *-algebra generated by the $\{V_i\}_{i \in I}$ is $\mathfrak{LC}(\mathscr{H})$. Therefore, $\Phi(\mathfrak{LC}(\mathscr{H}))$ is the norm-closed *-algebra generated by the $\{W_i\}_{i \in I}$ and hence each $\mathscr{K}_\alpha$ reduces $\Phi(\mathfrak{LC}(\mathscr{H}))$. If we define a mapping U_α from $\mathscr{H}$ to $\mathscr{K}_\alpha$ by $U_\alpha e_i = x_i^\alpha$, then U_α extends to an isometrical isomorphism and $\Phi(T)|\mathscr{K}_\alpha = U_\alpha T U_\alpha^*$. Therefore, Φ is spatially implemented on each $\mathscr{K}_\alpha$.

Lastly, since each $\mathscr{K}_\alpha$ reduces $\Phi(\mathfrak{LC}(\mathscr{H}))$, it follows that $\mathscr{K}_0$ also is a reducing subspace and since

$$\left(\sum_{i \in I} E_i\right) \Phi(T) \left(\sum_{i \in I} E_i\right) = \Phi(T) \qquad \text{for } T \text{ in } \mathfrak{LC}(\mathscr{H}),$$

while $I - \sum_{i \in I} E_i$ is the projection onto $\mathscr{K}_0$, we see that $\Phi(T)|\mathscr{K}_0 = 0$ for T in $\mathfrak{LC}(\mathscr{H})$. ∎

Such a result has a partial extension to a broader class of C^*-algebras.

5.41   Corollary   If $\mathfrak{A}$ is a C^*-algebra on $\mathscr{H}$ which contains $\mathfrak{LC}(\mathscr{H})$ and Φ is a *-homomorphism of $\mathfrak{A}$ into $\mathfrak{L}(\mathscr{K})$ such that $\Phi|\mathfrak{LC}(\mathscr{H})$ is not zero and $\Phi(\mathfrak{A})$ is irreducible, then there exists an isometrical isomorphism U from $\mathscr{H}$ onto $\mathscr{K}$ such that $\Phi(A) = UAU^*$ for A in $\mathfrak{A}$.

Proof   If $\Phi(\mathfrak{LC}(\mathscr{H}))$ is not irreducible, then by the preceding theorem there exists a proper closed subspace $\mathscr{K}'$ of $\mathscr{K}$ such that the projection P onto $\mathscr{K}'$ commutes with the operators $\Phi(K)$ for K in $\mathfrak{LC}(\mathscr{H})$, and there exists an isometrical isomorphism U from $\mathscr{H}$ onto $\mathscr{K}'$ such that $\Phi(K)|\mathscr{K}' = UKU^*$ for K in $\mathfrak{LC}(\mathscr{H})$. (The alternative leads to the conclusion that $\Phi|\mathfrak{LC}(\mathscr{H}) = 0$.) Then for A in $\mathfrak{A}$ and K in $\mathfrak{LC}(\mathscr{H})$, we have

$$[P\Phi(A) - \Phi(A)P]\Phi(K) = P\Phi(A)\Phi(K) - \Phi(A)\Phi(K)P$$
$$= P\Phi(AK) - \Phi(AK)P = 0,$$

since AK is in $\mathfrak{LC}(\mathscr{H})$. If $\{E_\alpha\}_{\alpha \in A}$ is a net of finite rank projections in $\mathscr{H}$ increasing to the identity, then $\{\Phi(E_\alpha)|\mathscr{K}'\}_{\alpha \in A}$ converges strongly to P, and thus we obtain $P\Phi(A)P = \Phi(A)P$ for every A in $\mathfrak{A}$. Since $\Phi(\mathfrak{A})$ is self-adjoint and is assumed to be irreducible, this implies $\mathscr{K}' = \mathscr{K}$. Lastly, for A in $\mathfrak{A}$ and K in $\mathfrak{LC}(\mathscr{H})$, we have

$$\Phi(K)[\Phi(A) - UAU^*] = \Phi(KA) - (UKU^*)(UAU^*)$$
$$= UKAU^* - UKAU^* = 0,$$

and again using a net of finite projections we obtain the fact that $\Phi(A) = UAU^*$ for A in $\mathfrak{A}$.   ■

These results enable us to determine the *-automorphisms of $\mathfrak{L}(\mathscr{H})$ and $\mathfrak{LC}(\mathscr{H})$.

5.42   Corollary   If $\mathscr{H}$ is a Hilbert space, then Φ is a *-automorphism of $\mathfrak{L}(\mathscr{H})$ if and only if there exists a unitary operator U in $\mathfrak{L}(\mathscr{H})$ such that $\Phi(A) = UAU^*$ for A in $\mathfrak{L}(\mathscr{H})$.

Proof   Immediate from the previous corollary.   ■

Such an automorphism is said to be inner and hence all *-automorphisms of $\mathfrak{L}(\mathscr{H})$ are inner. A similar but significantly different result holds for $\mathfrak{LC}(\mathscr{H})$.

5.43   Corollary   If $\mathscr{H}$ is a Hilbert space, then Φ is a *-automorphism of $\mathfrak{LC}(\mathscr{H})$ if and only if there exists a unitary operator U in $\mathfrak{L}(\mathscr{H})$ such that $\Phi(K) = UKU^*$ for K in $\mathfrak{LC}(\mathscr{H})$.

Proof   Again immediate.   ■

The difference in this case is that the unitary operator need not belong to the algebra and hence the automorphism need not be inner. The algebra $\mathfrak{LC}(\mathscr{H})$ has the property, however, that in each *-isomorphic image of the algebra every *-automorphism is spatially implemented by a unitary operator on the space.

We conclude with an observation concerning the Calkin algebra.

5.44   Theorem   If Φ is a *-isomorphism of the Calkin algebra $\mathfrak{L}(\mathscr{H})/\mathfrak{LC}(\mathscr{H})$ into $\mathfrak{L}(\mathscr{K})$, then $\Phi(\mathfrak{L}(\mathscr{H})/\mathfrak{LC}(\mathscr{H}))$ is not a W^*-algebra.

Proof   If $\Phi(\mathfrak{L}(\mathscr{H})/\mathfrak{LC}(\mathscr{H}))$ were a W^*-algebra, then the group of invertible elements would be connected by Proposition 5.29, thus contradicting Theorem 5.36.   ■

Notes

The earliest results on compact operators are implicit in the studies of Volterra and Fredholm on integral equations. The notion of compact operator is due to Hilbert, while it was F. Riesz who adopted an abstract point of view and formulated the so called "Fredholm alternative." Further study into certain classes of singular integral operators led Noether to introduce the notion of index and implicitly the class of Fredholm operators. The connection between this class and the Calkin algebra was made by Atkinson [4]. Finally, Gohberg and Kreĭn [48] systematized and extended the theory of Fredholm operators to approximately its present form. The connection between the components of the invertible elements in the Calkin algebra and the index was first established by Cordes and Labrouse [23] and Coburn and Lebow [22].

Further results including more detailed historical comments can be found in Riesz and Sz.-Nagy [92], Maurin [79], Goldberg [51], and the expository article of Gohberg and Kreĭn [48]. The reader can also consult Lang [74] or Palais [85] for a modern treatment of a slightly different flavor. Again the results on C^*-algebras can be found in Dixmier [28]. The proof of Theorem 5.38 is taken from Naimark [80], whereas the short and clever proof of Lemma 5.7 is due to Halmos [58].

Exercises

5.1   If $\mathcal{H}$ is a Hilbert space and T is in $\mathfrak{L}(\mathcal{H})$, then T is compact if and only if $(T^*T)^{\frac{1}{2}}$ is compact.

5.2   If T is a compact normal operator on $\mathcal{H}$, then there exists a sequence of complex numbers $\{\lambda_n\}_{n=1}^\infty$ and a sequence $\{E_n\}_{n=1}^\infty$ of pairwise orthogonal finite rank projections such that $\lim_{n\to\infty} \lambda_n = 0$ and

$$\lim_{N\to\infty} \left\| T - \sum_{n=1}^N \lambda_n E_n \right\| = 0.$$

5.3   If $\mathcal{H}$ is a Hilbert space, then $\mathfrak{LC}(\mathcal{H})$ is strongly dense in $\mathfrak{L}(\mathcal{H})$.

5.4   If $\mathcal{H}$ is a Hilbert space, then the commutator ideal of $\mathfrak{L}(\mathcal{H})$ is $\mathfrak{L}(\mathcal{H})$.

5.5   If K_1 and K_2 are complex functions in $L^2([0,1] \times [0,1])$ and T_1 and T_2 the integral operators on $L^2([0,1])$ with kernels K_1 and K_2, respectively, then show that T_1^* and $T_1 T_2$ are integral operators and determine their kernels.

5.6   Show that for every finite rank operator F on $L^2([0,1])$, there exists a kernel K in $L^2([0,1] \times [0,1])$ such that $F = T_K$.

5.7   If T_K is an integral operator on $L^2([0,1])$ with kernel K in

$$L^2([0,1] \times [0,1])$$

and $\{f_n\}_{n=1}^\infty$ is an orthonormal basis for $L^2([0,1])$, then the series

$$\sum_{n=1}^\infty |(T_K f_n, f_n)|^2$$

converges absolutely to $\int_0^1 \int_0^1 |K(x,y)|^2 \, dx \, dy$. (Hint: Consider the expansion of K as an element of $L^2([0,1] \times [0,1])$ in terms of the orthonormal basis $\{f_n(x) f_m(y)\}_{n,m=1}^\infty$.)

5.8   Show that not every compact operator on $L^2([0,1])$ is an integral operator with kernel belonging to $L^2([0,1] \times [0,1])$.

5.9   (*Weyl*) If T is a normal operator on the Hilbert space $\mathcal{H}$ and K is a compact operator on $\mathcal{H}$, then any λ in $\sigma(T)$ but not in $\sigma(T+K)$ is an isolated eigenvalue for T of finite multiplicity.

5.10   If T is a quasinilpotent operator on $\mathcal{H}$ for which $T+T^*$ is in $\mathfrak{LC}(\mathcal{H})$, then T is in $\mathfrak{LC}(\mathcal{H})$.

5.11   If T is an operator on $\mathscr{H}$ for which the algebraic dimension of the linear space $\mathscr{H}/\mathrm{ran}\, T$ is finite, then T has closed range.*

5.12   If T is an operator on $\mathscr{H}$, then the set of λ for which $T-\lambda$ is not Fredholm is compact and nonempty.

5.13   (*Gohberg*)   If T is a Fredholm operator on the Hilbert space $\mathscr{H}$, then there exists $\varepsilon > 0$ such that

$$\alpha = \dim \ker(T-\lambda)$$

is constant for $0 < |\lambda| < \varepsilon$ and $\alpha \leqslant \dim \ker T$.* (Hint: For sufficiently small λ, $(T-\lambda)|\mathscr{L}$ is right invertible, where $\mathscr{L} = \bigcap_{n \geqslant 0}^{\infty} T^n \mathscr{H}$ is closed, and $\ker(T-\lambda) \subset \mathscr{L}$.)

5.14   If T is an operator on $\mathscr{H}$, then the function $\dim \ker(T-\lambda)$ is locally constant on the open set on which $T-\lambda$ is Fredholm except for isolated points at which it is larger.

5.15   If H is a self-adjoint operator on $\mathscr{H}$ and K is a compact operator on $\mathscr{H}$, then $\sigma(H+K)\backslash\sigma(H)$ consists of isolated eigenvalues of finite multiplicity.

5.16   If $\mathscr{H}$ is a Hilbert space and π is the natural map from $\mathfrak{L}(\mathscr{H})$ to the Calkin algebra $\mathfrak{L}(\mathscr{H})/\mathfrak{LC}(\mathscr{H})$ and T is an operator on $\mathscr{H}$, then $\pi(T)$ is self-adjoint if and only if $T = H + iK$, where H is self-adjoint and K is compact. Further, $\pi(T)$ is unitary if and only if $T = V + K$, where K is compact and either V or V^* is an isometry for which $I - VV^*$ or $I - V^*V$ is finite rank. What, if anything, can be said if $\pi(T)$ is normal?**

5.17   (*Weyl–von Neumann*).   If H is a self-adjoint operator on the separable Hilbert space $\mathscr{H}$, then there exists a compact self-adjoint operator K on $\mathscr{H}$ such that $H+K$ has an orthonormal basis consisting of eigenvectors.* (Hint: Show for every vector x in $\mathscr{H}$ there exists a finite rank operator F of small norm such that $H+F$ has a finite-dimensional reducing subspace which almost contains x and proceed to exhaust the space.)

5.18   If U is a unitary operator on the separable Hilbert space $\mathscr{H}$, then there exists a compact operator K such that $U+K$ is unitary and $\mathscr{H}$ has an orthonormal basis consisting of eigenvalues for $U+K$.

5.19   If U_+ is the unilateral shift on $l^2(\mathbb{Z}_+)$, then for any unitary operator V on a separable Hilbert space $\mathscr{H}$, there exists a compact operator K on $l^2(\mathbb{Z}_+)$ such that $U_+ + K$ is unitarily equivalent to $U_+ \oplus V$ on $l^2(\mathbb{Z}_+)\oplus \mathscr{H}$.* (Hint: Consider the case of finite-dimensional $\mathscr{H}$ with the additional requirement that K have small norm and use the preceding result to handle the general case.)

5.20 If V_1 and V_2 are isometries on the separable Hilbert space $\mathcal{H}$ and at least one is not unitary, then there exists a compact perturbation of V_1 which is unitarily equivalent to V_2 if and only if $\dim \ker V_1{}^* = \dim \ker V_2{}^*$.

Definition If $\mathfrak{A}$ is a C^*-algebra, then a state φ on $\mathfrak{A}$ is a complex linear function which satisfies $\varphi(A^*A) \geqslant 0$ for A in $\mathfrak{A}$ and $\varphi(I) = 1$.

5.21 If φ is a state on the C^*-algebra $\mathfrak{A}$, then $(A, B) = \varphi(B^*A)$ has the properties of an inner product except $(A, A) = 0$ need not imply $A = 0$. Moreover, φ is continuous and has norm 1. (Hint: Use a generalization of the Cauchy–Schwarz inequality.)

5.22 If φ is a state on the C^*-algebra $\mathfrak{A}$, then $\mathfrak{N} = \{A \in \mathfrak{A} : \varphi(A^*A) = 0\}$ is a closed left ideal in $\mathfrak{A}$. Further, φ induces an inner product on the quotient space $\mathfrak{A}/\mathfrak{N}$, such that $\pi(B)(A + \mathfrak{N}) = BA + \mathfrak{N}$ defines a bounded operator for B in $\mathfrak{A}$. If we let $\pi_\varphi(B)$ denote the extension of this operator to the completion $\mathcal{H}_\varphi$ of $\mathfrak{A}/\mathfrak{N}$, then π_φ defines a *-homomorphism from $\mathfrak{A}$ into $\mathfrak{L}(\mathcal{H}_\varphi)$.

5.23 If $\mathfrak{A}$ is a C^*-algebra contained in $\mathfrak{L}(\mathcal{K})$ having the unit vector f as a cyclic vector, then $\varphi(A) = (Af, f)$ is a state on $\mathfrak{A}$. Moreover, if π_φ is the representation of $\mathfrak{A}$ given by φ on $\mathcal{H}_\varphi$, then there exists an isometrical isomorphism ψ from $\mathcal{H}_\varphi$ onto $\mathcal{K}$ such that $A = \psi \pi_\varphi(A) \psi^*$.

5.24 (*Kreĭn*) If $\mathscr{L}$ is a self-adjoint subspace of the C^*-algebra $\mathfrak{A}$ containing the identity and φ_0 is a positive linear functional on $\mathscr{L}$ (that is, $\varphi_0(A) \geqslant 0$ for $A \geqslant 0$), satisfying $\varphi_0(I) = 1$, then there exists a state φ on $\mathfrak{A}$ extending φ_0. (Hint: Use the Hahn–Banach theorem to extend φ_0 to φ and prove that φ is positive.)

5.25 If $\mathfrak{A}$ is a C^*-algebra and A is in $\mathfrak{A}$ then there exists a state ψ on $\mathfrak{A}$ such that $\varphi(A^*A) = \|A\|^2$. (Hint: Consider first the abelian subalgebra generated by A^*A.)

5.26 If $\mathfrak{A}$ is a C^*-algebra, then there exists a Hilbert space $\mathcal{H}$ and a *-isometrical isomorphism π from $\mathfrak{A}$ into $\mathfrak{L}(\mathcal{H})$. Moreover, if $\mathfrak{A}$ is separable in the norm topology, then $\mathcal{H}$ can be chosen to be separable. (Hint: Find a representation π_A of $\mathfrak{A}$ for which $\|\pi_A(A)\| = \|A\|$ for each A in $\mathfrak{A}$ and consider the direct sum.)

5.27 The collection of states on a C^*-algebra $\mathfrak{A}$ is a weak *-compact convex subset of the dual of $\mathfrak{A}$. Moreover, a state φ is an extreme point of the set of all states if and only if $\pi_\varphi(\mathfrak{A})$ is an irreducible subset of $\mathfrak{L}(\mathcal{H}_\varphi)$. Such states are called pure states.

5.28 Show that there are no proper closed two-sided ideals in $\mathfrak{LC}(\mathscr{H})$. (Hint: Assume $\mathfrak{I}$ were such an ideal and show that a representation of $\mathfrak{LC}(\mathscr{H})/\mathfrak{I}$ given by Exercise 5.26 contradicts Theorem 5.40.)

5.29 If $\mathfrak{A}$ is a Banach algebra with an involution, no identity, but satisfying $\|T^*T\| = \|T\|^2$ for T in $\mathfrak{A}$, then $\mathfrak{A} \oplus \mathbb{C}$ can be given a norm making it into a C^*-algebra. (Hint: Consider the operator norm of $\mathfrak{A} \oplus \mathbb{C}$ acting on $\mathfrak{A}$.)

6 *The Hardy Spaces*

6.1 In this chapter we study various properties of the spaces H^1, H^2, and H^∞ in preparation for our study of Toeplitz operators in the following chapter. Due to the availability of several excellent accounts of this subject (see Notes), we do not attempt a comprehensive treatment and proceed in the main using the techniques which we have already introduced.

We begin by recalling some pertinent definitions from earlier chapters. For n in $\mathbb{Z}$ let χ_n be the function on $\mathbb{T}$ defined by $\chi_n(e^{i\theta}) = e^{in\theta}$. For $p = 1, 2, \infty$, we define the Hardy space:

$$H^p = \left\{ f \subset L^p(\mathbb{T}) . \int_0^{2\pi} f(e^{i\theta}) \chi_n(e^{i\theta}) \, d\theta = 0 \qquad \text{for} \quad n > 0 \right\}.$$

It is easy to see that each H^p is a closed subspace of the corresponding $L^p(\mathbb{T})$, and hence is a Banach space. Moreover, since $\{\chi_n\}_{n \in \mathbb{Z}}$ is an orthonormal basis for $L^2(\mathbb{T})$, it follows that H^2 is the closure in the L^2-norm of the analytic trigonometric polynomials $\mathscr{P}_+$. The closure of $\mathscr{P}_+$ in $C(\mathbb{T})$ is the disk algebra A with maximal ideal space equal to the closed unit disk $\overline{\mathbb{D}}$. Lastly, recall the representation of $L^\infty(\mathbb{T})$ into $\mathfrak{L}(L^2(\mathbb{T}))$ given by the mapping $\varphi \to M_\varphi$, where M_φ is the multiplication operator defined by $M_\varphi f = \varphi f$ for f in $L^2(\mathbb{T})$.

We begin with the following result which we use to show that H^∞ is an algebra.

6.2   Proposition   If φ is in $L^\infty(\mathbb{T})$, then H^2 is an invariant subspace for M_φ if and only if φ is in H^∞.

Proof   If $M_\varphi H^2$ is contained in H^2, then $\varphi \cdot 1$ is in H^2, since 1 is in H^2, and hence φ is in H^∞. Conversely, if φ is in H^∞, then $\varphi \mathscr{P}_+$ is contained in H^2, since for $p = \sum_{j=0}^N \alpha_j \chi_j$ in $\mathscr{P}_+$, we have

$$\int_0^{2\pi} (\varphi p)\, \chi_n \, d\theta = \sum_{j=0}^N \alpha_j \int_0^{2\pi} \varphi \chi_{j+n} \, d\theta = 0 \qquad \text{for } n > 0.$$

Lastly, since H^2 is the closure of $\mathscr{P}_+$, we have φH^2 contained in H^2 which completes the proof. ∎

6.3   Corollary   The space H^∞ is an algebra.

Proof   If φ and ψ are in H^∞, then $M_{\varphi\psi} H^2 = M_\varphi(M_\psi H^2) \subset M_\varphi H^2 \subset H^2$ by the proposition, which then implies that $\varphi\psi$ is in H^∞. Thus H^∞ is an algebra. ∎

The following result is essentially the uniqueness of the Fourier–Stieltjes transform for measures on $\mathbb{T}$.

6.4   Theorem   If μ is in the space $M(\mathbb{T})$ of Borel measures on $\mathbb{T}$ and $\int_\mathbb{T} \chi_n \, d\mu = 0$ for n in $\mathbb{Z}$, then $\mu = 0$.

Proof   Since the linear span of the functions $\{\chi_n\}_{n \in \mathbb{Z}}$ is uniformly dense in $C(\mathbb{T})$ and $M(\mathbb{T})$ is the dual of $C(\mathbb{T})$, the measure μ represents the zero functional and hence must be the zero measure. ∎

6.5   Corollary   If f is a function in $L^1(\mathbb{T})$ such that

$$\int_0^{2\pi} f(e^{i\theta})\, \chi_n(e^{i\theta}) \, d\theta = 0 \qquad \text{for } n \text{ in } \mathbb{Z},$$

then $f = 0$ a.e.

Proof   If we define the measure μ on $\mathbb{T}$ such that $\mu(E) = \int_E f(e^{i\theta}) \, d\theta$, then our hypotheses become $\int_\mathbb{T} \chi_n \, d\mu = 0$ for n in $\mathbb{Z}$, and hence $\mu = 0$ by the preceding result. Therefore, $f = 0$ a.e. ∎

6.6   Corollary   If f is a real-valued function in H^1, then $f = \alpha$ a.e. for some α in $\mathbb{R}$.

Proof   If we set $\alpha = (1/2\pi) \int_0^{2\pi} f(e^{i\theta})\, d\theta$, then α is real and

$$\int_0^{2\pi} (f - \alpha)\chi_n\, d\theta = 0 \qquad \text{for} \quad n \geq 0.$$

Since $f - \alpha$ is real valued, taking the complex conjugate of the preceding equation yields

$$\int_0^{2\pi} (f - \alpha)\bar{\chi}_n\, d\theta = \int_0^{2\pi} (f - \alpha)\chi_{-n}\, d\theta = 0 \qquad \text{for} \quad n \geq 0.$$

Combining this with the previous identity yields

$$\int_0^{2\pi} (f - \alpha)\chi_n\, d\theta = 0 \qquad \text{for} \quad \text{all } n,$$

and hence $f = \alpha$ a.e.   ■

6.7   Corollary   If both f and $\bar{f}$ are in H^1, then $f = \alpha$ a.e. for some α in $\mathbb{C}$.

Proof   Apply the previous corollary to the real-valued functions $\frac{1}{2}(f + \bar{f})$ and $\frac{1}{2}(f - \bar{f})/i$ which are in H^1 by hypothesis.   ■

We now consider the characterization of the invariant subspaces of certain unitary operators. It was the results of Beurling on a special case of this problem which led to much of the modern work on function algebras and, in particular, to the recent interest in the Hardy spaces.

6.8   Theorem   If μ is a positive Borel measure on $\mathbb{T}$, then a closed subspace $\mathcal{M}$ of $L^2(\mu)$ satisfies $\chi_1 \mathcal{M} = \mathcal{M}$ if and only if there exists a Borel subset E of $\mathbb{T}$ such that

$$\mathcal{M} = L_E^2(\mu) = \{f \in L^2(\mu) : f(e^{it}) = 0 \text{ for } e^{it} \notin E\}.$$

Proof   If $\mathcal{M} = L_E^2(\mu)$, then clearly $\chi_1 \mathcal{M} = \mathcal{M}$. Conversely, if $\chi_1 \mathcal{M} = \mathcal{M}$, then it follows that $\mathcal{M} = \chi_{-1}\chi_1 \mathcal{M} = \chi_{-1}\mathcal{M}$ and hence $\mathcal{M}$ is a reducing subspace for the operator M_{χ_1} on $L^2(\mu)$. Therefore, if F denotes the projection onto $\mathcal{M}$, then F commutes with M_{χ_1} by Proposition 4.42 and hence with M_φ for φ in $C(\mathbb{T})$. Combining Corollary 4.53 with Propositions 4.22 and 4.51 allows us to conclude that F is of the form M_φ for some φ in $L^\infty(\mu)$, and hence the result follows.   ■

The role of H^2 in the general theory is established in the following description of the simply invariant subspaces for M_{χ_1}.

6.9   Theorem   If μ is a positive Borel measure on $\mathbb{T}$, then a nontrivial closed subspace $\mathcal{M}$ of $L^2(\mu)$ satisfies $\chi_1 \mathcal{M} \subset \mathcal{M}$ and $\bigcap_{n \geq 0} \chi_n \mathcal{M} = \{0\}$ if and only if there exists a Borel function φ such that $|\varphi|^2 \, d\mu = d\theta/2\pi$ and $\mathcal{M} = \varphi H^2$.

Proof   If φ is a Borel function satisfying $|\varphi|^2 \, d\mu = d\theta/2\pi$, then the function $\Psi f = \varphi f$ is μ-measurable for f in H^2 and

$$\|\Psi f\|_2^2 = \int_{\mathbb{T}} |\varphi f|^2 \, d\mu = \frac{1}{2\pi} \int_0^{2\pi} |f|^2 \, d\theta = \|f\|_2^2.$$

Thus the image $\mathcal{M}$ of H^2 under the isometry Ψ is a closed subspace of $L^2(\mu)$ and is invariant for M_{χ_1}, since $\chi_1(\Psi f) = \Psi(\chi_1 f)$. Lastly, we have

$$\bigcap_{n \geq 0} \chi_n \mathcal{M} = \Psi \left[\bigcap_{n \geq 0} \chi_n H^2 \right] = \{0\}$$

and hence $\mathcal{M}$ is a simply invariant subspace for M_{χ_1}.

Conversely, suppose $\mathcal{M}$ is a nontrivial closed invariant subspace for M_{χ_1} which satisfies $\bigcap_{n \geq 0} \chi_n \mathcal{M} = \{0\}$. Then $\mathcal{L} = \mathcal{M} \ominus \chi_1 \mathcal{M}$ is nontrivial and $\chi_n \mathcal{L} = \chi_n \mathcal{M} \ominus \chi_{n+1} \mathcal{M}$, since multiplication by χ_1 is an isometry on $L^2(\mu)$. Therefore, the subspace $\sum_{n=0}^{\infty} \oplus \chi_n \mathcal{L}$ is contained in $\mathcal{M}$, and an easy argument reveals $\mathcal{M} \ominus (\sum_{n=0}^{\infty} \oplus \chi_n \mathcal{L})$ to be $\bigcap_{n \geq 0} \chi_n \mathcal{M}$ and hence $\{0\}$.

If φ is a unit vector in $\mathcal{L}$, then φ is orthogonal to $\chi_n \mathcal{M}$ and hence to $\chi_n \varphi$ for $n > 0$, and thus we have

$$0 = (\varphi, \chi_n \varphi) = \int_{\mathbb{T}} |\varphi|^2 \chi_n \, d\mu \qquad \text{for} \quad n > 0.$$

Combining Theorem 6.4 and Corollary 6.6, we see that $|\varphi|^2 \, d\mu = d\theta/2\pi$. Now suppose $\mathcal{L}$ has dimension greater than one and φ' is a unit vector in $\mathcal{L}$ orthogonal to φ. In this case, we have

$$0 = (\chi_n \varphi, \chi_m \varphi') = \int_{\mathbb{T}} \varphi \bar{\varphi}' \chi_{n-m} \, d\mu \qquad \text{for} \quad n, m \geq 0,$$

and thus $\int_{\mathbb{T}} \chi_k \, d\nu = 0$ for k in $\mathbb{Z}$, where $d\nu = \varphi \bar{\varphi}' \, d\mu$. Therefore, $\varphi \bar{\varphi}' = 0 \ \mu$ a.e. Combining this with the fact that $|\varphi|^2 \, d\mu = |\varphi'|^2 \, d\mu$ leads to a contradiction, and hence $\mathcal{L}$ is one dimensional. Thus we obtain that $\varphi \mathcal{P}_+$ is dense in $\mathcal{M}$ and hence $\mathcal{M} = \varphi H^2$, which completes the proof.   ∎

The case of the preceding theorem considered by Beurling will be given after the following definition.

6.10   Definition   A function φ in H^∞ is an inner function if $|\varphi| = 1$ a.e.

6.11   Corollary (Beurling)   If $T_{\chi_1} = M_{\chi_1}|H^2$, then a nontrivial closed subspace $\mathcal{M}$ of H^2 is invariant for T_{χ_1} if and only if $\mathcal{M} = \varphi H^2$ for some inner function φ.

Proof   If φ is an inner function, then $\varphi \mathscr{P}_+$ is contained in H^∞, since the latter is an algebra, and is therefore contained in H^2. Since φH^2 is the closure of $\varphi \mathscr{P}_+$, we see that φH^2 is a closed invariant subspace for T_{χ_1}.

Conversely, if $\mathcal{M}$ is a nontrivial closed invariant subspace for T_{χ_1}, then $\mathcal{M}$ satisfies the hypotheses of the preceding theorem for $d\mu = d\theta/2\pi$, and hence there exists a measurable function φ such that $\mathcal{M} = \varphi H^2$ and

$$|\varphi|^2 \, d\theta/2\pi = d\theta/2\pi. \quad .$$

Therefore, $|\varphi| = 1$ a.e., and since 1 is in H^2 we see that $\varphi = \varphi \cdot 1$ is in H^2; thus φ is an inner function. ∎

A general invariant subspace for M_{χ_1} on $L^2(\mu)$ need not be of the form covered by either of the preceding two theorems. The following result enables us to reduce the general case to these, however.

6.12   Theorem   If μ is a positive Borel measure on $\mathbb{T}$, then a closed invariant subspace $\mathcal{M}$ for M_{χ_1} has a unique direct sum decomposition $\mathcal{M} = \mathcal{M}_1 \oplus \mathcal{M}_2$ such that each of $\mathcal{M}_1$ and $\mathcal{M}_2$ is invariant for M_{χ_1}, $\chi_1 \mathcal{M}_1 = \mathcal{M}_1$, and $\bigcap_{n \geq 0} \chi_n \mathcal{M}_2 = \{0\}$.

Proof   If we set $\mathcal{M}_1 = \bigcap_{n \geq 0} \chi_n \mathcal{M}$, then $\mathcal{M}_1$ is a closed invariant subspace for M_{χ_1} satisfying $\chi_1 \mathcal{M}_1 = \mathcal{M}_1$. To prove the latter statement observe that a function f is in $\mathcal{M}_1$ if and only if it can be written in the form $\chi_n g$ for some g in $\mathcal{M}$ for each $n > 0$. Now if we set $\mathcal{M}_2 = \mathcal{M} \ominus \mathcal{M}_1$, then a function f in $\mathcal{M}$ is in $\mathcal{M}_2$ if and only if $(f, g) = 0$ for all g in $\mathcal{M}_1$. Since $0 = (f, g) = (\chi_1 f, \chi_1 g)$ and $\chi_1 \mathcal{M}_1 = \mathcal{M}_1$, it follows that $\chi_1 f$ is in $\mathcal{M}_2$ and hence $\mathcal{M}_2$ is invariant for M_{χ_1}. If f is in $\bigcap_{n \geq 0} \chi_n \mathcal{M}_2$, then it is in $\mathcal{M}_1$ and hence $f = 0$. Thus the proof is complete. ∎

Although we could combine the three preceding theorems to obtain a complete description of the invariant subspaces for M_{χ_1}, the statement would be very unwieldy and hence we omit it.

The preceding theorems correspond to the multiplicity one case of certain structure theorems for isometries (see [66], [58]).

To illustrate the power of the preceding results we obtain as corollaries the following theorems which will be important in what follows.

6.13   Theorem (F. and M. Riesz)   If f is a nonzero function in H^2, then the set $\{e^{it} \in \mathbb{T}: f(e^{it}) = 0\}$ has measure zero.

Proof   Set $E = \{e^{it} \in \mathbb{T}: f(e^{it}) = 0\}$ and define

$$\mathcal{M} = \{g \in H^2: g(e^{it}) = 0 \text{ for } e^{it} \in E\}.$$

It is clear that $\mathcal{M}$ is a closed invariant subspace for T_{χ_1} which is nontrivial since f is in it. Hence, by Beurling's theorem there exists an inner function φ such that $\mathcal{M} = \varphi H^2$. Since 1 is in H^2, it follows that φ is in $\mathcal{M}$ and hence that E is contained in $\{e^{it} \in \mathbb{T}: \varphi(e^{it}) = 0\}$. Since $|\varphi| = 1$ a.e., the result follows.   ∎

6.14   Theorem (F. and M. Riesz)   If v is a Borel measure on $\mathbb{T}$ such that $\int_{\mathbb{T}} \chi_n \, dv = 0$ for $n > 0$, then v is absolutely continuous and there exists f in H^1 such that $dv = f \, d\theta$.

Proof   If μ denotes the total variation of v, then there exists a Borel function ψ such that $dv = \psi \, d\mu$ and $|\psi| = 1$ a.e. with respect to μ. If $\mathcal{M}$ denotes the closed subspace of $L^2(\mu)$ spanned by $\{\chi_n : n > 0\}$, then

$$(\chi_n, \bar{\psi}) = \int_{\mathbb{T}} \chi_n \psi \, d\mu = \int_{\mathbb{T}} \chi_n \, dv = 0,$$

and hence $\bar{\psi}$ is orthogonal to $\mathcal{M}$ in $L^2(\mu)$.

Suppose $\mathcal{M} = \mathcal{M}_1 \oplus \mathcal{M}_2$ is the decomposition given by Theorem 6.12. If E is the Borel subset of $\mathbb{T}$ given by Theorem 6.8 such that $\mathcal{M}_1 = L_E^2(\mu)$, then we have

$$\mu(E) = \int_{\mathbb{T}} |\psi|^2 I_E \, d\mu = (\bar{\psi}, \bar{\psi} I_E) = 0,$$

since $\bar{\psi} I_E$ is in $\mathcal{M}_1$ and $\bar{\psi}$ is orthogonal to $\mathcal{M}$. Therefore, $\mathcal{M}_1 = \{0\}$ and hence there exists a μ-measurable function φ such that $\mathcal{M} = \varphi H^2$ and $|\varphi|^2 \, d\mu = d\theta/2\pi$ by Theorem 6.9. Since χ_1 is in $\mathcal{M}$, it follows that there exists g in H^2 such that $\chi_1 = \varphi g$ a.e. with respect to μ, and since $\varphi \neq 0$ μ a.e., we have that μ is mutually absolutely continuous with Lebesgue measure. If f is a function in $L^1(\mathbb{T})$ such that $dv = f \, d\theta$, then the hypotheses imply that f is in H^1, and hence the proof is complete.   ∎

Actually, the statements of the preceding two theorems can be combined into one: an analytic measure is mutually absolutely continuous with respect to Lebesgue measure.

6.15 We now turn to the investigation of the maximal ideal space M_∞ of the commutative Banach algebra H^∞. We begin by imbedding the open unit disk $\mathbb{D}$ in M_∞. For z in $\mathbb{D}$ define the bounded linear functional φ_z on H^1 such that

$$\varphi_z(f) = \frac{1}{2\pi} \int_0^{2\pi} \frac{f(e^{i\theta})}{1 - ze^{-i\theta}} \, d\theta \qquad \text{for } f \text{ in } H^1.$$

Since the function $1/(1 - ze^{-i\theta})$ is in $L^\infty(\mathbb{T})$ and H^1 is contained in $L^1(\mathbb{T})$, it follows that φ_z is a bounded linear functional on H^1. Moreover, since $1/(1 - ze^{-i\theta}) = \sum_{n=0}^\infty e^{-in\theta} z^n$ and the latter series converges absolutely, we see that

$$\varphi_z(f) = \sum_{k=0}^\infty z^k \left(\frac{1}{2\pi} \int_0^{2\pi} f\bar{\chi}_k \, d\theta \right).$$

Thus, if p is an analytic trigonometric polynomial, then $\varphi_z(p) = p(z)$ and hence φ_z is a multiplicative linear functional on $\mathscr{P}_+$. To show that φ_z is multiplicative on H^∞ we proceed as follows.

6.16 Lemma If f and g are in H^2 and z is in $\mathbb{D}$, then fg is in H^1 and $\varphi_z(fg) = \varphi_z(f)\varphi_z(g)$.

Proof Let $\{p_n\}_{n=1}^\infty$ and $\{q_n\}_{n=1}^\infty$ be sequences of analytic trigonometric polynomials such that

$$\lim_{n\to\infty} \|f - p_n\|_2 = \lim_{n\to\infty} \|g - q_n\|_2 = 0.$$

Since the product of two L^2-functions is in L^1, we have

$$\|fg - p_n q_n\|_1 \leqslant \|fg - p_n g\|_1 + \|p_n g - p_n q_n\|_1$$

$$\leqslant \|f - p_n\|_2 \|g\|_2 + \|p_n\|_2 \|g - q_n\|_2,$$

and hence $\lim_{n\to\infty} \|fg - p_n q_n\|_1 = 0$. Therefore, since each $p_n q_n$ is in H^1 we have fg in H^1. Moreover, since φ_z is continuous, we have

$$\varphi_z(fg) = \lim_{n\to\infty} \varphi_z(p_n q_n) = \lim_{n\to\infty} \varphi_z(p_n) \lim_{n\to\infty} \varphi_z(q_n) = \varphi_z(f)\varphi_z(g). \quad \blacksquare$$

With these preliminary considerations taken care of, we can now imbed $\mathbb{D}$ in M_∞.

6.17 Theorem For z in $\mathbb{D}$ the restriction of φ_z to H^∞ is a multiplicative linear functional on H^∞. Moreover, the mapping F from $\mathbb{D}$ into M_∞ defined by $F(z) = \varphi_z$ is a homeomorphism.

Proof That φ_z restricted to H^∞ is a multiplicative linear functional follows from the preceding lemma.

Since for a fixed f in H^1, the function $\varphi_z(f)$ is analytic in z, it follows that F is continuous. Moreover, since $\varphi_z(\chi_1) = z$, it follows that F is one-to-one. Lastly, if $\{\varphi_{z_\alpha}\}_{\alpha \in A}$ is a net in M_∞ converging to φ_z, then

$$\lim_{\alpha \in A} z_\alpha = \lim_{\alpha \in A} \varphi_{z_\alpha}(\chi_1) = \varphi_z(\chi_1) = z,$$

and hence F is a homeomorphism. ∎

From now on we shall simply identify $\mathbb{D}$ as a subset of M_∞. Further, we shall denote the Gelfand transform of a function f in H^∞ by $\hat{f}$. Note that $\hat{f}|\mathbb{D}$ is analytic. Moreover, for f in H^1 we shall let $\hat{f}$ denote the function defined on $\mathbb{D}$ by $\hat{f}(z) = \varphi_z(f)$. This dual use of the ^-notation should cause no confusion.

The maximal ideal space M_∞ is quite large and is extremely complex. The deepest result concerning M_∞ is the corona theorem of Carleson, stating that $\mathbb{D}$ is dense in M_∞. Although the proof of this result has been somewhat clarified (see [15], [39]) it is still quite difficult and we do not consider it in this book.

Due to the complexity of M_∞ it is not feasible to determine the spectrum of a function f in H^∞ using $\hat{f}$, but it follows from the corona theorem that the spectrum of f is equal to the closure of $\hat{f}(\mathbb{D})$. Fortunately, a direct proof of this result is not difficult.

6.18 **Theorem** If φ is a function in H^∞, then φ is invertible in H^∞ if and only if $\hat{\varphi}|\mathbb{D}$ is bounded away from zero.

Proof If φ is invertible in H^∞, then $\hat{\varphi}$ is nonvanishing on the compact space M_∞ and hence $|\hat{\varphi}(z)| \geq \varepsilon > 0$ for z in $\mathbb{D}$. Conversely, if $|\hat{\varphi}(z)| \geq \varepsilon > 0$ for z in $\mathbb{D}$ and we set $\psi(z) = 1/\hat{\varphi}(z)$, then ψ is analytic and bounded by $1/\varepsilon$ on $\mathbb{D}$. Thus ψ has a Taylor series expansion $\psi(z) = \sum_{n=0}^{\infty} a_n z^n$, which converges in $\mathbb{D}$. Since for $0 < r < 1$, we have

$$\sum_{n=0}^{\infty} |a_n|^2 r^{2n} = \frac{1}{2\pi} \int_0^{2\pi} |\psi(re^{it})|^2 \, dt \leq \frac{1}{\varepsilon^2},$$

it follows that $\sum_{n=0}^{\infty} |a_n|^2 \leq 1/\varepsilon^2$. Therefore, there exists a function f in H^2 such that $f = \sum_{n=0}^{\infty} a_n \chi_n$.

If $\varphi = \sum_{n=0}^{\infty} b_n \chi_n$ is the orthonormal expansion of φ as an element of H^2, then $\hat{\varphi}(z) = \sum_{n=0}^{\infty} b_n z^n$ for z in $\mathbb{D}$. Since $\hat{\varphi}(z)\psi(z) = 1$, it follows that

$(\sum_{n=0}^{\infty} b_n z^n)(\sum_{n=0}^{\infty} a_n z^n) = 1$ for z in $\mathbb{D}$. Therefore, $\sum_{n=0}^{\infty}(\sum_{k=0}^{n} b_k a_{n-k}) z^n = 1$ for z in $\mathbb{D}$, and hence the uniqueness of power series implies that

$$\sum_{k=0}^{n} b_k a_{n-k} = \begin{cases} 1 & \text{if } n = 0, \\ 0 & \text{if } n > 0. \end{cases}$$

Since

$$\lim_{N \to \infty} \left\| \varphi - \sum_{n=0}^{N} b_n \chi_n \right\|_2 = \lim_{M \to \infty} \left\| f - \sum_{m=0}^{M} a_m \chi_m \right\|_2 = 0,$$

we have that

$$\lim_{N \to \infty} \left\| \varphi f - \left(\sum_{n=0}^{N} b_n \chi_n \right) \left(\sum_{m=0}^{N} a_m \chi_m \right) \right\|_1 = 0,$$

which implies that

$$\lim_{N \to \infty} \left\| (\varphi f - 1) + \sum_{n=N+1}^{2N} c_n \chi_n \right\|_1 = 0 \qquad \text{for} \quad c_n = \sum_{k=n-N}^{N} a_k b_{n-k}.$$

Therefore,

$$\frac{1}{2\pi} \int_0^{2\pi} \varphi f \chi_k \, d\theta = \begin{cases} 1 & \text{if } k = 0, \\ 0 & \text{if } k \neq 0, \end{cases}$$

and hence $\varphi f = 1$ by Corollary 6.6. It remains only to show that f is in $L^\infty(\mathbb{T})$ and this follows from the fact that the functions $\{f_r\}_{r \in (0,1)}$ are uniformly bounded by $1/\varepsilon$, where $f_r(e^{it}) = \hat{f}(re^{it})$, and the fact that $\lim_{r \to 1} \|f - f_r\|_2 = 0$ Thus f is an inverse for φ which lies in H^∞. ∎

The preceding proof was complicated by the fact that we have not investigated the precise relation between the function $\hat{f}$ on $\mathbb{D}$ and the function f on $\mathbb{T}$. It can be shown that for f in H^1 we have $\lim_{r \to 1} \hat{f}(re^{it}) = f(e^{it})$ for almost all e^{it} in $\mathbb{T}$. We do not prove this but leave it as an exercise (see Exercise 6.23).

Observe that we proved in the last paragraph of the preceding proof that if f is in H^2 and $\hat{f}$ is bounded on $\mathbb{D}$, then f is in H^∞.

We give another characterization of invertibility for functions in H^∞ which will be used in the following chapter, but first we need a definition.

6.19 Definition   A function f in H^2 is an outer function if $\operatorname{clos}[f\mathscr{P}_+] = H^2$.

An alternate definition is that outer functions are those functions in H^2 which are cyclic vectors for the operator T_{χ_1} which is multiplication by χ_1 on H^2.

6.20   Proposition   A function φ in H^∞ is invertible in H^∞ if and only if φ is invertible in L^∞ and is an outer function.

Proof   If $1/\varphi$ is in H^∞, then obviously φ is invertible in L^∞. Moreover, since

$$\mathrm{clos}\,[\varphi \mathscr{P}_+] = \varphi H^2 \supset \varphi\left(\frac{1}{\varphi} H^2\right) = H^2,$$

it follows that φ is an outer function. Conversely, if $1/\varphi$ is in $L^\infty(\mathbb{T})$, and φ is an outer function, then $\varphi H^2 = \mathrm{clos}\,[\varphi \mathscr{P}_+] = H^2$. Therefore, there exists a function ψ in H^2 such that $\varphi\psi = 1$, and hence $1/\varphi = \psi$ is in H^2. Thus, $1/\varphi$ is in H^∞ and the proof is complete.   ■

Note, in particular, that by combining the last two results we see that an outer function can not vanish on $\mathbb{D}$. The property of being an outer function, however, is more subtle than this.

The following result shows one of the fundamental uses of inner and outer functions.

6.21   Proposition   If f is a nonzero function in H^2, then there exist inner and outer functions φ and g such that $f = \varphi g$. Moreover, f is in H^∞ if and only if g is in H^∞.

Proof   If we set $\mathscr{M} = \mathrm{clos}\,[f\mathscr{P}_+]$, then $\mathscr{M}$ is a nontrivial closed invariant subspace for T_{χ_1} and hence by Beurling's Theorem 6.11 is of the form φH^2 for some inner function φ. Since f is in $\mathscr{M}$, there must exist g in H^2 such that $f = \varphi g$. If we set $\mathscr{N} = \mathrm{clos}\,[g\mathscr{P}_+]$, then again there exists an inner function ψ such that $\mathscr{N} = \psi H^2$. Then the inclusion $f\mathscr{P}_+ = \varphi g \mathscr{P}_+ \subset \varphi\psi H^2$ implies $\varphi H^2 = \mathrm{clos}\,[f\mathscr{P}_+] \subset \varphi\psi H^2$, and hence there must exist h in H^2 such that $\varphi = \varphi\psi h$. Since φ and ψ are inner functions, it follows that $\bar{\psi} = h$ and therefore ψ is constant by Corollary 6.7. Hence, $\mathrm{clos}\,[g\mathscr{P}_+] = H^2$ and g is an outer function. Lastly, since $|f| = |g|$, we see that f is in H^∞ if and only if g is.   ■

We next show that the modulus of an outer function determines it up to a constant as a corollary to the following proposition.

6.22   Proposition   If g and h are functions in H^2 such that g is outer, then $|h| \leqslant |g|$ if and only if there exists a function k in H^2 such that $h = gk$ and $|k| \leqslant 1$.

Proof If $h = gk$ and $|k| \leqslant 1$, then clearly $|h| \leqslant |g|$. Conversely, if g is an outer function, then there exists a sequence of analytic trigonometric polynomials $\{p_n\}_{n=1}^{\infty}$ such that $\lim_{n \to \infty} \|1 - p_n g\|_2 = 0$. If $|h| \leqslant |g|$, then we have

$$\|hp_n - hp_m\|_2^{\ 2} = \frac{1}{2\pi} \int_0^{2\pi} |p_n - p_m|^2 |h|^2 \, d\theta \leqslant \frac{1}{2\pi} \int_0^{2\pi} |p_n - p_m|^2 |g|^2 \, d\theta$$

$$= \|gp_n - gp_m\|_2^{\ 2},$$

and hence $\{p_n h\}_{n=1}^{\infty}$ is a Cauchy sequence. Thus the sequence $\{p_n h\}_{n=1}^{\infty}$ converges to a function k in H^2, and

$$\|gk - h\|_1 \leqslant \lim_{n \to \infty} \|g\|_2 \|k - p_n h\|_2 + \lim_{n \to \infty} \|gp_n - 1\|_2 \|h\|_2 = 0.$$

Therefore, $gk = h$ and the proof is complete. ∎

6.23 Corollary If g_1 and g_2 are outer functions in H^2 such that $|g_1| = |g_2|$, then $g_1 = \lambda g_2$ for some complex number of modulus one.

Proof By the preceding result there exist functions h and k in H^2 such that $|h|$, $|k| \leqslant 1$, $g_1 = hg_2$, and $g_2 = kg_1$. This implies $g_1 = khg_1$, and by Theorem 6.13 we have that $kh = 1$. Thus $h = \bar{k}$, and hence both h and $\bar{h}$ are in H^2. Therefore, h is constant by Corollary 6.7 and the result follows. ∎

The question of which nonnegative functions in L^2 can be the modulus of a function in H^2 is interesting from several points of view. Although an elegant necessary and sufficient condition can be given, we leave this for the exercises and obtain only those results which we shall need. Our first result shows the equivalence of this question to another.

6.24 Theorem If f is a function in $L^2(\mathbb{T})$, then there exists an outer function g such that $|f| = |g|$ a.e. if and only if $\operatorname{clos}[f\mathscr{P}_+]$ is a simply invariant subspace for M_{χ_1}.

Proof If $|f| = |g|$ for some outer function g, then $f = \varphi g$ for some unimodular function φ in $L^{\infty}(\mathbb{T})$. Then

$$\operatorname{clos}[f\mathscr{P}_+] = \operatorname{clos}[\varphi g \mathscr{P}_+] = \varphi \operatorname{clos}[g\mathscr{P}_+] = \varphi H^2,$$

and hence $\operatorname{clos}[f\mathscr{P}_+]$ is simply invariant.

Conversely, if $\operatorname{clos}[f\mathscr{P}_+]$ is simply invariant for M_{χ_1}, then there exists a unimodular function φ in $L^{\infty}(\mathbb{T})$ by Theorem 6.9 such that $\operatorname{clos}[f\mathscr{P}_+] = \varphi H^2$. Since f is in $\operatorname{clos}[f\mathscr{P}_+]$, there must exist a function g in H^2 such that

$f = \varphi g$. The proof is concluded either by applying Proposition 6.21 to g or by showing that this g is outer.   ∎

6.25   Corollary   If f is a function in $L^2(\mathbb{T})$ such that $|f| \geq \varepsilon > 0$, then there exists an outer function g such that $|g| = |f|$.

Proof   If we set $\mathcal{M} = \text{clos}[f\mathcal{P}_+]$, then $M_{\chi_1}\mathcal{M}$ is the closure of

$$\{fp : p \in \mathcal{P}_+, p(0) = 0\}.$$

If we compute the distance from f to such an fp, we find that

$$\|f - fp\|_2^2 = \frac{1}{2\pi} \int_0^{2\pi} |f|^2 |1 - p|^2 \, d\theta \geq \frac{\varepsilon^2}{2\pi} \int_0^{2\pi} |1 - p|^2 \, d\theta \geq \varepsilon^2,$$

and hence f is not in $M_{\chi_1}\mathcal{M}$. Therefore, $\mathcal{M}$ is simply invariant and hence the outer function exists by the preceding theorem.   ∎

We can also use the theorem to establish the following relation between functions in H^2 and H^1.

6.26   Corollary   If f is a function in H^1, then there exists g in H^2 such that $|g|^2 = |f|$ a.e.

Proof   If $f = 0$, then take $g = 0$. If f is a nonzero function in H^1, then there exists h in L^2 such that $|h|^2 = |f|$. It is sufficient in view of the theorem to show that $\text{clos}[h\mathcal{P}_+]$ is a simply invariant subspace for M_{χ_1}. Suppose it is not. Then $\chi_{-N}h$ is in $\text{clos}[h\mathcal{P}_+]$ for $N > 0$, and hence there exists a sequence of analytic trigonometric polynomials $\{p_n\}_{n=1}^{\infty}$ such that

$$\lim_{n \to \infty} \|p_n h - \chi_{-N}h\|_2 = 0.$$

Since

$$\|\chi_{-N}h - p_n h\|_2^2 = \frac{1}{2\pi} \int_0^{2\pi} |h^2 \chi_{-2N} - 2h^2 p_n \chi_{-N} + h^2 p_n^2| \, d\theta$$

$$= \frac{1}{2\pi} \int_0^{2\pi} |h^2 \chi_{-N} - h^2 (2p_n - p_n^2 \chi_N)| \, d\theta$$

$$= \|h^2 \chi_{-N} - h^2 (2p_n - p_n^2 \chi_N)\|_1,$$

we see that $h^2 \chi_{-N}$ is in the closure, $\text{clos}_1[h^2\mathcal{P}_+]$, of $h^2\mathcal{P}_+$ in $L^1(\mathbb{T})$. Since there exists a unimodular function φ such that $f = \varphi h^2$, we see that the function $\chi_{-N}f = \varphi(\chi_{-N}h^2)$ is in $\varphi\,\text{clos}_1[h^2\mathcal{P}_+] = \text{clos}_1[f\mathcal{P}_+] \subseteq H^1$ for $N > 0$. This implies $f \equiv 0$, which is a contradiction. Thus $\text{clos}[h\mathcal{P}_+]$ is simply invariant and the proof is complete.   ∎

6.27   Corollary   If f is a function in H^1, then there exist functions g_1 and g_2 in H^2 such that $|g_1| = |g_2| = (|f|)^{\frac{1}{2}}$ and $f = g_1 g_2$.

Proof   If g is an outer function such that $|g| = (|f|)^{\frac{1}{2}}$, then there exists a sequence of analytic trigonometric polynomials $\{p_n\}_{n=1}^\infty$ such that

$$\lim_{n \to \infty} \|gp_n - 1\|_2 = 0.$$

Thus we have

$$\|fp_n^2 - fp_m^2\|_1 \leqslant \|f(p_n - p_m)p_n\|_1 + \|f(p_n - p_m)p_m\|_1$$
$$\leqslant \|gp_n\|_2 \|g(p_n - p_m)\|_2 + \|gp_m\|_2 \|g(p_n - p_m)\|_2,$$

and hence the sequence $\{fp_n^2\}_{n=1}^\infty$ is Cauchy in the $L^1(\mathbb{T})$ norm and therefore converges to some function φ in H^1. Extracting a subsequence, if necessary, such that $\lim_{n \to \infty} (fp_n^2)(e^{it}) = \varphi(e^{it})$ a.e. and $\lim_{n \to \infty} (gp_n)(e^{it}) = 1$ a.e., we see that $\varphi g^2 = f$. Since $|g^2| = |f|$ a.e., we see that $|\varphi| = 1$ a.e., and thus the functions $g_1 = \varphi g$ and $g_2 = g$ are in H^2 and satisfy $f = g_1 g_2$ and $|g_1| = |g_2| = (|f|)^{\frac{1}{2}}$.   ∎

6.28   Corollary   The closure of $\mathscr{P}_+$ in $L^1(\mathbb{T})$ is H^1.

Proof   If f is in H^1, then $f = g_1 g_2$ with g_1 and g_2 in H^2. If $\{p_n\}_{n=1}^\infty$ and $\{q_n\}_{n=1}^\infty$ are sequences of analytic trigonometric polynomials chosen such that $\lim_{n \to \infty} \|g_1 - p_n\|_2 = \lim_{n \to \infty} \|g_2 - q_n\|_2 = 0$, then $\{p_n q_n\}_{n=1}^\infty$ is a sequence of analytic trigonometric polynomials such that $\lim_{n \to \infty} \|f - p_n q_n\|_1 = 0$.   ∎

With this corollary we can determine the dual of the Banach space H^1. Before stating this result we recall that H_0^p denotes the closed subspace

$$\left\{ f \in H^p : \frac{1}{2\pi} \int_0^{2\pi} f \, d\theta = 0 \right\} \quad \text{of} \quad H^p \quad \text{for} \quad p = 1, 2, \infty.$$

6.29   Theorem   There is a natural isometrical isomorphism between $(H^1)^*$ and $L^\infty(\mathbb{T})/H_0^\infty$.

Proof   Since H^1 is contained in $L^1(\mathbb{T})$, we obtain a contractive mapping Ψ from $L^\infty(\mathbb{T})$ into $(H^1)^*$ such that

$$[\Psi(\varphi)](f) = \frac{1}{2\pi} \int_0^{2\pi} \varphi f \, d\theta \quad \text{for} \quad \varphi \text{ in } L^\infty(\mathbb{T}) \quad \text{and} \quad f \text{ in } H^1.$$

Moreover, from the Hahn–Banach theorem and the characterization of $L^1(\mathbb{T})^*$, it follows that given Φ in $(H^1)^*$ there exists a function φ in $L^\infty(\mathbb{T})$ such that $\|\varphi\|_\infty = \|\Phi\|$ and $\Psi(\varphi) = \Phi$. Thus the mapping Ψ is onto and induces an isometrical isomorphism of $L^\infty(\mathbb{T})/\ker \Psi$ onto $(H^1)^*$.

We must determine the kernel of Ψ. If φ is a function in $\ker \Psi$, then

$$\frac{1}{2\pi} \int_0^{2\pi} \varphi \chi_n \, d\theta = [\Psi(\varphi)](\chi_n) = 0 \qquad \text{for} \quad n \geqslant 0,$$

since each χ_n is in H^1 and hence φ is in H_0^∞. Conversely, if φ lies in H_0^∞, then $[\Psi(\varphi)](p) = 0$ for each p in $\mathscr{P}_+$, and hence φ is in $\ker \Psi$ by the preceding corollary. ∎

Although $L^1(\mathbb{T})$ can be shown not to be a dual space, the subspace H^1 is.

6.30 Theorem There is a natural isometrical isomorphism between $(C(\mathbb{T})/A)^*$ and H_0^1.

Proof If φ is a function in H_0^1, then the linear functional defined

$$\Phi(f) = \frac{1}{2\pi} \int_0^{2\pi} f\varphi \, d\theta \qquad \text{for} \quad f \text{ in } C(\mathbb{T})$$

is bounded and vanishes on A. Therefore, the mapping

$$\Phi_0(f+A) = \Phi(f) = \frac{1}{2\pi} \int_0^{2\pi} f\varphi \, d\theta$$

is well defined on $C(\mathbb{T})/A$ and hence defines an element of $(C(\mathbb{T})/A)^*$. Moreover, the mapping $\Psi(\varphi) = \Phi_0$ is clearly a contractive homomorphism of H_0^1 into $(C(\mathbb{T})/A)^*$.

On the other hand, if Φ_0 is a bounded linear functional on $C(\mathbb{T})/A$, then the composition $\Phi_0 \circ \pi$, where π is the natural homomorphism of $C(\mathbb{T})$ onto $C(\mathbb{T})/A$, defines an element v of $C(\mathbb{T})^* = M(\mathbb{T})$ such that

$$\Phi_0(f+A) = \Phi(f) = \int_{\mathbb{T}} f \, dv \qquad \text{for} \quad f \text{ in } C(\mathbb{T})$$

and $\|v\| = \|\Phi_0\|$. Since this implies, in particular, that $\int_{\mathbb{T}} g \, dv = 0$ for g in A, it follows from the F. and M. Riesz theorem that there exists a function φ in H_0^1 such that

$$\Phi_0(f+A) = \frac{1}{2\pi} \int_0^{2\pi} f\varphi \, d\theta \quad \text{for } f \text{ in } C(\mathbb{T}) \quad \text{and} \quad \|\varphi\|_1 = \|v\| = \|\Phi_0\|.$$

Therefore, the mapping Ψ is an isometrical isomorphism of H_0^1 onto $(C(\mathbb{T})/A)^*$. ∎

6.31 Observe that the natural mapping i of $C(\mathbb{T})/A$ into its second dual $L^\infty(\mathbb{T})/H^\infty$ (see Exercise 1.15) is $i(f+A) = f+H^\infty$. Since the natural map is an isometry, it follows that $i[C(\mathbb{T})/A]$ is a closed subspace of $L^\infty(\mathbb{T})/H^\infty$. Hence, the inverse image of this latter subspace under the natural homomorphism of $L^\infty(\mathbb{T})$ onto $L^\infty(\mathbb{T})/H^\infty$ is closed, and therefore the linear span $H^\infty + C(\mathbb{T})$ is a closed subspace of $L^\infty(\mathbb{T})$. This proof that $H^\infty + C(\mathbb{T})$ is closed is due to Sarason [97].

The subspace $H^\infty + C(\mathbb{T})$ is actually an algebra and is just one of a large family of closed algebras which lie between H^∞ and $L^\infty(\mathbb{T})$. Much of the remainder of this chapter will be concerned with their study. We begin with the following approximation theorem.

6.32   Theorem   The collection $\mathscr{Q}$ of functions in $L^\infty(\mathbb{T})$ of the form $\psi\bar{\varphi}$ for ψ in H^∞ and φ an inner function forms a dense subalgebra of $L^\infty(\mathbb{T})$.

Proof   That $\mathscr{Q}$ is an algebra follows from the identities

$$(\psi_1\bar{\varphi}_1)(\psi_2\bar{\varphi}_2) = (\psi_1\psi_2)(\overline{\varphi_1\varphi_2})$$

and

$$\psi_1\bar{\varphi}_1 + \psi_2\bar{\varphi}_2 = (\psi_1\varphi_2 + \psi_2\varphi_1)(\overline{\varphi_1\varphi_2}).$$

Since $\mathscr{Q}$ is a linear space and the simple step functions are dense in L^∞, to conclude that $\mathscr{Q}$ is dense in $L^\infty(\mathbb{T})$ it suffices to show every characteristic function is in $\text{clos}_\infty[\mathscr{Q}]$. Thus let E be a measurable subset of $\mathbb{T}$ and let f be a function in H^2 such that

$$|f(e^{it})| = \begin{cases} \frac{1}{2} & \text{if } e^{it} \in E, \\ 2 & \text{if } e^{it} \notin E. \end{cases}$$

The existence of such a function follows from Corollary 6.25. Moreover, since f is bounded, it is in H^∞ and consequently so is $1+f^n$ for $n > 0$. If $1+f^n = \varphi_n g_n$ is a factorization given by Proposition 6.21, when φ_n is an inner and g_n is an outer function, then $|g_n| = |1+f^n| \geq \frac{1}{2}$ and hence $1/g_n$ is in H^∞ by Proposition 6.20. Therefore, the function $1/(1+f^n) = (1/g_n)\bar{\varphi}_n$ is in $\mathscr{Q}$, and since $\lim_{n\to\infty} \|I_E - 1/(1+f^n)\|_\infty = 0$, we see that I_E is in $\text{clos}_\infty[\mathscr{Q}]$. Thus, $\mathscr{Q}$ is dense in $L^\infty(\mathbb{T})$ by our previous remarks.   ∎

We next prove a certain uniqueness result.

6.33 Theorem (Gleason–Whitney) If Φ is a multiplicative linear functional on H^∞ and L_1 and L_2 are positive linear functionals on $L^\infty(\mathbb{T})$ such that $L_1|H^\infty = L_2|H^\infty = \Phi$, then $L_1 = L_2$.

Proof If u is a real-valued function in $L^\infty(\mathbb{T})$, then there exists an invertible function φ in H^∞ by Proposition 6.20 and Corollary 6.25 such that $|\varphi| = e^u$. Since L_1 and L_2 are positive, we have

$$|\Phi(\varphi)| = |L_1(\varphi)| \leqslant L_1(|\varphi|) = L_1(e^u)$$

and

$$\left|\Phi\left(\frac{1}{\varphi}\right)\right| = \left|L_2\left(\frac{1}{\varphi}\right)\right| \leqslant L_2\left(\frac{1}{|\varphi|}\right) = L_2(e^{-u}).$$

Multiplying, we obtain

$$1 = |\Phi(\varphi)| \left|\Phi\left(\frac{1}{\varphi}\right)\right| \leqslant L_1(e^u) L_2(e^{-u})$$

and hence the function $\Psi(t) = L_1(e^{tu}) L_2(e^{-tu})$ defined for all real t has an absolute minimum at $t = 0$. Since Ψ is a differentiable function of t by the linearity and continuity of L_1 and L_2, we obtain

$$\Psi'(t) = L_1(ue^{tu}) L_2(e^{-tu}) - L_1(e^{tu}) L_2(ue^{-tu}).$$

Substituting $t = 0$ yields $0 = \Psi'(0) = L_1(u) L_2(1) - L_1(1) L_2(u)$, and hence $L_1(u) = L_2(u)$ which completes the proof. ∎

6.34 Theorem If $\mathfrak{A}$ is a closed algebra satisfying $H^\infty \subset \mathfrak{A} \subset L^\infty(\mathbb{T})$, then the maximal ideal space $M_{\mathfrak{A}}$ of $\mathfrak{A}$ is naturally homeomorphic to a subset of M_∞.

Proof If Φ is a multiplicative linear functional on $\mathfrak{A}$, then $\Phi|H^\infty$ is a multiplicative linear functional on H^∞ and hence we have a continuous natural map η from $M_{\mathfrak{A}}$ into M_∞. Moreover, let Φ' denote any Hahn–Banach extension of Φ to $L^\infty(\mathbb{T})$. Since $L^\infty(\mathbb{T})$ is isometrically isomorphic to $C(M_{L^\infty})$ by Theorem 2.64, Φ' is integration with respect to a Borel measure ν on M_{L^∞} by the Riesz–Markov representation theorem (see Section 1.38). Since $\nu(M_{L^\infty}) = \Phi'(1) = 1 = \|\Phi'\| = |\nu|(M_{L^\infty})$, the functional Φ' is positive and hence uniquely determined by $\Phi|H^\infty$ by the previous result. Therefore, the mapping η is one-to-one and hence a homeomorphism. ∎

Observe that the maximal ideal space for $\mathfrak{A}$ contains the maximal ideal space for $L^\infty(\mathbb{T})$ and, in fact, as we indicate in the problems, the latter is the Šilov boundary of $\mathfrak{A}$.

We now introduce some concrete examples of algebras lying between H^∞ and $L^\infty(\mathbb{T})$.

6.35 Definition If Σ is a semigroup of inner functions containing the constant function 1, then the collection $\{\psi\bar\varphi: \psi \in H^\infty, \varphi \in \Sigma\}$ is a subalgebra of $L^\infty(\mathbb{T})$ and the closure is denoted $\mathfrak{A}_\Sigma$.

The argument that $\mathfrak{A}_\Sigma$ is an algebra is the same as was given in the proof of Theorem 6.32.

We next observe that $H^\infty + C(\mathbb{T})$ is one of these algebras.

6.36 Proposition If $\Sigma(\chi)$ denotes the semigroup of inner functions $\{\chi_n: n \geqslant 0\}$, then $\mathfrak{A}_{\Sigma(\chi)} = H^\infty + C(\mathbb{T})$.

Proof Since the linear span $H^\infty + C(\mathbb{T})$ is closed by Section 6.31, we have $H^\infty + C(\mathbb{T}) = \text{clos}_\infty[H^\infty + \mathscr{P}]$. Lastly, since

$$H^\infty + \mathscr{P} = \{\psi\bar\chi_n: \psi \in H^\infty, n \geqslant 0\},$$

the result follows. ∎

The maximal ideal space of $\mathfrak{A}_\Sigma$ can be identified as a closed subset of M_∞ by Theorem 6.34. The following more abstract result will enable us to identify the subset.

6.37 Proposition Let X be a compact Hausdorff space, $\mathfrak{A}$ be a function algebra contained in $C(X)$ with maximal ideal space M, and Σ be a semigroup of unimodular functions in $\mathfrak{A}$. If $\mathfrak{A}_\Sigma$ is the algebra

$$\text{clos}\{\psi\varphi: \psi \in \mathfrak{A}, \varphi \in \Sigma\}$$

and M_Σ is the maximal ideal space of $\mathfrak{A}_\Sigma$, then M_Σ can be identified with

$$\{m \in M: |\hat\varphi(m)| = 1 \text{ for } \varphi \in \Sigma\},$$

where $\hat\varphi$ denotes the Gelfand transform.

Proof If Ψ is a multiplicative linear functional on $\mathfrak{A}_\Sigma$, then $\Psi|\mathfrak{A}$ is a multiplicative linear functional on $\mathfrak{A}$, and hence $\eta(\Psi) = \Psi|\mathfrak{A}$ defines a continuous mapping from M_Σ into M. If Ψ_1 and Ψ_2 are elements of M_Σ such that $\eta(\Psi_1) = \eta(\Psi_2)$, then $\Psi_1|\mathfrak{A} = \Psi_2|\mathfrak{A}$. Further, for φ in Σ, we have

$$\Psi_1(\bar\varphi) = \Psi_1\left(\frac{1}{\varphi}\right) = \Psi_1(\varphi)^{-1} = \Psi_2(\varphi)^{-1} = \Psi_2(\bar\varphi)$$

and thus $\Psi_1 = \Psi_2$. Therefore, η is a homeomorphism of M_Σ into M. Moreover, since $|\Psi(\varphi)| \leqslant \|\varphi\| = 1$ and

$$\frac{1}{|\Psi(\varphi)|} = |\Psi(\bar\varphi)| \leqslant \|\bar\varphi\| = 1,$$

we have $|\Psi(\varphi)| = 1$ for Ψ in M_Σ and φ in Σ; therefore, the range of η is contained in

$$\{m \in M : |\hat\varphi(m)| = 1 \text{ for } \varphi \in \Sigma\}$$

and only the reverse inclusion remains.

Let m be a point in M such that $|\hat\varphi(m)| = 1$ for every φ in Σ. If we define Ψ on $\{\psi\bar\varphi : \psi \in \mathfrak{A}, \varphi \in \Sigma\}$ such that $\Psi(\psi\bar\varphi) = \hat\psi(m)\overline{\hat\varphi(m)}$, then Ψ can easily be shown to be multiplicative, and the inequality

$$|\Psi(\psi\bar\varphi)| = |\hat\psi(m)||\hat\varphi(m)| = |\hat\psi(m)| \leqslant \|\psi\| = \|\psi\bar\varphi\|$$

shows that Ψ can be extended to a multiplicative linear functional on $\mathfrak{A}_\Sigma$. Since $\eta(\Psi) = m$, the proof is complete. ∎

6.38 Corollary If Σ is a semigroup of inner functions, then the maximal ideal space M_Σ of $\mathfrak{A}_\Sigma$ can be identified with

$$\{m \in M_\infty : |\hat\varphi(m)| = 1 \text{ for } \varphi \in \Sigma\}.$$

Proof Since $L^\infty(\mathbb{T}) = C(X)$ for some compact Hausdorff space X, the result follows. ∎

Using the Gleason–Whitney theorem we can determine the Gelfand transform in the following sense.

6.39 Theorem There is a homeomorphism η from M_∞ into the unit ball of the dual of $L^\infty(\mathbb{T})$ such that $\hat\psi(m) = \eta(m)(\psi)$ for ψ in any algebra $\mathfrak{A}$ lying between H^∞ and $L^\infty(\mathbb{T})$ and m in $M_\mathfrak{A}$.

Proof For m in M_∞ let $\eta(m)$ denote the unique positive extension of m to $L^\infty(\mathbb{T})$ by Theorem 6.33. Since a multiplicative linear functional on $\mathfrak{A}$ extends to a positive extension of m on $L^\infty(\mathbb{T})$, we have $\hat\psi(m) = \eta(m)(\psi)$ for ψ in $\mathfrak{A}$. The only thing to prove is that η is a homeomorphism. Recall that the unit ball of $L^\infty(\mathbb{T})^*$ is w^*-compact. Thus if $\{m_\alpha\}_{\alpha \in A}$ is a net in M_∞ which converges to m, then any subnet of $\{\eta(m_\alpha)\}_{\alpha \in A}$ has a convergent subnet whose limit is a positive extension of m and hence equal to $\eta(m)$. Therefore, η is continuous and hence an into homeomorphism. ∎

We now adopt the notation $\hat{\varphi}(m) = \eta(m)(\varphi)$ for φ in $L^\infty(\mathbb{T})$. The restriction $\hat{\varphi}|\mathbb{D}$ will be shown to agree with the classical harmonic extension of a function in $L^\infty(\mathbb{T})$ into the disk. We illustrate the usefulness of the preceding by proving the following result showing the unique position occupied by $H^\infty + C(\mathbb{T})$ in the hierarchy of subalgebras of $L^\infty(\mathbb{T})$.

6.40 Corollary If $\mathfrak{A}$ is an algebra lying between H^∞ and $L^\infty(\mathbb{T})$, then either $\mathfrak{A} = H^\infty$ or $\mathfrak{A}$ contains $H^\infty + C(\mathbb{T})$.

Proof From Theorem 6.34 it follows that the maximal ideal space of $\mathfrak{A}$ can be identified as a subset $M_\mathfrak{A}$ of M_∞. If the origin in $\mathbb{D}$ is not in $M_\mathfrak{A}$, then χ_1 is invertible in $\mathfrak{A}$ ($\hat{\chi}_1 \neq 0$), and hence $C(\mathbb{T})$ is contained in $\mathfrak{A}$, whence the result follows. Thus suppose the origin in $\mathbb{D}$ is in $M_\mathfrak{A}$. Since

$$\varphi \to \frac{1}{2\pi} \int_0^{2\pi} \varphi \, dt$$

defines a positive extension of evaluation at 0, it follows that $\hat{\varphi}(0) = (1/2\pi)\int_0^{2\pi} \varphi \, dt$. If φ is contained in $\mathfrak{A}$ but not in H^∞, then $(1/2\pi)\int_0^{2\pi} \varphi \chi_n \, dt \neq 0$ for some $n > 0$, and hence $0 \neq \widehat{\varphi\chi_n}(0) = \hat{\varphi}(0)\,\hat{\chi}_n(0) = 0$. This contradiction completes the proof. ∎

One can also show that either $M_\mathfrak{A}$ is contained in $M_\infty \backslash \mathbb{D}$ or $\mathfrak{A} = H^\infty$.

Before we can apply this to $H^\infty + C(\mathbb{T})$ we need the following lemma on factoring out zeros.

6.41 Lemma If φ is in H^∞ and z is in $\mathbb{D}$ such that $\hat{\varphi}(z) = 0$, then there exists ψ in H^∞ such that $\varphi = (\chi_1 - z)\psi$.

Proof If θ is in H^∞, then $\widehat{\varphi\theta}(z) = \hat{\varphi}(z)\hat{\theta}(z) = 0$. If $\theta\varphi = \sum_{n=0}^\infty a_n \chi_n$ is the orthonormal expansion of $\theta\varphi$ viewed as an element of H^2, then

$$\sum_{n=0}^\infty a_n z^n = \widehat{\theta\varphi}(z) = 0$$

by Section 6.15, and hence

$$\left(\theta\varphi, \frac{1}{1 - \bar{z}\chi_1}\right) = \left(\sum_{n=0}^\infty a_n \chi_n, \sum_{n=0}^\infty \bar{z}^n \chi_n\right) = \sum_{n=0}^\infty a_n z^n = 0.$$

Therefore, we have

$$
\frac{1}{2\pi}\int_0^{2\pi}\chi_k\frac{\bar{\chi}_1\varphi}{1-z\bar{\chi}_1}\,d\theta = \left(\chi_k,\frac{\chi_1\bar{\varphi}}{1-\bar{z}\chi_1}\right) = \frac{1}{2\pi}\int_0^{2\pi}\frac{\varphi\chi_{k-1}}{1-z\bar{\chi}_1}\,dt
$$

$$
= \left(\chi_{k-1}\,\varphi,\frac{1}{1-\bar{z}\chi_1}\right) = 0
$$

$$
\text{for}\quad k = 1,2,3,...,
$$

and hence the function $\bar{\chi}_1\,\varphi/(1-z\bar{\chi}_1)$ is in H^∞. Thus setting $\psi = \bar{\chi}_1\varphi/(1-z\bar{\chi}_1)$, we obtain $(\chi_1-z)\psi = \varphi$. ∎

6.42   Corollary   The maximal ideal space of $H^\infty + C(\mathbb{T})$ can be identified with $M_\infty\backslash\mathbb{D}$.

Proof   From the preceding corollary, we have

$$
M_{H^\infty + C(\mathbb{T})} = \{m \in M_\infty : |\hat{\chi}_1(m)| = 1\}.
$$

It remains to show that this latter set is $M_\infty\backslash\mathbb{D}$. Let m be in M_∞ such that $|\hat{\chi}_1(m)| < 1$ and set $\hat{\chi}_1(m) = z$. If φ is in H^∞, then $\varphi - \hat{\varphi}(z)1$ vanishes at z, and hence by the preceding lemma we have $\varphi - \hat{\varphi}(z)1 = (\chi_1-z)\psi$ for some ψ in H^∞. Evaluating at m in M_∞, we have

$$
\hat{\varphi}(m) - \hat{\varphi}(z) = (\hat{\chi}_1(m)-z)\hat{\psi}(z) = 0,
$$

and hence $\hat{\varphi}(m) = \hat{\varphi}(z)$. Therefore, $m = z$ and the proof is complete. ∎

In the next chapter we shall be interested in determining when functions in $H^\infty + C(\mathbb{T})$ are invertible. From this point of view, the preceding result seems somewhat unfortunate since the only portion of the maximal ideal space of H^∞ over which we have some control, namely $\mathbb{D}$, has disappeared. We shall show, however, that the question of invertibility of functions in $H^\infty + C(\mathbb{T})$ can be answered by considering the harmonic extension of the function on $\mathbb{D}$. Our motivation for introducing the harmonic extension is quite different from that considered classically. We begin by determining a more explicit representation for $\hat{\varphi}$ on $\mathbb{D}$.

6.43   Lemma   If $z = re^{i\theta}$ is in $\mathbb{D}$ and φ in $L^\infty(\mathbb{T})$, then

$$
\hat{\varphi}(z) = \sum_{n=-\infty}^{\infty} a_n r^{|n|} e^{in\theta} = \frac{1}{2\pi}\int_0^{2\pi}\varphi(e^{it})k_r(\theta-t)\,dt,
$$

where

$$k_r(t) = \frac{1-r^2}{1-2r\cos t+r^2} \quad \text{and} \quad a_n = \frac{1}{2\pi}\int_0^{2\pi}\varphi\chi_{-n}\,dt.$$

Moreover, $\|\hat{\varphi}\|_\infty \leqslant \|\varphi\|_\infty$.

Proof The function k_r is positive and continuous; moreover, since

$$k_r(t) = \mathrm{Re}\left\{\frac{1+re^{it}}{1-re^{it}}\right\} = \sum_{n=-\infty}^{\infty} r^{|n|}e^{int},$$

it follows that

$$\|k_r\|_1 = \frac{1}{2\pi}\int_0^{2\pi}k_r(t)\,dt = 1.$$

Therefore, we have

$$\left|\frac{1}{2\pi}\int_0^{2\pi}\varphi(e^{it})\,k_r(\theta-t)\,dt\right| \leqslant \|\varphi\|_\infty\|k_r\|_1 = \|\varphi\|_\infty.$$

Lastly, since $\hat{\varphi}(z) = \sum_{n=-\infty}^{\infty}a_n r^{|n|}e^{in\theta}$ where $z = re^{i\theta}$, for φ in H^∞ it follows that this defines a positive extension of evaluation at z. The uniqueness of the latter by the Gleason–Whitney theorem completes the proof. ∎

6.44 Lemma The mapping from $H^\infty + C(\mathbb{T})$ to $C(\mathbb{D})$ defined by $\varphi \to \hat{\varphi}|\mathbb{D}$ is asymptotically multiplicative, that is, for φ and ψ in $H^\infty + C(\mathbb{T})$ and $\varepsilon > 0$, there exists K compact in $\mathbb{D}$ such that

$$|\hat{\varphi}(z)\hat{\psi}(z)-\widehat{\varphi\psi}(z)| < \varepsilon \quad \text{for} \quad z \text{ in } \mathbb{D}\backslash K.$$

Proof Since $H^\infty + C(\mathbb{T}) = \mathrm{clos}\left[\bigcup_{n\geqslant 0}\chi_{-n}H^\infty\right]$, it is sufficient to establish the result for functions of the form $\bar{\chi}_n\varphi$ for φ in H^m. If $\psi = \sum_{n=0}^{m}a_n\chi_n$ is the Fourier expansion of φ, then for $z = re^{it}$, we have

$$|\widehat{(\chi_{-n}\varphi)}(z) - \hat{\chi}_{-n}(z)\hat{\varphi}(z)|$$

$$\leqslant \sum_{k=0}^{n}|r^{|k-n|} - r^{k+n}||a_k| + \left(\frac{1}{r^n} - r^n\right)\left\|\sum_{k=n+1}^{\infty}a_k\chi_k\right\|_\infty.$$

Thus, if $|1-r| < \delta$, then $|\widehat{(\chi_{-n}\varphi)}(z)-\hat{\chi}_{-n}(z)\hat{\varphi}(z)| < \varepsilon$.

Since for φ_1 and φ_2 in H^∞ and z in $\mathbb{D}$, we have

$$|\widehat{(\chi_{-n}\varphi_1)}(z)\widehat{(\chi_{-m}\varphi_2)}(z) - \widehat{(\chi_{-n-m}\varphi_1\varphi_2)}(z)|$$

$$\leqslant |\widehat{(\chi_{-n}\varphi_1)}(z)\widehat{(\chi_{-m}\varphi_2)}(z) - \hat{\chi}_{-n}(z)\hat{\varphi}_1(z)\hat{\chi}_{-m}(z)\hat{\varphi}_2(z)|$$

$$+ \, |\hat{\chi}_{-n}(z)\,\hat{\varphi}_1(z)\,\hat{\chi}_{-m}(z)\,\hat{\varphi}_2(z) - \hat{\chi}_{-n-m}(z)\,\widehat{\varphi_1\varphi_2}(z)|$$

$$+ \, |\hat{\chi}_{-n-m}(z)\,\hat{\varphi}_1\,\hat{\varphi}_2(z) - (\widehat{\chi_{-n-m}\varphi_1\varphi_2})(z)|,$$

the result follows. ∎

An abstract proof could have been given for the preceding lemma, where the compact set K is replaced by a set $\{z \in \mathbb{D}: |\hat{\chi}_1(z)| \leqslant 1-\delta\}$. Moreover, a similar result holds for the algebras $\mathfrak{A}_\Sigma$ and can be used to state an invertibility criteria for functions in $\mathfrak{A}_\Sigma$ in terms of their harmonic extension on $\mathbb{D}$ (see [30], [31]).

6.45   Theorem   If φ is in $H^\infty + C(\mathbb{T})$, then φ is invertible if and only if there exist $\delta, \varepsilon > 0$ such that

$$|\hat{\varphi}(re^{it})| \geqslant \varepsilon \qquad \text{for} \quad 1 - \delta < r < 1.$$

Proof   Using the preceding lemma for $\varepsilon > 0$ there exists $\delta > 0$ such that for $1 - \delta < r < 1$, we have

$$\left| \hat{\varphi}(re^{it}) \, \widehat{\frac{1}{\varphi}}(re^{it}) - 1 \right| < \varepsilon$$

whence the implication follows one way if ε is chosen sufficiently small.

Conversely, let φ be a function in $H^\infty + C(\mathbb{T})$ such that

$$|\hat{\varphi}(re^{it})| \geqslant \varepsilon > 0 \qquad \text{for} \quad 1 - \delta < r < 1.$$

Choose ψ in H^∞ and an integer N such that $\|\varphi - \chi_{-N}\psi\|_\infty < \varepsilon/3$. Then there exists $\delta_1 > 0$ such that for $1 - \delta_1 < r < 1$, we have

$$|\widehat{\chi_{-N}\psi}(re^{it}) - \hat{\chi}_{-N}(re^{it})\hat{\psi}(re^{it})| < \frac{\varepsilon}{3}.$$

Therefore, for $1 - \delta_1 < r < 1$ we have using Lemma 6.43 that

$$|\hat{\varphi}(re^{it}) - r^N e^{-iNt}\hat{\psi}(re^{it})| < \frac{2\varepsilon}{3},$$

and hence $|\hat{\psi}(re^{it})| \geqslant \varepsilon/3$ if we also assume $r > 1 - \delta$. Let $z_1, \ldots, z_N$ be the zeros of the analytic function $\hat{\psi}(z)$ on $\mathbb{D}$ counting multiplicities. (Since $\hat{\psi}(z)$ is not zero near the boundary, the number is finite.) Using Lemma 6.41 repeatedly we can find a function θ in H^∞ such that $\psi = p\theta$, where

$$p = (\chi_1 - z_1)(\chi_1 - z_2)\cdots(\chi_1 - z_n).$$

Since $\hat{\psi} = \hat{p}\hat{\theta}$, we conclude that $\hat{\theta}$ does not vanish on $\mathbb{D}$ and is bounded away from zero in a neighborhood of the boundary. Therefore, θ is invertible in H^∞ by Theorem 6.18. Since p is invertible in $C(\mathbb{T})$, it follows that $\psi = p\theta$, and hence also $\chi_{-N}\psi$ is invertible in $H^\infty + C(\mathbb{T})$.

Lastly, since $\lim_{r \to 1^-} \|\varphi - \varphi_r\|_2 = 0$, where $\varphi_r(e^{it}) = \hat{\varphi}(re^{it})$, we have $|\varphi(e^{it})| \geq \varepsilon$ a.e., and hence $|(\chi_{-N}\psi)(e^{it})| \geq 2\varepsilon/3$ a.e. Therefore,

$$\|(\chi_{-N}\psi)^{-1}\| < \frac{3}{2\varepsilon}$$

and hence φ is invertible in $H^\infty + C(\mathbb{T})$ by Proposition 2.7. ∎

We conclude this chapter by showing that the harmonic extension of a continuous function on $\mathbb{T}$ solves the classical Dirichlet problem.

6.46 Theorem If φ is a continuous function on $\mathbb{T}$, then the function $\tilde{\varphi}$ defined on the closed disk to be $\hat{\varphi}$ on $\mathbb{D}$ and φ on $\partial\mathbb{D} = \mathbb{T}$ is continuous.

Proof If p is a trigonometric polynomial, then the result is obvious. If φ is a continuous function on $\mathbb{T}$ and $\{p_n\}_{n=1}^\infty$ is a sequence of trigonometric polynomials such that $\lim_{n \to \infty} \|\varphi - p_n\| = 0$, then $\lim_{n \to \infty} \|\tilde{\varphi} - \tilde{p}_n\| = 0$, since $|\tilde{\varphi}(z) - \tilde{p}_n(z)| \leq \|\varphi - p_n\|_\infty$ by Lemma 6.43. Therefore, $\tilde{\varphi}$ is continuous on $\overline{\mathbb{D}}$. ∎

Notes

The classical literature on analytic functions in the Hardy spaces is quite extensive and no attempt will be made to summarize it here. Some of the earliest and most important results are due to F. Riesz [90] and F. and M. Riesz [91]. The proofs we have presented are, however, quite different from the classical proofs and largely stem from the work of Helson and Lowdenslager [62]. The best references on this subject are the books of Hoffman [66] and Duren [39]; in addition, the books of Helson [61] and Gamelin [40] should be mentioned.

As we suggested in the text, the interest of the functional analyst in the Hardy spaces is due largely to Beurling [6], who pointed out their role in the study of the unilateral shift. The F. and M. Riesz theorem occurs in [91], but our proof stems from ideas of Lowdenslager (unpublished) and Sarason [99]. Besides the work of Helson and Lowdenslager already mentioned, the study of the unilateral shift was extended by Lax [75], [76], Halmos [56], and more recently by Helson [61] and Sz.-Nagy and Foiaş [107].

The study of H^∞ as a Banach algebra largely began in [100] and the deepest result that has been obtained is the corona theorem of Carleson [14], [15].

The material covered in Theorem 6.24 is closely connected with what is called prediction theory (see [54]). The algebra $H^\infty + C(\mathbb{T})$ first occurred in a problem in prediction theory [63] and most of the results presented here are due to Sarason [97]. The study of the algebras between H^∞ and $L^\infty(\mathbb{T})$ was suggested to the author [31] in studying the invertibility question for Toeplitz operators. Theorem 6.32 is taken from [34] in which it is shown that quotients of inner functions are uniformly dense in the measurable unimodular functions. The Gleason–Whitney theorem is in [43]. The role of the harmonic extension in discussing the invertibility of functions in $H^\infty + C(\mathbb{T})$ is established in [30].

Exercises

6.1   If φ and ψ are unimodular functions in $L^\infty(\mathbb{T})$, then $\varphi H^2 = \psi H^2$ if and only if $\varphi = \lambda\psi$ for some λ in $\mathbb{C}$.

Definition   A function f in H^1 is an outer function if $\mathrm{clos}_1[f\mathscr{P}_+] = H^1$.

6.2   A function f in H^1 is outer if and only if it is the product of two outer functions in H^2.

6.3   A closed subspace $\mathscr{M}$ of H^1 satisfies $\chi_1\mathscr{M} \subset \mathscr{M}$ if and only if $\mathscr{M} = \varphi H^1$ for some inner function φ. State and prove the corresponding criteria for subspaces of $L^1(\mathbb{T})$.*

6.4   If f is a function in H^1, then $f = \varphi g$ for some inner function φ and outer function g in H^1.

6.5   If f is a nonzero function in H^1, then the set $\{e^{it} \in \mathbb{T} : f(e^{it}) = 0\}$ has measure zero.

6.6   An analytic polynomial is an outer function in either H^1 or H^2 if and only if it does not vanish on the interior of $\mathbb{D}$.

6.7   Show that rotation on $\mathbb{D}$ induces a natural representation of the circle group in $\mathrm{Aut}(H^\infty) = \mathrm{Hom}(M_\infty)$. Show that the orbit of a point in M_∞ under this group of homeomorphisms is closed if and only if it lies in $\mathbb{D}$.*

Definition   If $\hat{\chi}_1$ is the Gelfand transform of χ_1 on M_∞, then the fiber F_λ of M_∞ over λ in $\mathbb{T}$ is defined by $F_\lambda = \{m \in M_\infty : \hat{\chi}_1(m) = \lambda\}$.

6.8 The fibers F_λ of M_∞ are compact and homeomorphic.

6.9 If φ is in H^∞ and λ is in $\mathbb{T}$, then φ is bounded away from zero on a neighborhood of λ in $\{\lambda\} \cup \mathbb{D}$ if and only if $\hat{\varphi}$ does not vanish on F_λ.

6.10 If φ is in H^∞ and α is in the range of $\hat{\varphi}|F_\lambda$ for some λ in $\mathbb{T}$, then there exists a sequence $\{z_n\}_{n=1}^\infty$ in $\mathbb{D}$ such that $\lim_{n\to\infty} z_n = \lambda$ and $\lim_{n\to\infty} \hat{\varphi}(z_n) = \alpha$.

6.11 Show that the density of $\mathbb{D}$ in M_∞ is equivalent to the following statement: For $\varphi_1, \varphi_2, ..., \varphi_N$ in H^∞ satisfying $\sum_{i=1}^N |\hat{\varphi}_i(z)| \geq \varepsilon$ for z in $\mathbb{D}$, there exists $\psi_1, \psi_2, ..., \psi_N$ in H^∞ such that $\sum_{i=1}^N \varphi_i \psi_i = 1$. Prove this statement under the additional assumption that the $\bar{\varphi}_i$ are in $H^\infty + C(\mathbb{T})$.

6.12 If $\mathfrak{A}$ is a closed algebra satisfying $H^\infty \subset \mathfrak{A} \subset L^\infty(\mathbb{T})$, then the maximal ideal space of L^∞ is naturally embedded in $M_\mathfrak{A}$ as the Šilov boundary of $\mathfrak{A}$.

6.13 (*Newman*) Show that the closure of $\mathbb{D}$ in M_∞ contains the Šilov boundary.

Definition An isometry U on the Hilbert space $\mathcal{H}$ is pure if $\bigcap_{n\geq 0} U^n \mathcal{H} = \{0\}$. The multiplicity of U is $\dim \ker U^*$.

6.14 A pure isometry of multiplicity one is unitarily equivalent to T_{χ_1} on H^2.

6.15 A pure isometry of multiplicity N is unitarily equivalent to $\sum_{1\leq i\leq N} \oplus T_{\chi_1}$ on $\sum_{1\leq i\leq N} \oplus H^2$.

6.16 (*von Neumann–Wold*) If U is an isometry on the Hilbert space $\mathcal{H}$, then $\mathcal{H} = \mathcal{H}_1 \oplus \mathcal{H}_2$ such that $\mathcal{H}_1$ and $\mathcal{H}_2$ reduce U, $U|\mathcal{H}_1$ is a pure isometry, and $U|\mathcal{H}_2$ is unitary.

6.17 If U is an isometry on the Hilbert space $\mathcal{H}$, then there exists a unitary operator W on a Hilbert space $\mathcal{K}$ containing $\mathcal{H}$ such that $W\mathcal{H} \subset \mathcal{H}$ and $W|\mathcal{H} = U$.

6.18 (*Sz.-Nagy*) If T is a contraction on the Hilbert space $\mathcal{H}$, then there exists a unitary operator W on a Hilbert space $\mathcal{K}$ containing $\mathcal{H}$ such that $T^N = P_\mathcal{H} W^N|\mathcal{H}$ for N in $\mathbb{Z}_+$. (Hint: Choose operators B and C such that $\begin{pmatrix} T & B \\ 0 & C \end{pmatrix}$ is a coisometry and then apply the previous result to the adjoint.)

6.19 (*von Neumann*) If T is a contraction on the Hilbert space $\mathcal{H}$, then the mapping defined $\Psi(p) = p(T)$ for each analytic polynomial p extends to a contractive homomorphism from the disk algebra to $\mathfrak{L}(\mathcal{H})$. (Hint: Use the preceding exercise.)

6.20 Show that the w^*-topology on the unit ball of $L^\infty(\mathbb{T})$ coincides with the topology of uniform convergence of the harmonic extensions on compact subsets of $\mathbb{D}$. (Hint: Evaluation at a point of $\mathbb{D}$ is a w^*-continuous functional.)

6.21 If f is a function in H^1 and $f_r(e^{it}) = \hat{f}(re^{it})$ for $0 < r < 1$ and e^{it} in $\mathbb{T}$, then $\lim_{r \to 1} \|f - f_r\|_1 = 0$. (Hint: Imitate the proof of Theorem 6.46.)

6.22 If f is a function in $L^1(\mathbb{T})$ for which

$$\frac{1}{2\pi} \int_0^{2\pi} f(e^{it}) \, dt = 0 \quad \text{and} \quad F(t) = \int_{-\pi}^t f(e^{i\theta}) \, d\theta \,,$$

then the harmonic extension $\hat{f}$ satisfies

$$\hat{f}(re^{i\theta}) = \frac{1}{2\pi} \int_{-\pi}^{\pi} k_r(t) \, F(\theta - t) \, dt \,,$$

where

$$k_r(t) = \frac{1 - r^2}{1 + r^2 - 2r \cos t} \,.$$

(Hint: Observe that

$$\hat{f}(re^{i\theta}) = \frac{1}{2\pi} \int_{-\pi}^{\pi} k_r(\theta - t) \, dF(t)$$

and use integration by parts.)

6.23 *(Fatou)* If f is a function in $L^1(\mathbb{T})$ and $\hat{f}$ is its harmonic extension of f to $\mathbb{D}$, then $\lim_{r \to 1} \hat{f}(re^{it}) = f(e^{it})$ a.e.* (Hint: Show that

$$\hat{f}(re^{i\theta}) = \frac{1}{2\pi} \int_{-\pi}^{2\pi} [-t k_r(t)] \left\{ \frac{F(\theta + t) - F(\theta - t)}{2t} \right\} dt$$

and that $\lim_{r \to 1} \hat{f}(re^{i\theta})$ exists and is equal to $F'(\theta)$, whenever the latter derivative exists.)

6.24 If φ is a function in H^∞ and $\hat{\varphi}$ is its Gelfand transform on M_∞, then $|\hat{\varphi}|$ is subharmonic, that is, if $\hat{\psi}$ is the harmonic extension of a real-valued function to M_∞ such that $\psi \geq |\varphi|$ on $\mathbb{T}$, then $\hat{\psi} \geq |\hat{\varphi}|$ on M_∞.

6.25 A function f in H^1 is an outer function if and only if the inequality $|f| \geq |g|$ on $\mathbb{T}$ implies $|\hat{f}| \geq |\hat{g}|$ on M_∞ for every g in H^1.

6.26 *(Jensen's Inequality)* If f is a function in H^1, then

$$\log |\hat{f}(0)| \leq \frac{1}{2\pi} \int_0^{2\pi} \log |f(e^{it})| \, dt. *$$

(Hint: Assume that f is in A and approximate $\log(|f|+\varepsilon)$ by the real part u of a function g in A; show that $\log|\hat{f}(0)| - \hat{u}(0) < \varepsilon$ and let ε tend to zero.)

6.27 (*Kolmogorov–Kreĭn*) If μ is a positive measure on $\mathbb{T}$ and μ_a is the absolutely continuous part of μ, then

$$\inf_{f \in A_0} \int_{\mathbb{T}} |1-f|^2 \, d\mu = \inf_{f \in A_0} \int_{\mathbb{T}} |1-f|^2 \, d\mu_a.*$$

(Hint: Show that if F is the projection of 1 onto the closure of A_0 in $L^2(\mu)$, then $1 - F = 0$ a.e. μ_s, where μ_s is the singular part of μ.)

6.28 A function f in H^2 is outer if and only if

$$\inf_{h \in A_0} \frac{1}{2\pi} \int_0^{2\pi} |1-h|^2 \, |f|^2 \, d\theta = |\hat{f}(0)|^2.*$$

(Hint: Show that f is outer if and only if $1 - \hat{f}(0)/f$ is the projection of 1 on the closure of A_0 in $L^2(|f|^2 \, d\theta)$.)

6.29 (*Szegö*) If μ is a positive measure on $\mathbb{T}$, then

$$\inf_{\in A_0} \int_{\mathbb{T}} |1-f|^2 \, d\mu = \exp\left(\frac{1}{2\pi} \int_0^{2\pi} \log h \, d\theta\right),$$

where h is the Radon–Nikodym derivative of μ with respect to Lebesgue measure.* (Hint: Use Exercise 6.27 to reduce it to μ of the form $w \, d\theta$; use the geometric–arithmetic mean inequality for one direction and reduce to the case $|h|^2 \, d\theta$ for h an outer function in the other.)

6.30 If $\mathfrak{A}$ is a closed algebra satisfying $H^\infty \subset \mathfrak{A} \subset L^\infty(\mathbb{T})$, then $\mathfrak{A}$ is generated by H^∞ together with the unimodular functions u for which both u and $\bar{u}$ are in $\mathfrak{A}$. (Hint: If f is in $\mathfrak{A}$, then $f + 2\|f\| = ug$, where u is unimodular and g is outer.)

6.31 If Γ is a group of unimodular functions in $L^\infty(\mathbb{T})$, then the maximal ideal space M_Γ of the subalgebra $\mathfrak{A}_\Gamma$ of $L^\infty(\mathbb{T})$ generated by H^∞ and Γ can be identified by

$$M_\Gamma = \{m \in M_\infty : |\hat{u}(m)| = 1 \text{ for } u \in \Gamma\}.$$

6.32 Is every $\mathfrak{A}_\Gamma$ of the form $\mathfrak{A}_\Sigma$ for some semigroup Σ of inner functions?**

6.33 Show that the closure of $H^\infty + \bar{H}^\infty$ is not equal to $L^\infty(\mathbb{T})$.* (Hint: If $\arg z$ were in the closure of $H^\infty + \bar{H}^\infty$, then there would exist φ in H^∞ such that ze^φ would be invertible in H^∞.)

6.34 If m is in F_λ and μ is the unique positive measure on M_{L^∞} such that

$$\hat\varphi(m) = \int_{M_{L^\infty}} \hat\varphi \, d\mu \qquad \text{for} \quad \varphi \text{ in } L^\infty(\mathbb{T}),$$

then μ is supported on $M_{L^\infty} \cap F_\lambda$.* (Hint: Show that the maximum of $|\hat\varphi|$ on F_λ is achieved on $M_{L^\infty} \cap F_\lambda$ for φ in H^∞.)

6.35 If φ is a continuous function and ψ is a function in $L^\infty(\mathbb{T})$, then $\widehat{\varphi\psi}$ and $\hat\varphi\hat\psi$ are asymptotically equal on $\mathbb{D}$ and equal on $M_\infty \backslash \mathbb{D}$.

6.36 If φ is a function in $L^\infty(\mathbb{T})$, then the linear functional on H^1 defined by $Lf = (1/2\pi)\int_0^{2\pi} f\varphi \, d\theta$ is continuous in the w^*-topology on H^1 if and only if φ is in $H^\infty + C(\mathbb{T})$.

6.37 Show that the collection PC of right-continuous functions on $\mathbb{T}$ possessing a limit from the left at every point of $\mathbb{T}$ is a uniformly closed self-adjoint subalgebra of $L^\infty(\mathbb{T})$. Show that the piecewise continuous functions form a dense subalgebra of PC. Show that the maximal ideal space of PC can be identified with two copies of $\mathbb{T}$ given an exotic topology.

6.38 If φ is a function in PC, then the range of the harmonic extension of φ on F_λ is the closed line segment joining the limits of φ from the left and right at λ.

6.39 Show that $QC = [H^\infty + C(\mathbb{T})] \cap \overline{[H^\infty + C(\mathbb{T})]}$ is a uniformly closed self-adjoint subalgebra of $L^\infty(\mathbb{T})$ which properly contains $C(\mathbb{T})$. Show that every inner function in QC is continuous but that $QC \cap H^\infty \neq A$.* (Hint: There exists a real function φ in $C(\mathbb{T})$ not in $\operatorname{Re} A$; if ψ is a real function in $L^2(\mathbb{T})$ such that $\varphi + i\psi$ is in H^2 then $e^{\varphi + i\psi}$ is in H^∞ and $e^{i\psi}$ is in QC.)

6.40 If u is a unimodular function in QC, then $|\hat u| = 1$ on $M_\infty \backslash \mathbb{D}$. Is the converse true?**

6.41 Show that $M_\infty \backslash \mathbb{D}$ is the maximal ideal space of the algebra generated by H^∞ and the functions u in $L^\infty(\mathbb{T})$ for which $|\hat u|$ has a continuous extension to $\overline{\mathbb{D}}$. Is this algebra $H^\infty + C(\mathbb{T})$?**

6.42 Show that $PC \cap QC = C(\mathbb{T})$. (Hint: Consider the unimodular functions in the intersection and use Exercises 6.38 and 6.40.)

6.43 Show that there is a natural isometrical isomorphism between H^∞ and $(L^1(\mathbb{T})/H_0^1)^*$. Show that the analytic trigonometric polynomials $\mathscr{P}_+$ are w^*-dense in H^∞.

7 *Toeplitz Operators*

7.1 Despite considerable effort there are few classes of operators on Hilbert space which one can declare are fully understood. Except for the self-adjoint operators and a few other examples, very little is known about the detailed structure of any class of operators. In fact, in most cases even the appropriate questions are not clear. In this chapter we study a class of operators about which much is known and even more remains to be known. Although the results we obtain would seem to fully justify their study, the occurrence of this class of operators in other areas of mathematics suggests they play a larger role in operator theory than would at first be obvious.

We begin with the definition of this class of operators.

7.2 Definition Let P be the projection of $L^2(\mathbb{T})$ onto H^2. For φ in $L^\infty(\mathbb{T})$ the Toeplitz operator T_φ on H^2 is defined by $T_\varphi f = P(\varphi f)$ for f in H^2.

7.3 The original context in which Toeplitz operators were studied was not that of the Hardy spaces but rather as operators on $l^2(\mathbb{Z}_+)$. Consider the orthonormal basis $\{\chi_n : n \in \mathbb{Z}_+\}$ for H^2, and the matrix for a Toeplitz operator with respect to it. If φ is a function in $L^\infty(\mathbb{T})$ with Fourier coefficients $\hat{\varphi}(n) = (1/2\pi)\int_0^{2\pi} \varphi \chi_{-n}\, dt$, then the matrix $\{a_{m,n}\}_{m,n\in\mathbb{Z}_+}$ for T_φ with respect to $\{\chi_n : n \in \mathbb{Z}_+\}$ is

$$a_{m,n} = (T_\varphi \chi_n, \chi_m) = \frac{1}{2\pi}\int_0^{2\pi} \varphi \chi_{n-m}\, dt = \hat{\varphi}(m-n).$$

Thus the matrix for T_φ is constant on diagonals; such a matrix is called a Toeplitz matrix, and it can be shown that if the matrix defines a bounded operator, then its diagonal entries are the Fourier coefficients of a function in $L^\infty(\mathbb{T})$ (see [11]).

We begin our study of Toeplitz operators by considering some elementary properties of the mapping ξ from $L^\infty(\mathbb{T})$ to $\mathfrak{L}(H^2)$ defined by $\xi(\varphi) = T_\varphi$.

7.4   Proposition   The mapping ξ is a contractive *-linear mapping from $L^\infty(\mathbb{T})$ into $\mathfrak{L}(H^2)$.

Proof   That ξ is contractive and linear is obvious. To show that $\xi(\varphi)^* = \xi(\bar\varphi)$, let f and g be in H^2. Then we have

$$(T_{\bar\varphi} f, g) = (P(\bar\varphi f), g) = (f, \varphi g) = (f, P(\varphi g)) = (f, T_\varphi g) = (T_\varphi^* f, g),$$

and hence $\xi(\varphi)^* = T_\varphi^* = T_{\bar\varphi} = \xi(\bar\varphi)$. ∎

The mapping ξ is not multiplicative, and hence ξ is not a homomorphism. We see later that ξ is actually an isometric cross section for a *-homomorphism from the C^*-algebra generated by $\{T_\varphi \colon \varphi \in L^\infty(\mathbb{T})\}$ onto $L^\infty(\mathbb{T})$, that is, if α is the *-homomorphism, then $\alpha \circ \xi$ is the identity on $L^\infty(\mathbb{T})$.

In special cases, ξ is multiplicative, and this will be important in what follows.

7.5   Proposition   If φ is in $L^\infty(\mathbb{T})$ and ψ and $\bar\theta$ are functions in H^∞, then $T_\varphi T_\psi = T_{\varphi\psi}$ and $T_\theta T_\varphi = T_{\theta\varphi}$.

Proof   If f is in H^2, then ψf is in H^2 by Proposition 6.2 and hence $T_\psi f = P(\psi f) = \psi f$. Thus

$$T_\varphi T_\psi f = T_\varphi(\psi f) = P(\varphi\psi f) = T_{\varphi\psi} f \qquad \text{and} \qquad T_\varphi T_\psi = T_{\varphi\psi}.$$

Taking adjoints reduces the second part to the first. ∎

The converse of this proposition is also true [11] but will not be needed in what follows.

Next we consider a basic result which will enable us to show that ξ is an isometry.

7.6   Proposition   If φ is a function in $L^\infty(\mathbb{T})$ such that T_φ is invertible, then φ is invertible in $L^\infty(\mathbb{T})$.

Proof   Using Corollary 4.24 it is sufficient to show that M_φ is an invertible operator if T_φ is. If T_φ is invertible, then there exists $\varepsilon > 0$ such that

$\|T_\varphi f\| \geqslant \varepsilon \|f\|$ for f in H^2. Thus for each n in $\mathbb{Z}$ and f in H^2, we have

$$\|M_\varphi(\chi_n f)\| = \|\varphi \chi_n f\| = \|\varphi f\| \geqslant \|P(\varphi f)\| = \|T_\varphi f\| \geqslant \varepsilon \|f\| = \varepsilon \|\chi_n f\|.$$

Since the collection of functions $\{\chi_n f : f \in H^2, n \in \mathbb{Z}\}$ is dense in $L^2(\mathbb{T})$, it follows that $\|M_\varphi g\| \geqslant \varepsilon \|g\|$ for g in $L^2(\mathbb{T})$. Similarly, $\|M_{\bar\varphi} f\| \geqslant \varepsilon \|f\|$, since $T_{\bar\varphi} = T_\varphi^*$ is also invertible, and thus M_φ is invertible by Corollary 4.9, which completes the proof. ∎

As a corollary we obtain the spectral inclusion theorem.

7.7 Corollary (Hartman–Wintner) If φ is in $L^\infty(\mathbb{T})$, then $\mathcal{R}(\varphi) = \sigma(M_\varphi) \subset \sigma(T_\varphi)$.

Proof Since $T_\varphi - \lambda = T_{\varphi-\lambda}$ for λ in $\mathbb{C}$, we see by the preceding proposition that $\sigma(M_\varphi) \subset \sigma(T_\varphi)$. Since the identity $\mathcal{R}(\varphi) = \sigma(M_\varphi)$ was established in Corollary 4.24, the proof is complete. ∎

This result enables us to complete the elementary properties of ξ.

7.8 Corollary The mapping ξ is an isometry from $L^\infty(\mathbb{T})$ into $\mathfrak{L}(H^2)$.

Proof Using Proposition 2.28 and Corollaries 4.24 and 7.7, we have for φ in $L^\infty(\mathbb{T})$ that

$$\|\varphi\|_\infty \geqslant \|T_\varphi\| \geqslant r(T_\varphi) = \sup\{|\lambda| : \lambda \in \sigma(T_\varphi)\}$$
$$\geqslant \sup\{|\lambda| : \lambda \in \mathcal{R}(\varphi)\} = \|\varphi\|_\infty,$$

and hence ξ is isometric. ∎

7.9 Certain additional properties of the correspondence are now obvious. If T_φ is quasinilpotent, then $\mathcal{R}(\varphi) \subset \sigma(T_\varphi) = \{0\}$, and hence $T_\varphi = 0$. If T_φ is self-adjoint, then $\mathcal{R}(\varphi) \subset \sigma(T_\varphi) \subset \mathbb{R}$ and hence φ is a real-valued function.

We now exhibit the homomorphism for which ξ is a cross section.

7.10 Definition If S is a subset of $L^\infty(\mathbb{T})$, then $\mathfrak{T}(S)$ is the smallest closed subalgebra of $\mathfrak{L}(H^2)$ containing $\{T_\varphi : \varphi \in S\}$.

7.11 Theorem If $\mathfrak{C}$ is the commutator ideal in $\mathfrak{T}(L^\infty(\mathbb{T}))$, then the mapping ξ_c induced from $L^\infty(\mathbb{T})$ to $\mathfrak{T}(L^\infty(\mathbb{T}))/\mathfrak{C}$ by ξ is a *-isometrical isomorphism. Thus there is a short exact sequence

$$(0) \to \mathfrak{C} \to \mathfrak{T}(L^\infty(\mathbb{T})) \xrightarrow{\rho} L^\infty(\mathbb{T}) \to (0)$$

for which ξ is an isometrical cross section.

Proof The mapping ξ_c is obviously linear and contractive. To show that ξ_c is multiplicative, observe for inner functions φ_1 and φ_2 and functions ψ_1 and ψ_2 in H^∞, that we have

$$\xi(\psi_1\bar\varphi_1)\xi(\psi_2\bar\varphi_2) - \xi(\psi_1\bar\varphi_1\psi_2\bar\varphi_2) = T_{\psi_1\bar\varphi_1}T_{\psi_2\bar\varphi_2} - T_{\psi_1\bar\varphi_1\psi_2\bar\varphi_2}$$
$$= T_{\varphi_1}^*(T_{\psi_1}T_{\varphi_2}^* - T_{\psi_1\bar\varphi_2})T_{\psi_2}$$
$$= T_{\varphi_1}^*(T_{\psi_1}T_{\varphi_2}^* - T_{\varphi_2}^*T_{\psi_1})T_{\psi_2}.$$

Since $T_{\psi_1}T_{\varphi_2}^* - T_{\varphi_2}^*T_{\psi_1}$ is a commutator and $\mathfrak{C}$ is an ideal, it follows that the latter operator lies in $\mathfrak{C}$. Thus ξ_c is multiplicative on the subalgebra

$$\mathcal{Q} = \{\psi\bar\varphi \colon \psi \in H^\infty, \ \varphi \text{ an inner function}\}$$

of $L^\infty(\mathbb{T})$ and the density of $\mathcal{Q}$ in $L^\infty(\mathbb{T})$ by Theorem 6.32 implies that ξ_c is a *-homomorphism.

To complete the proof we show that $\|T_\varphi + K\| \geqslant \|T_\varphi\|$ for φ in $L^\infty(\mathbb{T})$ and K in $\mathfrak{C}$ and hence that ξ_c is an isometry. A dense subset of operators in $\mathfrak{C}$ can be written in the form

$$K = \sum_{i=1}^{n} A_i[T_{\psi_i\bar\varphi_i}, T_{\psi_i'\bar\varphi_i'}]\prod_{j=1}^{m} T_{\alpha_{ij}\bar\beta_{ij}},$$

where A_i is in $\mathfrak{T}(L^\infty(\mathbb{T}))$, the functions φ_i, φ_i', and β_{ij} are inner functions, the functions ψ_i, ψ_i', and α_{ij} are in H^∞, and square brackets denote commutator. If we set

$$\theta = \prod_{\substack{i=1 \\ j=1}}^{n,m} \beta_{ij}\varphi_i\varphi_i',$$

then θ is an inner function and $K(\theta f) = 0$ for f in H^2.

Fix $\varepsilon > 0$ and let f be a function in H^2 chosen such that $\|f\| = 1$ and $\|T_\varphi f\| \geqslant \|T_\varphi\| - \varepsilon$. If $\varphi f = g_1 + g_2$, where g_1 is in H^2, and g_2 is orthogonal to H^2, then since θ is inner we have that θg_1 is in H^2 and orthogonal to θg_2. Thus

$$\|(T_\varphi + K)(\theta f)\| = \|T_\varphi(\theta f)\| = \|P(\varphi\theta f)\|$$
$$\geqslant \|\theta g_1\| = \|g_1\| = \|T_\varphi f\| \geqslant \|T_\varphi\| - \varepsilon,$$

and therefore $\|T_\varphi + K\| \geqslant \|T_\varphi\|$, which completes the proof. ∎

A direct proof of this result which avoids Theorem 6.32 can be given based on a theorem due to Bunce [12]. In this case the spectral inclusion theorem is then a corollary.

The C^*-algebra $\mathfrak{T}(L^\infty(\mathbb{T}))$ is a very interesting one; the preceding result shows that its study largely reduces to that of the commutator ideal $\mathfrak{C}$ about which very little is known. We can show that $\mathfrak{C}$ contains the compact operators, from which several important corollaries follow.

7.12 Proposition The commutator ideal in the C^*-algebra $\mathfrak{T}(C(\mathbb{T}))$ is $\mathfrak{LC}(H^2)$. Moreover, the commutator ideal of $\mathfrak{T}(L^\infty(\mathbb{T}))$ contains $\mathfrak{LC}(H^2)$.

Proof Since the operator T_{χ_1} is the unilateral shift, we see that the commutator ideal of $\mathfrak{T}(C(\mathbb{T}))$ contains the nonzero rank one operator $T_{\chi_1}^* T_{\chi_1} - T_{\chi_1} T_{\chi_1}^*$. Moreover, the algebra $\mathfrak{T}(C(\mathbb{T}))$ is irreducible since T_{χ_1} has no proper reducing subspaces by Beurling's theorem. Therefore, $\mathfrak{T}(C(\mathbb{T}))$ contains $\mathfrak{LC}(H^2)$ by Theorem 5.39.

Lastly, since the image of T_{χ_1} in $\mathfrak{T}(C(\mathbb{T}))/\mathfrak{LC}(H^2)$ is normal and generates this algebra, it follows that $\mathfrak{T}(C(\mathbb{T}))/\mathfrak{LC}(H^2)$ is commutative, and hence $\mathfrak{LC}(H^2)$ contains the commutator ideal of $\mathfrak{T}(C(\mathbb{T}))$. To complete the identification of $\mathfrak{LC}(H^2)$ as the commutator ideal in $\mathfrak{T}(C(\mathbb{T}))$, it is sufficient to show that $\mathfrak{LC}(H^2)$ contains no proper closed ideal. If $\mathfrak{I}$ were such an ideal, then it would contain a self-adjoint compact operator H. Multiplying H by the projection onto the subspace spanned by a nonzero eigenvector, we obtain a rank one projection in $\mathfrak{I}$. Now the argument used in the last paragraph of the proof of Theorem 5.39 can be applied, and hence $\mathfrak{LC}(H^2)$ is the commutator ideal in $\mathfrak{T}(C(\mathbb{T}))$. Obviously, $\mathfrak{LC}(H^2)$ is contained in the commutator ideal of $\mathfrak{T}(L^\infty(\mathbb{T}))$. ■

7.13 Corollary There exists a *-homomorphism ζ from the quotient algebra $\mathfrak{T}(L^\infty(\mathbb{T}))/\mathfrak{LC}(H^2)$ onto $L^\infty(\mathbb{T})$ such that the diagram

$$\mathfrak{T}(L^\infty(\mathbb{T})) \xrightarrow{\pi} \mathfrak{T}(L^\infty(\mathbb{T}))/\mathfrak{LC}(H^2)$$
$$\searrow{\scriptstyle\rho} \qquad {\scriptstyle\zeta}\swarrow$$
$$L^\infty(\mathbb{T})$$

commutes.

Proof Immediate from Theorem 7.11 and the preceding proposition. ■

7.14 Corollary If φ is a function in $L^\infty(\mathbb{T})$ such that T_φ is a Fredholm operator, then φ is invertible in $L^\infty(\mathbb{T})$.

Proof If T_φ is a Fredholm operator, then $\pi(T_\varphi)$ is invertible in

$$\mathfrak{T}(L^\infty(\mathbb{T}))/\mathfrak{LC}(H^2)$$

by Definition 5.14, and hence $\varphi = (\zeta \circ \pi)(T_\varphi)$ is invertible in $L^\infty(\mathbb{T})$. ■

7.15   Certain other results follow from this circle of ideas. In particular, it follows from Corollary 7.13 that $\|T_\varphi + K\| \geqslant \|T_\varphi\|$ for φ in $L^\infty(\mathbb{T})$ and K in $\mathfrak{LC}(H^2)$, and hence the only compact Toeplitz operator is 0.

Let us consider again the Toeplitz operator T_{χ_1}. Since the spectrum of T_{χ_1} is the closed unit disk we see, in general, that the spectrum of a Toeplitz operator T_φ is larger than the essential range of its symbol φ. It is this phenomenon which shall largely concern us. In particular, we are interested in determining criteria for a Toeplitz operator to be invertible and, in addition, for obtaining the spectrum. The deepest and perhaps the most striking result along these lines is due to Widom and states that the spectrum of a Toeplitz operator is a connected subset of $\mathbb{C}$. This will be proved at the end of this chapter.

We now show that the spectrum of a Toeplitz operator cannot be too much larger than the essential range of its symbol. We begin by recalling an elementary definition and lemma concerning convex sets.

7.16   Definition   If E is a subset of $\mathbb{C}$, then the closed convex hull of E, denoted $h(E)$, is the intersection of all closed convex subsets of $\mathbb{C}$ which contain E.

7.17   Lemma   If E is a subset of $\mathbb{C}$, then $h(E)$ is the intersection of the open half planes which contain E.

Proof   Elementary plane geometry.   ■

The lemma and the following result combine to show that $\sigma(T_\varphi)$ is contained in $h(\mathscr{R}(\varphi))$.

7.18   Proposition   If φ is an invertible function in $L^\infty(\mathbb{T})$ whose essential range is contained in the open right half-plane, then T_φ is invertible.

Proof   If Δ denotes the subset $\{z \in \mathbb{C}: |z-1| < 1\}$, then there exists an $\varepsilon > 0$ such that $\varepsilon \mathscr{R}(\varphi) = \{\varepsilon z: z \in \mathscr{R}(\varphi)\} \subset \Delta$. Hence we have $\|\varepsilon\varphi - 1\| < 1$ which implies $\|I - T_{\varepsilon\varphi}\| < 1$ by Corollary 7.8, and thus $T_{\varepsilon\varphi} = \varepsilon T_\varphi$ is invertible by Proposition 2.5.   ■

7.19   Corollary (Brown–Halmos)   If φ is a function in $L^\infty(\mathbb{T})$, then $\sigma(T_\varphi) \subset h(\mathscr{R}(\varphi))$.

Proof   By virtue of Lemma 7.17 it is sufficient to show that every open half-plane containing $\mathscr{R}(\varphi)$ also contains $\sigma(T_\varphi)$. This follows from the proposition after a translation and rotation of the open half-plane to co-incide with the open right half-plane. ■

We now obtain various results on the invertibility and spectrum of certain classes of Toeplitz operators. We begin with the self-adjoint operators.

7.20   Theorem (Hartman–Wintner)   If φ is a real-valued function in $L^{\infty}(\mathbb{T})$, then

$$\sigma(T_{\varphi}) = [\text{ess inf } \varphi, \text{ ess sup } \varphi].$$

Proof   Since the spectrum of T_{φ} is real it is sufficient to show that $T_{\varphi} - \lambda$ invertible implies that either $\varphi - \lambda \geqslant 0$ for almost all e^{it} in $\mathbb{T}$ or $\varphi - \lambda \leqslant 0$ for almost all e^{it} in $\mathbb{T}$. If $T_{\varphi} - \lambda$ is invertible for λ real, then there exists g in H^{2} such that $(T_{\varphi} - \lambda)g = 1$. Thus there exists h in $H_0{}^2$ such that $(\varphi - \lambda)g = 1 + \bar{h}$. Since $(\varphi - \lambda)\bar{g} = 1 + h$ is in H^{2}, we have $(\varphi - \lambda)|g|^{2} = (1 + h)g$ is in H^{1}, and therefore $(\varphi - \lambda)|g|^{2} = \alpha$ for some α in $\mathbb{R}$ by Corollary 6.6. Since $g \neq 0$ a.e. by the F. and M. Riesz theorem, it follows that $\varphi - \lambda$ has the sign of α and the result follows. ■

Actually much more is known about the self-adjoint Toeplitz operators. In particular, a spectral resolution is known for such operators up to unitary equivalence (see Ismagilov [68], Rosenblum [94], and Pincus [86]).

The Toeplitz operators with analytic symbol are particularly amenable to study. If φ is in H^{∞}, then the operator T_{φ} is the restriction of the normal operator M_{φ} on $L^{2}(\mathbb{T})$ to the invariant subspace H^{2} and hence is what is called a subnormal operator.

7.21   Theorem (Wintner)   If φ is a function in H^{∞}, then T_{φ} is invertible if and only if φ is invertible in H^{∞}. Moreover, if $\hat{\varphi}$ is the Gelfand transform of φ, then $\sigma(T_{\varphi}) = \text{clos}[\hat{\varphi}(\mathbb{D})]$.

Proof   If φ is invertible in H^{∞}, then there exists ψ in H^{∞} such that $\varphi\psi = 1$. Hence $I = T_{\psi}T_{\varphi} = T_{\varphi}T_{\psi}$ by Proposition 7.5. Conversely, if T_{φ} is invertible, then φ is invertible in $L^{\infty}(\mathbb{T})$ by Proposition 7.6. If ψ is $1/\varphi$, then $T_{\psi}T_{\varphi} = T_{\psi\varphi} = I$ by Proposition 7.5, and hence T_{ψ} is a left inverse for T_{φ}. Thus $T_{\psi} = T_{\varphi}^{-1}$ and therefore $1 = T_{\varphi}T_{\psi}1 = \varphi P(\psi)$. Multiplying both sides by $1/\varphi = \psi$, we obtain $P(\psi) = \psi$ implying that ψ is in H^{∞} and completing the proof that φ is invertible in H^{∞}. The fact that $\sigma(T_{\varphi}) = \text{clos}[\hat{\varphi}(\mathbb{D})]$ follows from Theorem 6.18. ■

This result yields especially nice answers to the questions of when T_φ is invertible and what its spectrum is for analytic φ. It is answers along these lines that we seek to determine. We next investigate the Toeplitz operators with continuous symbol and find in this case that an additional ingredient of a different nature enters into the answer. Our results on these operators depend on an analysis of the C^*-algebra $\mathfrak{T}(C(\mathbb{T}))$.

We begin by showing the ξ is almost multiplicative if the symbol of one of the factors is continuous.

7.22 Proposition If φ is in $C(\mathbb{T})$ and ψ is in $L^\infty(\mathbb{T})$, then $T_\varphi T_\psi - T_{\varphi\psi}$ and $T_\psi T_\varphi - T_{\psi\varphi}$ are compact.

Proof If ψ is in $L^\infty(\mathbb{T})$ and f is in H^2, then

$$T_\psi T_{\chi_{-1}} f = T_\psi P(\chi_{-1} f) = PM_\psi(\chi_{-1} f - (f, 1)\chi_{-1})$$
$$= P(\psi \chi_{-1} f) - (f, 1) P(\psi \chi_{-1})$$
$$= T_{\psi \chi_{-1}} f - (f, 1) P(\psi \chi_{-1}),$$

and hence $T_\psi T_{\chi_{-1}} - T_{\psi\chi_{-1}}$ is a rank one operator.

Suppose $T_\psi T_{\chi_{-n}} - T_{\psi\chi_{-n}}$ has been shown to be compact for every ψ in $L^\infty(\mathbb{T})$ and $-N \leq n < 0$. Then we have

$$T_\psi T_{\chi_{-N-1}} - T_{\psi\chi_{-N-1}} = (T_\psi T_{\chi_{-N}} - T_{\psi\chi_{-N}}) T_{\chi_{-1}} + (T_{\psi\chi_{-N}} T_{\chi_{-1}} - T_{(\psi\chi_{-N})\chi_{-1}})$$

and hence is compact. Since $T_\psi T_{\chi_n} = T_{\psi\chi_n}$ for $n \geq 0$ by Proposition 7.5, it follows that $T_\psi T_p - T_{\psi p}$ is compact for every trigonometric polynomial p. The density of the trigonometric polynomials in $C(\mathbb{T})$ and the fact that ξ is isometric complete the proof that $T_\psi T_\varphi - T_{\psi\varphi}$ is compact for ψ in $L^\infty(\mathbb{T})$ and φ in $C(\mathbb{T})$. Lastly, since

$$(T_\varphi T_\psi - T_{\varphi\psi})^* = T_{\bar\psi} T_{\bar\varphi} - T_{\bar\psi\bar\varphi}$$

we see that this operator is also compact. ∎

The basic facts about $\mathfrak{T}(C(\mathbb{T}))$ are contained in the following theorem due to Coburn.

7.23 Theorem The C^*-algebra $\mathfrak{T}(C(\mathbb{T}))$ contains $\mathfrak{LC}(H^2)$ as its commutator and the sequence

$$(0) \to \mathfrak{LC}(H^2) \to \mathfrak{T}(C(\mathbb{T})) \to C(\mathbb{T}) \to (0)$$

is short exact; that is, the quotient algebra $\mathfrak{T}(C(\mathbb{T}))/\mathfrak{LC}(H^2)$ is *-isometrically isomorphic to $C(\mathbb{T})$.

Proof It follows from the preceding proposition that the mapping ζ of Corollary 7.13 restricted to $\mathfrak{T}(C(\mathbb{T}))/\mathfrak{LC}(H^2)$ is a *-isometrical isomorphism onto $C(\mathbb{T})$, and hence the result follows. ■

Combining this with the following proposition yields the spectrum of a Toeplitz operator with continuous symbol.

7.24 Proposition (Coburn) If φ is a function in $L^\infty(\mathbb{T})$ not almost everywhere zero, then either $\ker T_\varphi = \{0\}$ or $\ker T_\varphi^* = \{0\}$.

Proof If f is in $\ker T_\varphi$ and g is in $\ker T_\varphi^*$, then $\bar\varphi \bar f$ and $\varphi \bar g$ are in H_0^2. Thus $\varphi f \bar g$ and $\bar\varphi f g$ are in H_0^1 by Lemma 6.16, and therefore $\varphi f \bar g$ is 0 by Corollary 6.7. If neither f nor g is the zero vector, then it follows from the F. and M. Riesz theorem that φ must vanish for almost all e^{it} in $\mathbb{T}$, which is a contradiction. ■

7.25 Corollary If φ is a function in $L^\infty(\mathbb{T})$ such that T_φ is a Fredholm operator, then T_φ is invertible if and only if $j(T_\varphi) = 0$.

Proof Immediate from the proposition. ■

Thus the problem of determining when a Toeplitz operator is invertible has been replaced by that of determining when it is a Fredholm operator and what is its index. If φ is continuous, then this is readily done. The result is due to a number of authors including Kreĭn, Widom, and Devinatz.

7.26 Theorem If φ is a continuous function on $\mathbb{T}$, then the operator T_φ is a Fredholm operator if and only if φ does not vanish and in this case $j(T_\varphi)$ is equal to minus the winding number of the curve traced out by φ with respect to the origin.

Proof First, T_φ is a Fredholm operator if and only if φ is invertible in $C(\mathbb{T})$ by Theorem 7.23. To determine the index of T_φ we first observe that $j(T_\varphi) = j(T_\psi)$ if φ and ψ determine homotopic curves in $\mathbb{C}\backslash\{0\}$. To see this, let Φ be a continuous map from $[0,1] \times \mathbb{T}$ to $\mathbb{C}\backslash\{0\}$ such that $\Phi(0, e^{it}) = \varphi(e^{it})$ and $\Phi(1, e^{it}) = \psi(e^{it})$ for e^{it} in $\mathbb{T}$. Then the mapping $t \to T_{\Phi_t}$ is norm continuous and each T_{Φ_t} is a Fredholm operator. Since $j(T_{\Phi_t})$ is continuous and integer valued, we see that $j(T_\varphi) = j(T_\psi)$.

If n is the winding number of the curve determined by φ, then φ is homotopic in $\mathbb{C}\backslash\{0\}$ to χ_n. Since $j(T_{\chi_n}) = -n$, we have $j(T_\varphi) = -n$ and the result is completely proved. ■

7.27 Corollary If φ is a continuous function on $\mathbb{T}$, then T_φ is invertible if and only if φ does not vanish and the winding number of the curve determined by φ with respect to the origin is zero.

Proof Combine the previous theorem and Corollary 7.25. ∎

7.28 Corollary If φ is a continuous function on $\mathbb{T}$, then

$$\sigma(T_\varphi) = \mathcal{R}(\varphi) \cup \{\lambda \in \mathbb{C} : i_t(\varphi, \lambda) \neq 0\},$$

where $i_t(\varphi, \lambda)$ is the winding number of the curve determined by φ with respect to λ.

In particular, the spectrum of T_φ is seen to be connected since it is formed from the union of $\mathcal{R}(\varphi)$ and certain components of the complement.

In this case the invertibility of the Toeplitz operator T_φ depended on φ being invertible in the appropriate Banach algebra $C(\mathbb{T})$ along with a topological criteria. In particular, the condition on the winding number amounts to requiring φ to lie in the connected component of the identity in $C(\mathbb{T})$ or that φ represent the identity in the abstract index group for $C(\mathbb{T})$. Although we shall extend the above to the larger algebra $H^\infty + C(\mathbb{T})$, these techniques are not adequate to treat the general case of a bounded measurable function.

We begin again by identifying the commutator ideal in $\mathfrak{T}(H^\infty + C(\mathbb{T}))$ and the corresponding quotient algebra.

7.29 Theorem The commutator ideal in $\mathfrak{T}(H^\infty + C(\mathbb{T}))$ is $\mathfrak{LC}(H^2)$ and the mapping $\xi_K = \pi \circ \xi$ from $H^\infty + C(\mathbb{T})$ to $\mathfrak{T}(H^\infty + C(\mathbb{T}))/\mathfrak{LC}(H^2)$ is an isometrical isomorphism.

Proof The algebra $\mathfrak{T}(H^\infty + C(\mathbb{T}))$ contains $\mathfrak{LC}(H^2)$, since it contains $\mathfrak{T}(C(\mathbb{T}))$ and thus the mapping ξ_K is well defined and isometric by Corollary 7.13 and the comments in Section 7.15. If φ and ψ are functions in H^∞ and f and g are continuous, then

$$T_{\varphi+f} T_{\psi+g} - T_{(\varphi+f)(\psi+g)} = T_{\varphi+f} T_\psi - T_{(\varphi+f)\psi} + T_{\varphi+f} T_g - T_{(\varphi+f)g}$$
$$= T_{\varphi+f} T_g - T_{(\varphi+f)g}$$

by Proposition 7.5, and the latter operator is compact by Proposition 7.22. Thus the commutator ideal is contained in $\mathfrak{LC}(H^2)$ and hence is equal to it. Thus ξ_K is multiplicative, since

$$\xi_K(\varphi+f)\xi_K(\psi+g) - \xi_K((\varphi+f)(\psi+g)) = \pi[T_{\varphi+f} T_{\psi+g} - T_{(\varphi+f)(\psi+g)}] = 0.$$

Therefore, ξ_K is an isometrical isomorphism, which completes the proof. ■

Note that $\mathfrak{T}(H^\infty + C(\mathbb{T}))$ is not a C^*-algebra, and hence the fact that T_φ is a Fredholm operator and $\pi(T_\varphi)$ has an inverse in $\mathfrak{L}(H^2)/\mathfrak{LC}(H^2)$ does not automatically imply that $\pi(T_\varphi)$ has an inverse in $\mathfrak{T}(H^\infty + C(\mathbb{T}))/\mathfrak{LC}(H^2)$. To show this we must first prove an invertibility criteria due to Widom and based on a result of Helson and Szëgo from prediction theory.

7.30   Theorem   If φ is a unimodular in $L^\infty(\mathbb{T})$, then the operator T_φ is left invertible if and only if dist$(\varphi, H^\infty) < 1$.

Proof   If dist$(\varphi, H^\infty) < 1$, then there exists a function ψ in H^∞ such that $\|\varphi - \psi\|_\infty < 1$. This implies that $\|1 - \bar\varphi\psi\|_\infty < 1$ and hence $\|I - T_{\bar\varphi\psi}\| < 1$. Thus $T_\psi^* T_\varphi = T_{\bar\psi\varphi}$ is invertible in $\mathfrak{L}(H^2)$ by Proposition 2.5 and therefore T_φ is left invertible.

Conversely, if T_φ is left invertible, then there exists $\varepsilon > 0$ such that $\|T_\varphi f\| \geqslant \varepsilon \|f\|$ for f in H^2. Thus $\|P(\varphi f)\| \geqslant \varepsilon \|f\| = \varepsilon \|\varphi f\|$, and hence

$$\|\varphi f\|^2 = \|(I - P)(\varphi f)\|^2 + \|P(\varphi f)\|^2 \geqslant \|(I - P)(\varphi f)\|^2 + \varepsilon \|\varphi f\|^2.$$

Therefore, we have $\|(I - P)(\varphi f)\| \leqslant (1 - \delta) \|f\|$ for f in H^2, where $\delta = 1 - (1 - \varepsilon)^{1/2} > 0$. If f is in H^2 and $\bar g$ is in H_0^2, then

$$\left| \frac{1}{2\pi} \int_0^{2\pi} \varphi f \bar g \, dt \right| = |(\varphi f, g)| = |((I - P)(\varphi f), g)| \leqslant (1 - \delta) \|f\| \|g\|.$$

If for h in H_0^1 we choose f in H^2 and $\bar g$ in H_0^2 by Corollary 6.27, such that $h = f\bar g$ and $\|h\|_1 = \|f\|_2 \|g\|_2$, then we obtain

$$\left| \frac{1}{2\pi} \int_0^{2\pi} \varphi h \, dt \right| \leqslant (1 - \delta) \|h\|_1.$$

Thus, the linear functional defined by $\Phi(h) = (1/2\pi)\int_0^{2\pi} h\varphi \, dt$ for h in H_0^1 has norm less than one. Therefore, it follows from Theorem 6.29 (note that we are using $(H_0^1)^* = L^\infty(\mathbb{T})/H^\infty$), that there exists ψ in H^∞ such that $\|\varphi - \psi\|_\infty < 1$, which completes the proof. ■

7.31   Corollary   If φ is a unimodular function in $L^\infty(\mathbb{T})$, then T_φ is invertible if and only if there exists an outer function ψ such that $\|\varphi - \psi\|_\infty < 1$.

Proof   If $\|\varphi - \psi\|_\infty < 1$, then $|\psi| \geqslant \varepsilon > 0$ for $\varepsilon = 1 - \|\varphi - \psi\|_\infty$ and hence T_ψ is invertible by Theorem 7.21 and Proposition 6.20. Since $T_\psi^* T_\varphi$ is invertible, we see that T_φ is invertible. Conversely, if T_φ is invertible and ψ

is a function in H^∞ such that $\|\varphi - \psi\|_\infty < 1$, then T_ψ is invertible since $T_\psi^* T_\varphi$ is, and hence ψ is an outer function. ■

As a result of this we obtain a very general spectral inclusion theorem.

7.32   Theorem   If $\mathfrak{T}$ is a closed subalgebra of $\mathfrak{L}(H^2)$ containing $\mathfrak{T}(H^\infty)$ and T_φ is in $\mathfrak{T}$ for some φ in $L^\infty(\mathbb{T})$, then $\sigma(T_\varphi) = \sigma_\mathfrak{T}(T_\varphi)$.

Proof   Obviously $\sigma(T_\varphi)$ is contained in $\sigma_\mathfrak{T}(T_\varphi)$. To prove the reverse inclusion suppose T_φ is invertible. Using Corollary 6.25, we can write $\varphi = u\psi$, where u is a unimodular function and ψ is an outer function. Consequently, ψ is invertible in $L^\infty(\mathbb{T})$, and by Proposition 6.20 we obtain that T_ψ^{-1} is in $\mathfrak{T}$. Thus $T_u = T_\varphi T_\psi^{-1}$ is in $\mathfrak{T}$, moreover, T_u and hence $T_{\bar u}$ is invertible in $\mathfrak{L}(H^2)$. Employing the preceding corollary there exists an outer function θ such that $\|\theta - \bar u\|_\infty < 1$. Since $T_u T_\theta$ is in $\mathfrak{T}$ and $\|I - T_u T_\theta\| = \|1 - u\theta\|_\infty < 1$, we see that $(T_u T_\theta)^{-1}$ is in $\mathfrak{T}$ by Proposition 2.5. Since

$$T_\varphi^{-1} = T_\psi^{-1} T_u^{-1} = T_\psi^{-1} T_\theta (T_u T_\theta)^{-1}$$

is in $\mathfrak{T}$, the proof is complete. ■

This leads to a necessary condition for an operator to be Fredholm.

7.33   Corollary   If $\mathfrak{A}$ is a closed subalgebra of $L^\infty(\mathbb{T})$ containing $H^\infty + C(\mathbb{I})$ and φ is a function in $\mathfrak{A}$ for which T_φ is a Fredholm operator, then φ is invertible in $\mathfrak{A}$.

Proof   If T_φ is a Fredholm operator having index n, then $T_{\chi_n \varphi}$ is invertible by Propositions 7.5 and 7.24, and the function $\chi_n \varphi$ is also in $\mathfrak{A}$. Therefore, by the preceding result, $T_{\chi_n \varphi}^{-1}$ is in $\mathfrak{T}(\mathfrak{A})$, and hence using Theorem 7.11 we see that $\chi_n \varphi$ is invertible in $\rho[\mathfrak{T}(\mathfrak{A})] = \mathfrak{A}$. Thus φ is invertible in $\mathfrak{A}$, which completes the proof. ■

The condition is also sufficient for $H^\infty + C(\mathbb{T})$.

7.34   Corollary   If φ is a function in $H^\infty + C(\mathbb{T})$, then T_φ is a Fredholm operator if and only if φ is invertible in $H^\infty + C(\mathbb{T})$.

Proof   The result follows by combining the previous result and Theorem 7.29. ■

We need one more lemma to determine the spectrum of T_φ for φ in $H^\infty + C(\mathbb{T})$. Although this lemma is usually obtained as a corollary to the structure theory for inner functions, it is interesting to note that it also follows from our previous methods.

7.35 Lemma If φ is an inner function, then $H^2 \ominus \varphi H^2$ is finite dimensional if and only if φ is continuous.

Proof If φ is continuous, then T_φ is a Fredholm operator by Theorem 7.26, and hence $H^2 \ominus \varphi H^2 = \ker(T_\varphi{}^*)$ is finite dimensional. Conversely, if $H^2 \ominus \varphi H^2$ is finite dimensional, then T_φ is a Fredholm operator and hence φ is invertible in $H^\infty + C(\mathbb{T})$. We need to show that this implies that φ is continuous.

By Theorem 6.45 there exist $\varepsilon, \delta > 0$ such that $|\hat{\varphi}(re^{it})| \geqslant \varepsilon$ for $1 - \delta < r < 1$, and hence $\hat{\varphi}$ has at most finitely many zeros $z_1, z_2, ..., z_N$ in $\mathbb{D}$ counted according to multiplicity. Using Lemma 6.41 we obtain a function ψ in H^∞ such that $\psi \prod_{j=1}^{N}(\chi_1 - z_j) = \varphi$. Thus ψ does not vanish on $\mathbb{D}$ and is bounded away from zero on a neighborhood of the boundary. Therefore, ψ is invertible in H^∞ by Theorem 6.18. Moreover, since

$$|e^{it} - z_j| = |1 - \bar{z}_j e^{it}| \quad \text{for} \quad e^{it} \text{ in } \mathbb{T},$$

we have

$$\left| \frac{e^{it} - z_j}{1 - \bar{z}_j e^{it}} \right| = 1.$$

Thus the function $\prod_{j=1}^{N}(\chi_1 - z_j)/(1 - \bar{z}_j \chi_1)$ is a continuous inner function, and $\theta = \psi \prod_{j=1}^{N}(1 - \bar{z}_j \chi_1)$ has modulus one on $\mathbb{T}$. Since θ is invertible in H^∞, it follows that $\bar{\theta} = \theta^{-1}$ is in H^∞ and hence that θ is constant. Thus

$$\varphi = \theta \prod_{=1}^{N} \left(\frac{\chi_1 - z_j}{1 - \bar{z}_j \chi_1} \right)$$

and therefore is continuous. ■

7.36 Theorem If φ is a function in $H^\infty + C(\mathbb{T})$, then T_φ is a Fredholm operator if and only if there exist $\delta, \varepsilon > 0$ such that $|\hat{\varphi}(re^{it})| \geqslant \varepsilon$ for $1 - \delta < r < 1$, where $\hat{\varphi}$ is the harmonic extension of φ to $\mathbb{D}$. Moreover, in this case the index of T_φ is the negative of the winding number with respect to the origin of the curve $\hat{\varphi}(re^{it})$ for $1 - \delta < r < 1$.

Proof The first statement is obtained by combining Corollary 7.34 and Theorem 6.45.

If φ is invertible in $H^\infty + C(\mathbb{T})$ and $\varepsilon, \delta > 0$ are chosen such that

$$|\hat{\varphi}(re^{it})| \geqslant \varepsilon$$

for $1 - \delta < r < 1$, then choose ψ in H^∞ and a negative integer N such that $\|\varphi - \chi_N \psi\|_\infty < \varepsilon/3$. If we set $\varphi_\lambda = \lambda \varphi + (1 - \lambda)\chi_N \psi$ for $0 \leqslant \lambda \leqslant 1$, then each φ_λ is in $H^\infty + C(\mathbb{T})$ and $\|\varphi - \varphi_\lambda\|_\infty < \varepsilon/3$. Therefore, $|\hat{\varphi}_\lambda(re^{it})| \geqslant 2\varepsilon/3$ for $1 - \delta < r < 1$, and hence each φ_λ is invertible in $H^\infty + C(\mathbb{T})$ by Theorem 6.45. Hence each T_{φ_λ} is a Fredholm operator and the winding number of the curve $\varphi_\lambda(re^{it})$ is independent of λ and r for $0 \leqslant \lambda \leqslant 1$ and $1 - \delta < r < 1$. Thus it is sufficient to show that the index of $T_{\varphi_0} = T_{\chi_N \psi}$ is equal to the negative of the winding number of the curve $\widehat{(\chi_N \psi)}(re^{it})$.

Since $T_{\chi_N \psi} = T_{\chi_N} T_\psi$ we see that T_ψ is a Fredholm operator. If we write $\psi = \psi_o \psi_i$, where ψ_o is an outer function and ψ_i is an inner function, then T_{ψ_o} is invertible and hence T_{ψ_i} is a Fredholm operator. Thus ψ_i is continuous by the previous lemma which implies $\chi_N \psi_i$ is in $C(\mathbb{T})$ and we conclude by Theorem 7.26 that

$$j(T_{\chi_N \psi}) = j(T_{\chi_N \psi_i}) = i_t(\chi_N \psi_i).$$

Since there exists by Lemma 6.44 a $\delta_1 > 0$ such that $\delta > \delta_1 > 0$ and

$$|\widehat{\chi_N \psi}(re^{it}) - \widehat{\chi_N \psi_i}(re^{it})\widehat{\psi_o}(re^{it})| < \frac{\varepsilon}{3} \qquad \text{for} \quad 1 - \delta_1 < r < 1,$$

it follows that

$$i_t(\widehat{(\chi_N \psi)}(re^{it})) = i_t(\widehat{(\chi_N \psi_i)}(re^{it})) + i_t(\widehat{\psi_o}(re^{it})) \qquad \text{for} \quad 1 - \delta_1 < r < 1.$$

Using the fact that $\widehat{\psi_o}$ does not vanish on $\mathbb{D}$ and Theorem 6.46, we obtain the desired result. ■

We conclude our results for T_φ with φ in $H^\infty + C(\mathbb{T})$ by showing that the essential spectrum is connected. Although we shall show this for arbitrary φ, a direct proof would seem to be of interest.

7.37 Corollary If φ is a function in $H^\infty + C(\mathbb{T})$, then the essential spectrum of T_φ is connected.

Proof From the theorem it follows that λ is in the essential spectrum of T_φ for φ in $H^\infty + C(\mathbb{T})$ if and only if $\varphi - \lambda$ is not invertible in $H^\infty + C(\mathbb{T})$. Hence, by Theorem 6.46 we have that λ is in the essential spectrum of T_φ if and only if λ is in

$$\mathrm{clos}\{\hat{\varphi}(re^{it}): 1 > r > 1 - \delta\} \qquad \text{for each} \quad 1 > \delta > 0.$$

Since each of these sets is connected, the result follows. ∎

We now take up the proof of the connectedness of the essential spectrum for an arbitrary function in $L^\infty(\mathbb{T})$. The proof is considerably harder in this case and we begin with the following lemma.

7.38   Lemma   If φ is an invertible function in $L^\infty(\mathbb{T})$, then T_φ is a Fredholm operator if and only if $T_{1/\varphi}$ is and moreover, in this case, $j(T_\varphi) = -j(T_{1/\varphi})$.

Proof   If we set $\varphi = u\psi$ by Corollary 6.25, where ψ is an outer function and u is unimodular, then $T_{1/\varphi} = T_{\bar{u}} T_{1/\psi} = T_{1/\psi}^* T_\varphi^* T_{1/\psi}$ by Proposition 7.5. Since $T_{1/\psi}$ is invertible by Theorem 7.21, the result follows. ∎

The proof of connectedness is based on the analysis of the solutions f_λ and g_λ of the equations $T_{\varphi-\lambda} f_\lambda = 1$ and $T_{1/(\varphi-\lambda)} g_\lambda = 1$. Since we want to consider λ for which these operators are not invertible, the precise definition of f_λ and g_λ is slightly more complicated.

7.39   Definition   If φ is a function in $L^\infty(\mathbb{T})$, then the essential resolvent $\rho_e(T_\varphi)$ for T_φ is the open set of those λ in $\mathbb{C}$ for which $T_{\varphi-\lambda}$ is a Fredholm operator. If λ is in $\rho_e(T_\varphi)$ and $j(T_{\varphi-\lambda}) = n$, then

$$f_\lambda = T_{\chi_n(\varphi-\lambda)}^{-1} 1 \qquad \text{and} \qquad g_\lambda = T_{\chi_{-n}/(\varphi-\lambda)}^{-1} 1.$$

The basic result concerning the f_λ and g_λ is contained in the following.

7.40   Proposition   If φ is a function in $L^\infty(\mathbb{T})$ and λ is in $\rho_e(T_\varphi)$, then $f_\lambda g_\lambda = 1$ and

$$\frac{d}{d\lambda} \hat{f}_\lambda(z) = \hat{f}_\lambda(z) \cdot \widehat{P\left\{\frac{1}{\varphi-\lambda}\right\}}(z) \qquad \text{for} \quad z \text{ in } \mathbb{D}.$$

Proof   There exist functions u_λ and v_λ in H_0^2 such that $\chi_n(\varphi-\lambda)f_\lambda = 1 + \bar{u}_\lambda$ and $\chi_{-n} g_\lambda/(\varphi-\lambda) = 1 + \bar{v}_\lambda$. Multiplying these two identities, we obtain

$$f_\lambda g_\lambda = 1 + \overline{(u_\lambda + v_\lambda + u_\lambda v_\lambda)},$$

where $f_\lambda g_\lambda$ is in H^1 and $u_\lambda + v_\lambda + u_\lambda v_\lambda$ is in H_0^1 and thus $f_\lambda g_\lambda = 1$. We also have $\hat{f}_\lambda(z) \hat{g}_\lambda(z) = 1$ by Lemma 6.16.

Since the functions $\lambda \rightarrow \chi_n/(\varphi-\lambda)$ and $\lambda \rightarrow \chi_{-n}/(\varphi-\lambda)$ are analytic L^∞-valued functions, it follows from Corollary 7.8 that f_λ and g_λ are analytic H^2-valued functions. Differentiating the identity $\chi_n(\varphi-\lambda)f_\lambda = 1 + \bar{u}_\lambda$ with respect to λ yields $-\chi_n f_\lambda + \chi_n(\varphi-\lambda)f_\lambda' = \bar{u}_\lambda'$ and hence $f_\lambda = (\varphi-\lambda)f_\lambda' - \chi_{-n}\bar{u}_\lambda'$, where u_λ' lies in H_0^2. Multiplying both sides of this equation by $g_\lambda/(\varphi-\lambda)$, we obtain

$$\frac{1}{\varphi-\lambda} = \frac{1}{\varphi-\lambda}f_\lambda g_\lambda = f_\lambda' g_\lambda - (\chi_{-n}\bar{u}_\lambda')(\chi_n(1+\bar{v}_\lambda))$$

$$= f_\lambda' g_\lambda - \bar{u}_\lambda'(1+\bar{v}_\lambda).$$

Since $f_\lambda' g_\lambda$ is in H^1 and $\bar{u}_\lambda'(1+\bar{v}_\lambda)$ is in $\bar{H}_0^1$, we have $P\{1/(\varphi-\lambda)\} = f_\lambda' g_\lambda$. Finally, again using the identity proved in the first paragraph, we reach the equation $f_\lambda' = f_\lambda P\{1/(\varphi-\lambda)\}$ and observing that evaluation at z in $\mathbb{D}$ com-'mutes with differentiation with respect to λ we obtain the desired results. ∎

It is possible to solve this latter equation to relate the f_λ which lie in the same component of $\rho_e(T_\varphi)$.

7.41 Corollary If φ is a function in $L^\infty(\mathbb{T})$ and λ and λ_o are endpoints of a rectifiable curve Γ lying in $\rho_e(T_\varphi)$, then

$$\hat{f}_\lambda(z) = \hat{f}_{\lambda_o}(z) \exp\left\{\int_\Gamma \widehat{P\left\{\frac{1}{\varphi-\mu}\right\}}(z)\,d\mu\right\}$$

and

$$\hat{g}_\lambda(z) = \hat{g}_{\lambda_o}(z) \exp\left\{-\int_\Gamma \widehat{P\left\{\frac{1}{\varphi-\mu}\right\}}(z)\,d\mu\right\}$$

for z in $\mathbb{D}$.

Proof For each fixed z in $\mathbb{D}$ we are solving the ordinary first order linear differential equation $dx(\lambda)/d\lambda = F(\lambda)x(\lambda)$, where $F(\lambda)$ is an analytic function in λ. Hence the result follows for f_λ and the corresponding result for g_λ is obtained by using the identity $\hat{f}_\lambda(z)\hat{g}_\lambda(z) = 1$. ∎

We now show that no curve lying in $\rho_e(T_\varphi)$ can disconnect $\mathcal{R}(\varphi)$.

7.42 Proposition If φ is a function in $L^\infty(\mathbb{T})$ and C is a rectifiable simple closed curve lying in $\rho_e(T_\varphi)$, then $\mathcal{R}(\varphi)$ lies either entirely inside or entirely outside of C.

Proof Consider the analytic function defined by

$$F(z) = \frac{1}{2\pi i} \int_C P\widehat{\left\{\frac{1}{\varphi - \mu}\right\}}(z)\, d\mu \qquad \text{for each } z \text{ in } \mathbb{D}.$$

If λ_o is a fixed point on C, then it follows from the preceding corollary that

$$\hat{f}_{\lambda_o}(z) = \hat{f}_{\lambda_o}(z_o) \exp\{2\pi i F(z)\} \qquad \text{for } z \text{ in } \mathbb{D}.$$

Therefore, $\exp\{2\pi i F(z)\} = 1$ whenever $\hat{f}_{\lambda_o}(z) \neq 0$, and hence for all z in $\mathbb{D}$, since $\hat{f}_{\lambda_o}$ is analytic. Thus the function $F(z)$ is integervalued and hence equal to some constant N.

Now for each e^{it} in $\mathbb{T}$, the integral

$$\frac{1}{2\pi i} \int_C \frac{1}{\varphi(e^{it}) - \mu}\, d\mu$$

equals the winding number of the curve C with respect to the point $\varphi(e^{it})$. Thus the function defined by

$$\psi(e^{it}) = \frac{1}{2\pi i} \int_C \frac{1}{\varphi(e^{it}) - \mu}\, d\mu$$

is real valued. Since

$$P\psi = P\left\{\frac{1}{2\pi i} \int_C \frac{1}{\varphi - \mu}\, d\mu\right\} = \frac{1}{2\pi i} \int_C P\left\{\frac{1}{\varphi - \mu}\right\}\, d\mu = F$$

is constant, we see that ψ is constant a.e., and hence the winding number of C with respect to $\mathcal{R}(\varphi)$ is constant. Hence $\mathcal{R}(\varphi)$ lies either entirely inside or entirely outside of C. ∎

The remainder of the proof consists in showing that we can analytically continue solutions to any component of $\mathbb{C}\backslash\rho_e(T_\varphi)$ which does not contain $\mathcal{R}(\varphi)$. Before doing this we need to relate the inverse of $T_{\chi_n(\varphi - \lambda)}$ to the functions f_λ and g_λ.

7.43 Lemma If φ is a function in $L^\infty(\mathbb{T})$, λ is in $\rho_e(T_\varphi)$, k is in H^∞, and $h_\lambda = f_\lambda P\{\chi_{-n} g_\lambda k/(\varphi - \lambda)\}$, then h_λ is in H^2 and $T_{\chi_n(\varphi - \lambda)} h_\lambda = k$.

Proof If we set $h_\lambda = T^{-1}_{\chi_n(\varphi - \lambda)} k$, then there exists l in H_0^2 such that $\chi_n(\varphi - \lambda) h_\lambda = k + l$. Multiplying by $\chi_{-n} g_\lambda/(\varphi - \lambda) = 1 + \bar{v}_\lambda$, where v_λ is in H_0^2, we obtain

$$g_\lambda h_\lambda = \frac{1}{\varphi - \lambda} \chi_{-n} g_\lambda k + l(1 + \bar{v}_\lambda)$$

and hence $g_\lambda h_\lambda = P\{\chi_{-n} g_\lambda k/(\varphi - \lambda)\}$, since $l(1+v_\lambda)$ is in $H_0^{\,1}$. We now obtain the desired result upon multiplying both sides by f_λ and again using the identity $f_\lambda g_\lambda = 1$. ∎

We need one more lemma before proving the connectedness result.

7.44 Lemma If Ω is a simply connected open subset of $\mathbb{C}$, C is a rectifiable simple closed curve lying in Ω, and $F_\lambda(z)$ is a complex function on $\Omega \times \mathbb{D}$ such that $F_\lambda(z_0)$ is analytic on Ω for z_0 in $\mathbb{D}$, $F_{\lambda_0}(z)$ is analytic on $\mathbb{D}$ for λ_0 in Ω, and F_{λ_0} is in H^2 for λ_0 in C, then F_λ is in H^2 for λ in the interior of C and $\|F_\lambda\|_2 \leqslant \sup_{\lambda_0 \in C} \|F_{\lambda_0}\|_2$.

Proof An analytic function ψ on $\mathbb{D}$ is in H^2 if and only if

$$\sum_{n=0}^{\infty} |\psi^{(n)}(0)|^2/(n!)^2$$

is finite, since $f = \sum_{n=0}^{\infty} \psi^{(n)}(0)(n!)^{-1}\chi_n$ is in H^2 and $\hat{f} = \psi$; moreover $\|f\|_2 = (\sum_{n=0}^{\infty} |\psi^{(n)}(0)|^2/(n!)^{-2})^{1/2}$.

If $\alpha_0, \alpha_1, \ldots, \alpha_N$ are arbitrary complex numbers, then

$$\sum_{k=0}^{N} \frac{F_\lambda^{(k)}(0)}{k!}\bar{\alpha}_k = \frac{1}{2\pi} \int_0^{2\pi} F_\lambda(re^{it}) \overline{\sum_{k=0}^{N} \frac{1}{r^k}e^{ikt}\alpha_k}\,dt$$

is an analytic function of λ for $0 < r < 1$. Moreover, since

$$\left| \sum_{k=0}^{N} \frac{F_\lambda^{(k)}(0)}{k!}\bar{\alpha}_k \right| \leqslant \sup_{\lambda_0 \in C} \left| \sum_{k=0}^{N} \frac{F_{\lambda_0}^{(k)}(0)}{k!}\bar{\alpha}_k \right| \leqslant \sup_{\lambda_0 \in C} \|F_{\lambda_0}\|_2 \left(\sum_{k=0}^{N} |\alpha_k|^2 \right)^{1/2},$$

it follows that

$$\left(\sum_{k=0}^{N} \left| \frac{F_\lambda^{(k)}(0)}{k!} \right|^2 \right)^{1/2} \leqslant \sup_{\lambda_0 \in C} \|F_{\lambda_0}\|_2,$$

and hence the result follows. ∎

7.45 Theorem If φ is a function in $L^\infty(\mathbb{T})$, then the essential spectrum of T_φ is a connected subset of $\mathbb{C}$.

Proof It is sufficient to prove that if C is a rectifiable simple closed curve lying in $\rho_e(T_\varphi)$ with $\mathscr{R}(\varphi)$ lying outside, then $T_{\varphi-\lambda}$ is a Fredholm operator for λ in the interior of C. Let Ω be a simply connected open set which contains C and the interior of C and no point of the essential spectrum $\sigma_e(T_\varphi)$ exterior to C lies in Ω. We want to show that $\sigma_e(T_\varphi)$ and Ω are disjoint.

Fix λ_o in C. For each λ in Ω let Γ be a rectifiable simple arc lying in Ω with endpoints λ_0 and λ and define

$$F_\lambda(z) = \hat{f}_{\lambda_o}(z) \exp\left\{ \int_\Gamma \widehat{P\left\{\frac{1}{\varphi - \mu}\right\}}(z)\, d\mu \right\}$$

and

$$G_\lambda(z) = \hat{g}_{\lambda_o}(z) \exp\left\{ -\int_\Gamma \widehat{P\left\{\frac{1}{\varphi - \mu}\right\}}(z)\, d\mu \right\}$$

for z in $\mathbb{D}$. Since $\widehat{P\{1/(\varphi - \mu)\}}(z)$ is analytic on the simply connected region Ω, it follows that $F_\lambda(z)$ and $G_\lambda(z)$ are well defined. Moreover, since $F_\lambda(z) = \hat{f}_\lambda(z)$ for λ in some neighborhood of C by Corollary 7.41, it follows from the preceding lemma that F_λ is in H^2 for all λ in Ω. Similarly, G_λ is in H^2 for each λ in Ω.

If we consider the function $\psi(\lambda) = T_{\chi_n(\varphi - \lambda)} F_\lambda - 1$, then $\psi(\lambda)$ lies in H^2 for λ in Ω, is analytic, and vanishes on the exterior of C in Ω. Thus

$$T_{\chi_n(\varphi - \lambda)} F_\lambda = 1$$

for λ in Ω, and in a similar manner we see that $T_{\chi_{-n}(\varphi - \lambda)^{-1}} G_\lambda = 1$ for λ in Ω. If k is a function in H^∞, then

$$H_\lambda(z) = F_\lambda(z) \widehat{P\left\{\frac{1}{\varphi - \lambda} \chi_{-n} G_\lambda k\right\}}(z)$$

defines a function satisfying the hypotheses of the preceding lemma. Thus, H_λ is in H^2 and $T_{\chi_n(\varphi - \lambda)} H_\lambda = k$ for λ in Ω. Moreover, we have

$$\|H_\lambda\|_2 \leq \sup_{\lambda_0 \in C} \|H_{\lambda_o}\|_2 = \sup_{\lambda_0 \in C} \|T^{-1}_{\chi_n(\varphi - \lambda_0)}\| \, \|k\|_2.$$

Therefore, if k is in H^2 and $\{k_j\}_{j=1}^\infty$ is a sequence of functions in H^∞ such that $\lim_{j \to \infty} \|k - k_j\|_2 = 0$, then the corresponding functions $\{H_\lambda^j\}_{j=1}^\infty$ for a fixed λ in Ω form a Cauchy sequence and hence converge to a function H_λ such that $T_{\chi_n(\varphi - \lambda)} H_\lambda = k$. Thus we see that $T_{\chi_n(\varphi - \lambda)}$ is onto for λ in Ω. Since the same argument applied to $\bar{\lambda}$ on $\bar{\Omega}$ yields that $T_{\chi_{-n}(\bar{\varphi} - \bar{\lambda})}$ is onto for λ in Ω, we see that $T_{\chi_n(\varphi - \lambda)}$ is invertible, and hence that $T_\varphi - \lambda$ is a Fredholm operator, which completes the proof. ∎

7.46 Corollary (Widom) If φ is a function in $L^\infty(\mathbb{T})$, then $\sigma(T_\varphi)$ is a connected subset of $\mathbb{C}$.

Proof By virtue of Proposition 7.24, the spectrum of T_φ is formed from the union of the essential spectrum plus the λ for which $T_\varphi - \lambda$ is a Fredholm operator having index different from zero. Since $T_\varphi - \lambda$ is a Fredholm operator for λ in each component of the complement of the essential spectrum and the index is constant, it follows that the spectrum is obtained by taking the union of a compact connected set and some of the components in the complement—and hence is connected. ∎

Despite the elegance of the preceding proof of connectedness, we view it as not completely satisfactory for two reasons: First, the proof gives us no hint as to why the result is true. Second, the proof seems to depend on showing that the set of some kind of singularities for a function of two complex variables is connected, and it would be desirable to state it in these terms.

We conclude this chapter with a result of a completely different nature but of a kind which we believe will be important in the further study of Toeplitz operators. It involves a notion of "localization" and suggests that in order to understand certain phenomena concerning Toeplitz operators it is necessary to consider other representations of the C^*-algebra $\mathfrak{T}(L^\infty(\mathbb{T}))$.

We begin with a result concerning C^*-algebras having a nontrivial center. The center of an algebra is the commutative subalgebra consisting of those elements which commute with all the other elements in the algebra. In the proof we make use of the fact that an abstract C^*-algebras has a *-isometric isomorphic representation as an algebra of operators on some Hilbert space. Also we need to know that every *-isomorphism on $C(X)$ can be extended to the Borel functions on X. Although we have not proved these results in the text, outlines of the proofs were given in the Exercises in Chapter 4 and 5.

7.47   Theorem If $\mathfrak{T}$ is a C^*-algebra, $\mathfrak{A}$ is a C^*-algebra contained in the center of $\mathfrak{T}$ having maximal ideal space $M_\mathfrak{A}$, and for x in $M_\mathfrak{A}$, $\mathfrak{I}_x$ is the closed ideal in $\mathfrak{T}$ generated by the maximal ideal $\{A \in \mathfrak{A}: \hat{A}(x) = 0\}$ in $\mathfrak{A}$, then

$$\bigcap_{x \in M_\mathfrak{A}} \mathfrak{I}_x = \{0\}.$$

In particular, if Φ_x is the *-homomorphism from $\mathfrak{T}$ onto $\mathfrak{T}/\mathfrak{I}_x$, then $\sum_{x \in M_\mathfrak{A}} \oplus \Phi_x$ is a *-isomorphism of $\mathfrak{T}$ into $\sum_{x \in M_\mathfrak{A}} \oplus \mathfrak{T}/\mathfrak{I}_x$. Moreover, T is invertible in $\mathfrak{T}$ if and only if $\Phi_x(T)$ is invertible in $\mathfrak{T}/\mathfrak{I}_x$ for x in $M_\mathfrak{A}$.

Proof   By Proposition 4.67 it is sufficient to show that $\sup_{x \in M_\mathfrak{A}} \|\Phi_x(T)\| = \|T\|$ for T in $\mathfrak{T}$. Fix T in $\mathfrak{T}$ and suppose that $\|T\| - \sup_{x \in M_\mathfrak{A}} \|\Phi_x(T)\| = \varepsilon > 0$. For x_0 in $M_\mathfrak{A}$ let O_{x_0} denote the collection of products SA, where S is in $\mathfrak{T}$ and A is in $\mathfrak{A}$ such that $\hat{A}$ vanishes on a neighborhood of x_0. Then O_{x_0}

is an ideal in $\mathfrak{T}$ contained in $\mathfrak{J}_{x_0}$, and since the closure of O_{x_0} contains

$$\{A \in \mathfrak{A} : \hat{A}(x_0) = 0\},$$

we see that $\mathfrak{J}_{x_0} = \operatorname{clos}[O_{x_0}]$. Thus there exists S in $\mathfrak{T}$ and A in $\mathfrak{A}$ such that $\hat{A}$ vanishes on an open set U_{x_0} containing x_0 and $\|\Phi_{x_0}(T)\| + \varepsilon/3 > \|T + SA\|$. Choose a finite subcover $\{U_{x_i}\}_{i=1}^{N}$ of $M_{\mathfrak{A}}$ and the corresponding operators $\{S_i\}_{i=1}^{N}$ in $\mathfrak{T}$ and $\{A_i\}_{i=1}^{N}$ in $\mathfrak{A}$ such that

$$\|T\| - \frac{\varepsilon}{2} \geqslant \|\Phi_{x_i}(T)\| + \frac{\varepsilon}{2} \geqslant \|T + S_i A_i\|$$

and $\hat{A}_i$ vanishes on U_{x_i}.

Let Φ be a *-isomorphism of $\mathfrak{T}$ into $\mathfrak{L}(\mathscr{H})$ for some Hilbert space $\mathscr{H}$ and let ψ_0 be the corresponding *-isomorphism of $C(M_{\mathfrak{A}})$ into $\mathfrak{L}(\mathscr{H})$ defined by $\psi_0 = \Phi \circ \Gamma^{-1}$, where Γ is the Gelfand isomorphism of $\mathfrak{A}$ onto $C(M_{\mathfrak{A}})$. Since $\Phi(\mathfrak{A})$ is contained in the commutant of $\Phi(\mathfrak{T})$, it follows that $\psi_0(C(M_{\mathfrak{A}}))$ is contained in the commutant of $\Phi(\mathfrak{T})$, and hence there exists a *-homomorphism ψ from the algebra of bounded Borel functions on $M_{\mathfrak{A}}$ into the commutant of $\Phi(\mathfrak{T})$ which extends ψ_0.

Let $\{\Delta_i\}_{i=1}^{N}$ be a partition of $M_{\mathfrak{A}}$ by Borel sets such that Δ_i is contained in U_{x_i} for $i = 1, 2, \ldots, N$. Then

$$\Phi(T + S_i A_i)\psi(I_{\Delta_i}) = \Phi(T)\psi(I_{\Delta_i})$$

and hence

$$\|\Phi(T)\psi(I_{\Delta_i})\| \leqslant \|\Phi(T + S_i A_i)\| \, \|\psi(I_{\Delta_i})\| \leqslant \|T\| - \frac{\varepsilon}{2} \quad \text{for each } 1 \leqslant i \leqslant N.$$

Since the $\{\psi(I_{\Delta_i})\}_{i=1}^{N}$ are a family of commuting projections which reduce $\Phi(T)$ and such that $\sum_{i=1}^{N} \psi(I_{\Delta_i}) = I_{\mathscr{H}}$, it follows that

$$\|\Phi(T)\| = \sup_{1 \leqslant i \leqslant N} \|\Phi(T)\psi(I_{\Delta_i})\|$$

and hence $\|\Phi(T)\| \leqslant \|T\| - \varepsilon/2$. Since Φ is an isometry, this is a contradiction.

If T is invertible in $\mathfrak{T}$, then clearly $\Phi_x(T)$ is invertible in $\mathfrak{T}/\mathfrak{J}_x$ for x in $M_{\mathfrak{A}}$. Hence suppose $\Phi_x(T)$ is invertible in $\mathfrak{T}/\mathfrak{J}_x$ for each x in $M_{\mathfrak{A}}$. For x_0 in $M_{\mathfrak{A}}$ there exists S in $\mathfrak{T}$ by Theorem 4.28 such that $\Phi_{x_0}(ST - I) = 0$. Repeating the argument of the first paragraph, we obtain a neighborhood O_{x_0} on which $\|\Phi_x(ST - I)\| < \frac{1}{2}$ for x in O_{x_0}. From Proposition 2.5 we obtain that $\Phi_x(ST)$ is invertible and $\|\Phi_x(ST)^{-1}\| < 2$ for x in O_{x_0}. Since $\Phi_x(T)$ is invertible, $\Phi_x(ST)^{-1}\Phi_x(S)$ is its inverse, and hence $\|\Phi_x(T)^{-1}\|$ is bounded for x in

O_{x_0}. A standard compactness argument shows that $\sup_{x \in M_{\mathfrak{A}}} \|\Phi_x(T)^{-1}\| < \infty$. Therefore $\sum_{x \in M_{\mathfrak{A}}} \oplus \Phi_x(T)$ is invertible in $\sum_{x \in M_{\mathfrak{A}}} \oplus \mathfrak{T}/\mathfrak{T}_x$, and hence T is invertible in $\mathfrak{T}$ by Theorem 4.28. ∎

7.48 The context in which we want to apply this result is the following. Although the center of the C^*-algebra $\mathfrak{T}(L^\infty(\mathbb{T}))$ is equal to the scalars, the quotient algebra $\mathfrak{T}(L^\infty(\mathbb{T}))/\mathfrak{LC}(H^2)$ has a nontrivial center which contains $\mathfrak{T}(C(\mathbb{T}))/\mathfrak{LC}(H^2) = C(\mathbb{T})$ by Proposition 7.22. Thus we can "localize" the algebra $\mathfrak{T}(L^\infty(\mathbb{T}))/\mathfrak{LC}(H^2)$ to the points of $\mathbb{T}$.

For λ in $\mathbb{T}$ let $\mathfrak{T}_\lambda$ be the closed ideal in $\mathfrak{T}(L^\infty(\mathbb{T}))$ generated by

$$\{T_\varphi : \varphi \in C(\mathbb{T}), \varphi(\lambda) = 0\},$$

and let $\mathfrak{T}_\lambda$ be the quotient algebra $\mathfrak{T}(L^\infty(\mathbb{T}))/\mathfrak{T}_\lambda$, and Φ_λ be the natural *-homomorphism from $\mathfrak{T}(L^\infty(\mathbb{T}))$ to $\mathfrak{T}_\lambda$.

7.49 Theorem The C^*-algebras $\mathfrak{T}_\lambda$ are all *-isomorphic. If Φ is the *-homomorphism defined by $\Phi = \sum_{\lambda \in \mathbb{T}} \oplus \Phi_\lambda$ from $\mathfrak{T}(L^\infty(\mathbb{T}))$ to $\sum_{\lambda \in \mathbb{T}} \oplus \mathfrak{T}_\lambda$, then the sequence

$$(0) \longrightarrow \mathfrak{LC}(H^2) \longrightarrow \mathfrak{T}(L^\infty(\mathbb{T})) \overset{\Phi}{\longrightarrow} \sum_{\lambda \in \mathbb{T}} \oplus \mathfrak{T}_\lambda$$

is exact at $\mathfrak{T}(L^\infty(\mathbb{T}))$.

Proof Since $\mathfrak{T}(L^\infty(\mathbb{T}))$ is an irreducible algebra in $\mathfrak{L}(H^2)$, every nonzero closed ideal contains $\mathfrak{LC}(H^2)$ by Theorem 5.39. Thus $\mathfrak{LC}(H^2)$ is contained in the kernel of Φ and hence Φ induces a *-homomorphism Φ_c from the quotient algebra $\mathfrak{T}(L^\infty(\mathbb{T}))/\mathfrak{LC}(H^2)$ into $\sum_{\lambda \in \mathbb{T}} \oplus \mathfrak{T}_\lambda$. Since

$$\mathfrak{T}(C(\mathbb{T}))/\mathfrak{LC}(H^2)$$

is contained in the center of $\mathfrak{T}(L^\infty(T))/\mathfrak{LC}(H^2)$, the preceding theorem applies, and we conclude that Φ_c is a *-isomorphism. Rotation by λ on $\mathbb{T}$ obviously induces an automorphism on $\mathfrak{T}(L^\infty(\mathbb{T}))$ taking $\mathfrak{T}_0$ onto $\mathfrak{T}_\lambda$, and hence there exists a *-isomorphism from $\mathfrak{T}_0$ onto $\mathfrak{T}_\lambda$. ∎

The usefulness of this result lies in the fact that it reduces all questions concerning operators in $\mathfrak{T}(L^\infty(\mathbb{T}))$ modulo the compacts to questions concerning $\mathfrak{T}_\lambda$. Unfortunately, we do not know very much about the algebras $\mathfrak{T}_\lambda$. The following proposition shows that the operators in $\mathfrak{T}_\lambda$ depend only on local properties of the defining functions.

Recall that $M_\infty \backslash \mathbb{D}$ is fibered by the circle such that

$$F_\lambda = \{m \in M_\infty : \hat{\chi}_1(m) = \lambda\}$$

and that the Šilov boundary of H^∞ can be identified with the maximal ideal space M_{L^∞} of $L^\infty(\mathbb{T})$. Let us denote the intersection $F_\lambda \cap M_{L^\infty}$ by ∂F_λ.

7.50 Proposition If $\{\varphi_{ij}\}_{i,j=1}^N$ are functions in $L^\infty(\mathbb{T})$ with Gelfand transform $\hat{\varphi}_{ij}$ on M_{L^∞} and λ is in $\mathbb{T}$, then $\Phi_\lambda(\sum_{i=1}^N \prod_{j=1}^N T_{\varphi_{ij}})$ depends only on the functions $\{\hat{\varphi}_{ij}|\partial F_\lambda\}_{i,j=1}^N$.

Proof It is sufficient to show that $\Phi_\lambda(T_\varphi) = 0$ for φ in $L^\infty(\mathbb{T})$ such that $\hat{\varphi}|\partial F_\lambda \equiv 0$. By continuity and compactness, it follows that for $\varepsilon > 0$ there exists an open arc U of $\mathbb{T}$ containing λ such that $\|\varphi I_U\|_\infty < \varepsilon$. If ψ is a continuous function on $\mathbb{T}$ which equals 1 on the complement of U and vanishes at λ, then

$$T_\varphi = T_{\varphi I_U} + T_{\varphi I_{\mathbb{T}\setminus U}(1-\psi)} + T_{\varphi I_{\mathbb{T}\setminus U}\psi},$$

where

$$\|\Phi_\lambda(T_{\varphi I_U})\| \leqslant \|T_{\varphi I_U}\| < \varepsilon, \qquad \Phi_\lambda(T_{\varphi I_{\mathbb{T}\setminus U}(1-\psi)}) = 0,$$

since

$$I_{\mathbb{T}\setminus U}(1-\psi) = 0 \qquad \text{and} \qquad \Phi_\lambda(T_{\Phi I_{\mathbb{T}\setminus U}\psi}) = \Phi_\lambda(T_{\Phi I_{\mathbb{T}\setminus U}})\Phi_\lambda(T_\psi) = 0,$$

since T_ψ is in $\mathfrak{I}_\lambda$. Therefore, we have $\|\Phi_\lambda(T_\varphi)\| < \varepsilon$ for $\varepsilon > 0$ and hence $\Phi_\lambda(T_\varphi) = 0$. ∎

As corollaries, we obtain the following results proved originally by "localizing" in $H^\infty + C(\mathbb{T})$.

7.51 Corollary If φ is a function in $L^\infty(\mathbb{T})$, then T_φ is a Fredholm operator if and only if for each λ in $\mathbb{T}$ there exists ψ in $L^\infty(\mathbb{T})$ such that T_ψ is a Fredholm operator and $\varphi = \psi$ on ∂F_λ.

Proof From Theorem 7.49 and the definition, it follows that T_φ is a Fredholm operator if and only if $\Phi_\lambda(T_\varphi)$ is invertible for each λ in $\mathbb{T}$. If for λ in $\mathbb{T}$ there exists ψ in $L^\infty(\mathbb{T})$ such that T_ψ is a Fredholm operator and $\hat{\varphi} = \hat{\psi}$ on ∂F_λ, then $\Phi_\lambda(T_\psi)$ is invertible in $\mathfrak{I}_\lambda$ and equal to $\Phi_\lambda(T_\varphi)$ by the previous proposition. Thus the result follows from Theorem 7.47. ∎

7.52 Corollary If φ and ψ are functions in $L^\infty(\mathbb{T})$ such that for each λ in $\mathbb{T}$ either $\hat{\varphi} - \hat{\theta}_1|\partial F_\lambda \equiv 0$ for some θ_1 in $\bar{H}^\infty$ or $\hat{\psi} - \hat{\theta}_2|\partial F_\lambda \equiv 0$ for some θ_2 in H^∞, then $T_\varphi T_\psi - T_{\varphi\psi}$ is compact.

Proof Since $T_\varphi T_\psi - T_{\varphi\psi}$ is compact if and only if $\Phi_\lambda(T_\varphi)\Phi_\lambda(T_\psi) = \Phi_\lambda(T_{\varphi\psi})$ for each λ in $\mathbb{T}$ by Theorem 7.49, the result is seen to follow from Proposition 7.5. ∎

Notes

In an early paper [109] Toeplitz investigated finite matrices which are constant on diagonals and their relation to the corresponding one- and two-sided infinite matrices. The fundamental theorem in this line of study was proved by Szëgo (see [54]), and most of the early work concerned this type of question. In [116] Wintner determined the spectrum of analytic Toeplitz matrices, and he and Hartman set the tone for much of the work in this chapter in two papers [59] and [60] some twenty years later. The first systematic study of Toeplitz operators emphasizing the mapping $\varphi \to T_\varphi$ was made by Brown and Halmos in [11]. What might be called the algebra approach to these problems was first made explicit in [29] and [30] and was based on the earlier papers [17] and [18] of Coburn.

A vast literature exists for Wiener–Hopf operators beginning with the fundamental paper of Wiener and Hopf [115]. Most of the early work concerns the study of explicit operators, and a good exposition of that along with a bibliography can be found in Kreïn [72]. The studies of Toeplitz operators and Wiener–Hopf operators had parallel developments until Rosenblum observed [94] using Laguerre polynomials that the two classes of operators were unitarily equivalent. Subsequently, Devinatz showed in [25] that the canonical conformal mapping of the unit disk onto the upper half-plane establishes the unitary equivalence between a Toeplitz operator and the Fourier transform of a Wiener–Hopf operator. Thus a given result can be stated in the context of either Toeplitz operators or of Wiener–Hopf operators.

The spectral inclusion theorem is due to Hartman and Wintner [60], although the proof given here of Proposition 7.6 first occurs in [110]. Corollaries 7.8 and 7.19 as well as the remarks following 7.9 are due to Brown and Halmos [11]. The existence of the homomorphism in Theorem 7.11 as well as its role in these questions is established in [31]. In [103] Stampfli observed that a proof of Coburn in [17] actually yields Proposition 7.22. The analysis of the C^*-algebra $\mathfrak{T}(C(\mathbb{T}))$ was made by Coburn in [17] and [18] while its applicability to the invertibility problem for Toeplitz operators with continuous symbol was observed in [29]. Proposition 7.24 was proved by Coburn in [16]. The content of Corollary 7.27 is the culmination of several authors including Kreïn [72], Calderon, Spitzer, and Widom [13], Widom [110], and Devinatz [24]. The proof given here first appears in [29] and independently in [3], where Atiyah used the matrix analog in a proof of the periodicity theorem. A related proof was given by Gohberg and Fel'dman [45]. The study of Toeplitz operators with symbol in $H^\infty + C(\mathbb{T})$ was made in [30] with complements in [77] and [103].

The invertibility criteria stated in Theorem 7.30 and its corollary were given independently by Widom [111] and Devinatz [24] and are based on a study in prediction theory by Helson and Szëgo [64]. The spectral inclusion theorem given in Theorem 7.32 is based on an extension by Lee and Sarason [77] of a result of the author [31]. The question of the connectedness of the spectrum of a Toeplitz operator was posed by Halmos in [57] and answered by Widom in [113] and [114]. The proof of Theorem 7.45 is a slight adaptation of that in [114] to cover the essential spectrum and avoiding certain measure theoretic considerations as well as the use of the harmonic conjugate. The possibility of the essential spectrum being connected was suggested to the author by Abrahamse.

Theorem 7.47 is closely related to various central decompositions in C^*-algebra (see [96]) but its application to Toeplitz operators is new, and these results were suggested by the earlier Corollaries 7.51 and 7.52. The first corollary is due to Simonenko [102] and independently to Douglas and Sarason [35] and extends a result of Douglas and Widom [38]. The second corollary is due to Sarason [98].

Several further developments and additional topics should be mentioned. The invertibility problem for symbols in the algebra of piecewise continuous functions has been considered by Widom [110], Devinatz [24], and from the algebra viewpoint by Gohberg and Krupnik [50]. The invertibility problem has also been considered for certain algebras of functions which appear more natural in the context of the line. The algebra of almost periodic functions was considered independently by Coburn and Douglas [19] and by Gohberg and Fel'dman [46], [47]. Actually, the latter authors considered the Fourier transforms of measures having no continuous singular part. The problem for the Fourier transform of an arbitrary measure has been considered by Douglas and Taylor [37] using a deep result [108] of the latter on the cohomology of the maximal ideal space of the convolution algebra of measures. Lastly, Lee and Sarason [77] and Douglas and Sarason [36] have considered the invertibility problem for certain functions of the form $\varphi\bar\psi$ for inner functions φ and ψ.

Generalizations of the notion of Toeplitz operator have been considered by many authors: Douglas and Pearcy studied the role of the F. and M. Riesz theorem in [33]; Devinatz studied Toeplitz operators on the H^2-space of a Dirichlet algebra [24]; Devinatz and Shinbrot [26] studied the invertibility of compressions of operators to subspaces; and Abrahamse studied Toeplitz operators defined on the H^2-space of a finitely connected region in the plane [1]. A different kind of generalization is obtained by considering Wiener–Hopf operators. A subsemigroup of an abelian group is prescribed, and convolution operators compressed to the corresponding L^2 space of functions supported on it are studied. Certain basic results for rather arbitrary semigroups are obtained by Coburn and Douglas in [20]. Earlier work involving half-spaces is due to Goldenstein and Gohberg [52] and [53]. The case of the quarter plane is of particular importance and coincides with Toeplitz operators defined on H^2 of the bidisk. Results on this have been obtained by Simonenko [101], Osher [84], Malyšev [78], Strang [106], and Douglas and Howe [32]. In particular, the latter authors obtain necessary and sufficient conditions for such a Toeplitz operator to be a Fredholm operator. They show, moreover, that while such an operator must have index zero, it need not be invertible.

In many of the preceding contexts, various topological invariants of the symbol enter into the determination of when the corresponding operator is invertible, and usually these invariants must be zero. In the case of a continuous symbol on the circle, a nonzero invariant corresponds to the operator being a Fredholm operator having a nonzero index. A start at establishing similar results for the other classes of operators has been made by Coburn, Douglas, Schaeffer, and Singer in [21] where a generalized notion of Fredholm operator due to Breuer [8], [9] is utilized. It is expected that questions of this kind will prove important in future developments.

Although attention in this book has been confined to the scalar case, the matrix case is perhaps of even greater importance. The function φ is allowed to have $n \times n$ matrices as values and to operate on a $\mathbb{C}^n$-valued H^2-space. Many of the techniques of this chapter can be carried over to this case using the device of the tensor product. More specifically, if $\mathfrak{A}$ is an algebra of scalar functions, $\mathfrak{A}_n$ is equal to $\mathfrak{A} \otimes M_n$, where M_n is the C^*-algebra of operators on the n-dimensional Hilbert space $\mathbb{C}^n$, and more importantly, $\mathfrak{T}(\mathfrak{A}_n)$ is isometrically isomorphic to $\mathfrak{T}(\mathfrak{A}) \otimes M_n$. In particular, applying this to one of the exact sequences considered in the chapter one obtains the exact sequence

$$(0) \to \mathfrak{LC}(H^2) \otimes M_n \to \mathfrak{T}(C(\mathbb{T})) \otimes M_n \to C(\mathbb{T}) \otimes M_n \to (0).$$

(The fact that the "scalar" sequence has a continuous cross section is also used.) Since $\mathfrak{LC}(H^2) \otimes M_n$ is equal to $\mathfrak{LC}(H^2 \otimes \mathbb{C}^n)$, one obtains that a matrix Toeplitz operator with continuous symbol is a Fredholm operator if and only if the determinant function does not vanish on the circle. Moreover, the index can be shown to equal minus the winding number of the determinant. (This latter argument uses the fact that every continuous mapping from $\mathbb{T}$ into the invertible $n \times n$ matrice is homotopic to a mapping from $\mathbb{T}$ to the invertible diagonal matrices.)

This latter result is due to Gohberg and Kreĭn [49]. A generalization to certain operator-valued analogs can be found in [31]. Additional results on the matrix and operator-valued case are in [87] and [88].

Lastly, although this book does not comment on it, the study of Toeplitz and Wiener–Hopf operators is important in various areas of physics and probability (see [54], [69], [83]) and in examining the convergence of certain differences schemes for solving partial differential equations (see [83]).

Exercises

7.1 If $\mathfrak{A}$ and $\mathfrak{B}$ are C^*-algebras and ρ is an isometric *-linear map from $\mathfrak{A}$ into $\mathfrak{B}$, then $\sigma(T) \subset \sigma(\rho(T))$ for T in $\mathfrak{A}$. (Hint: If T is not left invertible in $\mathfrak{A}$, then there exists $\{S_n\}_{n=1}^{\infty}$ in $\mathfrak{A}$ such that $\|S_n\| = 1$ and $\lim_{n \to \infty} \|S_n T\| = 0$.) If, in addition, $\mathfrak{A}$ is commutative, then $\sigma(\rho(T)) \subset h[\sigma(T)]$ for T in $\mathfrak{A}$.

7.2 If φ is a nonconstant real function in $L^{\infty}(\mathbb{T})$, then T_{φ} has no eigenvalues.

7.3 An operator T in $\mathfrak{L}(H^2)$ is a Toeplitz operator if and only if

$$T_{\chi_1}^* T T_{\chi_1} = T.$$

7.4 If φ is a nonzero function in $L^{\infty}(\mathbb{T})$, then M_{φ} and T_{φ} have no eigenvalues in common.

7.5 If φ is a real function, then T_{φ} is invertible if and only if the function 1 is in its range.

7.6 If φ is in $L^{\infty}(\mathbb{T})$, then $W(T_{\varphi}) = h[\mathscr{R}(\varphi)]$.

7.7 If φ_1, φ_2, and φ_3 are functions in $L^{\infty}(\mathbb{T})$ such that $T_{\varphi_1} T_{\varphi_2} - T_{\varphi_3}$ is compact, then $\varphi_1 \varphi_2 = \varphi_3$.

Definition If $\{\alpha_n\}_{n=0}^{\infty}$ is a bounded sequence of complex numbers, then the associated Hankel matrix $\{a_{ij}\}_{i,j=0}^{\infty}$ is defined by $a_{ij} = \alpha_{i+j}$ for i,j in $\mathbb{Z}^+$.

7.8 (*Nehari*) If $\{\alpha_n\}_{n=0}^{\infty}$ is a bounded sequence, then the Hankel operator H defined by the associated Hankel matrix is bounded if and only if there exists a function φ in $L^{\infty}(\mathbb{T})$ such that

$$\alpha_n = \frac{1}{2\pi} \int_0^{2\pi} \varphi(e^{it}) \chi_n(e^{it}) \, dt.$$

Moreover, the norm of H is equal to the infimum of the norm $\|\varphi\|_\infty$ for all such functions φ.* (Hint: Consider the linear functional L defined on H^2 by $L(f) = (Hg, h)$, where $f = gh$, and show that L is continuous if H is bounded.)

7.9   (*Hartman*) If $\{\alpha_n\}_{n=0}^\infty$ is a bounded sequence, then the associated Hankel operator is compact if and only if there exists a continuous function φ on $\mathbb{T}$ such that

$$\alpha_n = \frac{1}{2\pi} \int_0^{2\pi} \varphi(e^{it}) \chi_n(e^{it}) \, dt \qquad \text{for} \quad n \text{ in } R^+.*$$

(Hint: Use Exercise 1.33 to show that a certain linear functional L is w^*-continuous if and only if H is compact and apply the analog of Exercise 6.36.)

7.10   An operator H in $\mathfrak{L}(H^2)$ is a Hankel operator if and only if $T_{\chi_1}^* H = H T_{\chi_1}$.

7.11   If φ is in $L^\infty(\mathbb{T})$, then $\pi(T_\varphi)$ is unitary in $\mathfrak{T}(L^\infty(\mathbb{T}))/\mathfrak{LC}(H^2)$ if and only if φ is in QC.* (Hint: Show that the Hankel operator H_φ is compact if and only if $\pi(T_\varphi^*)$ is an isometry in $\mathfrak{T}(L^\infty(\mathbb{T}))/\mathfrak{LC}(H^2)$.)

7.12   If φ is in $L^\infty(\mathbb{T})$, then $\pi(T_\varphi)$ is in the center of $\mathfrak{T}(L^\infty(\mathbb{T}))/\mathfrak{LC}(\mathbb{T}^2)$ if and only if φ is in QC. Are there any other operators in the center?**

7.13   Show that $T_\chi T_\varphi - T_{\chi\varphi}$ is compact for every χ in $L^\infty(\mathbb{T})$ if and only if φ is in $H^\infty + C(\mathbb{T})$.*

7.14   If φ is in $H^\infty + C(\mathbb{T})$, then φ is invertible in $L^\infty(\mathbb{T})$ if and only if T_φ has closed range.

7.15   If φ is in $H^\infty + C(\mathbb{T})$ and λ is in $\sigma(T_\varphi)\backslash\mathscr{R}(\varphi)$, then either λ is an eigenvalue for T_φ or $\bar{\lambda}$ is an eigenvalue for T_φ.

7.16   (*Widom–Devinatz*) If φ is an invertible functions in $L^\infty(\mathbb{T})$, then T_φ is invertible if and only if there exists an outer function g such that

$$|\arg g\varphi| < \pi/2 - \delta$$

for $\delta > 0$.

7.17   Show that there exists a natural homomorphism γ of $\mathrm{Aut}[\mathfrak{T}(C(\mathbb{T}))]$ onto the group $\mathrm{Hom}_+(\mathbb{T})$ of orientation preserving homomorphisms of $\mathbb{T}$. (Hint: If φ is in $\mathrm{Hom}_+(\mathbb{T})$, then there exists K in $\mathfrak{LC}(H^2)$ such that $T_\varphi + K$ is a unilateral shift.)

7.18   Show that the connected component $\mathrm{Aut}_0[\mathfrak{T}(C(\mathbb{T}))]$ of the identity in $\mathrm{Aut}[\mathfrak{T}(C(\mathbb{T}))]$ is contained in the kernel of γ. Is it equal?**

7.19 If φ is a unimodular function in QC and K is appropriately chosen in $\mathfrak{LC}(H^2)$, then $\alpha(T) = (T_\varphi + K)^* T(T_\varphi + K)$ defines an automorphism on $\mathfrak{T}(C(\mathbb{T}))$ in the kernel of γ. Show that the automorphisms of this form do not exhaust $\ker \gamma$.* Is α in $\mathrm{Aut}_0[\mathfrak{T}(C(\mathbb{T}))]$?**

7.20 If V is a pure isometry on $\mathscr{H}$ and E_n is the projection onto $V^n \mathscr{H}$, then $U_t = \sum_{n=0}^\infty e^{int}(E_n - E_{n+1})$ is a unitary operator on $\mathscr{H}$ for e^{it} in $\mathbb{T}$. If we set $\beta_t(T) = U_t^* T U_t$ for T in the C^*-algebra $\mathfrak{C}_V$ generated by V, then the mapping $\Gamma(e^{it}) = \beta_t$ is a homomorphism from the circle group into $\mathrm{Aut}[\mathfrak{C}_V]$ such that $F(e^{it}) = \beta_t(T)$ is continuous for T in $\mathfrak{C}_V$. Is it norm continuous? Moreover, for $V = T_{\chi_1}$, the unitary operator U_t coincides with that induced by rotation by e^{it} on H^2.

7.21 Identify the fixed points $\mathfrak{F}_V$ for the β_t as a maximal abelian subalgebra of $\mathfrak{C}_V$. (Hint: Consider the case of $V = T_{\chi_1}$, acting on $l^2(\mathbb{Z}_+)$.)

7.22 Show that the mapping ρ defined by

$$\rho(T) = \frac{1}{2\pi} \int_0^{2\pi} \beta_t(T)\, dt \qquad \text{for} \quad T \text{ in } \mathfrak{C}_V$$

is a contractive positive map from $\mathfrak{C}_V$ onto $\mathfrak{F}_V$ which satisfies $\rho(TF) = \rho(T)F$ for T in $\mathfrak{C}_V$ and F in $\mathfrak{F}_V$.

7.23 If V_1 and V_2 are pure isometries on $\mathscr{H}_1$ and $\mathscr{H}_2$, respectively, such that there exists a *-homomorphism Φ from $\mathfrak{C}_{V_1}$ to $\mathfrak{C}_{V_2}$ with $\Phi(V_1) = V_2$, then Φ is an isomorphism. (Hint: Show that $\rho \circ \Phi = \Phi \circ \rho$ and that $\Phi|\mathfrak{F}_{V_1}$ is an isomorphism.)

7.24 If A and B are operators on $\mathscr{H}$ and $\mathscr{K}$, respectively, and Φ is a *-homomorphism from $\mathfrak{C}_A$ to $\mathfrak{C}_B$ with $\Phi(A) = B$, then there exists *-isomorphism Ψ from $\mathfrak{C}_{A \oplus B}$ to $\mathfrak{C}_A$ such that $\Psi(A \oplus B) = A$.

7.25 If V is a pure isometry on $\mathscr{H}$ and W is a unitary operator on $\mathscr{K}$, then there exists a *-isomorphism ψ from $\mathfrak{C}_{V_1}$ onto $\mathfrak{C}_{V_2}$ such that $\psi(V_1) = V_2$.

7.26 (*Coburn*) If V_1 and V_2 are nonunitary isometries on H_1 and H_2, respectively, then there exists a *-isomorphism Ψ from $\mathfrak{C}_{V_1}$ onto $\mathfrak{C}_{V_2}$ such that $\Psi(V_1) = V_2$.

Definition A C^*-algebra $\mathfrak{A}$ is said to be an extension of $\mathfrak{LC}$ by $C(\mathbb{T})$, if there exists a *-isomorphism Φ from $\mathfrak{LC}(\mathscr{H})$ into $\mathfrak{A}$ and a *-homomorphism Ψ from $\mathfrak{A}$ onto $C(\mathbb{T})$ such that the sequence

$$(0) \longrightarrow \mathfrak{LC}(\mathscr{H}) \overset{\Phi}{\longrightarrow} \mathfrak{A} \overset{\Psi}{\longrightarrow} C(\mathbb{T}) \longrightarrow (0)$$

is exact. Two extensions $\mathfrak{A}_1$ and $\mathfrak{A}_2$ will be said to be equivalent if there exists a *-isomorphism θ from $\mathfrak{A}_1$ onto $\mathfrak{A}_2$ and an automorphism α in $\mathrm{Aut}[\mathfrak{LC}(\mathscr{H})]$ such that the diagram

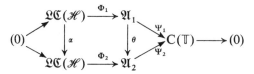

commutes.

7.27 If N is a positive integer and $\mathfrak{T}_N$ is the C^*-algebra generated by T_{χ_N} and $\mathfrak{LC}(H^2)$, then $\mathfrak{T}_N$ is an extension of $\mathfrak{LC}$ by $C(\mathbb{T})$ with $\Psi(T_{\chi_N}) = \chi_1$. Moreover, $\mathfrak{T}_N$ and $\mathfrak{T}_M$ are isomorphic C^*-algebras if and only if $N = M$. (Hint: Consider index in $\mathfrak{T}_N$.)

7.28 Let $K^2 = L^2(\mathbb{T}) \ominus H^2$, Q be the orthogonal projection onto K^2, and define $S_\varphi = QM_\varphi|K^2$. If N is a negative integer and $\mathfrak{S}_N$ is the C^*-algebra generated by S_{χ_N} and $\mathfrak{LC}(K^2)$, then $\mathfrak{S}_N$ is an extension of $\mathfrak{LC}$ by $C(\mathbb{T})$ with $\Psi(S_{\chi_N}) = \chi_1$. Moreover, the extensions $\mathfrak{T}_M$ and $\mathfrak{S}_N$ are all inequivalent despite the fact that the algebras $\mathfrak{T}_N$ and $\mathfrak{S}_N$ are isomorphic.

7.29 Let E be a closed perfect subset of $\mathbb{T}$ and μ be a probability measure on $\mathbb{T}$ such that the closed support of μ is $\mathbb{T}$ and the closed support of the restriction μ_E of μ to E is also E. If $\mathfrak{A}_E$ is the C^*-algebra generated by M_{χ_1} on $L^2(\mu)$ and $\mathfrak{LC}(L^2(\mu_E))$, then $\mathfrak{A}_E$ is an extension of $\mathfrak{LC}$ by $C(\mathbb{T})$ with $\Psi(M_{\chi_1}) = \chi_1$. Moreover, two of these extensions are equivalent if and only if the set E is the same.

7.30 If E is a closed subset of $\mathbb{T}$, then $E = E_0 \cup \{e^{it_n} : n \geqslant 1\}$, where E_0 is a closed perfect subset of $\mathbb{T}$. Let μ be a probability measure on $\mathbb{T}$ such that the closed support of μ is $\mathbb{T}$ and the closed support of the restriction μ_{E_0} of μ to E_0 is also E_0. If $\mathfrak{A}_E$ is the C^*-algebra generated by $W = M_{\chi_1} \oplus \sum_{n \geqslant 1} \oplus M_{e^{it_n}}$ on $L^2(\mu) \oplus \sum_{n \geqslant 1} \oplus l^2(\mathbb{Z})$ and $\mathfrak{LC}(L^2(\mu_{E_0}) \oplus \sum_{n \geqslant 1} \oplus l^2(\mathbb{Z}))$, then $\mathfrak{A}_E$ is an extension of $\mathfrak{LC}$ by $C(\mathbb{T})$ with $\Psi(W) = \chi_1$. Moreover, two of these extensions are equivalent if and only if the set E is the same.

7.31 Every extension $\mathfrak{A}$ of $\mathfrak{LC}$ by $C(\mathbb{T})$ is equivalent to exactly one extension of the form $\mathfrak{T}_N$, $\mathfrak{S}_N$, or $\mathfrak{A}_E$.* (Hint: Assume $\mathfrak{A}$ is contained in $\mathfrak{L}(\mathscr{H})$ and decompose the representation of $\mathfrak{LC}$ on $\mathscr{H}$. Use Exercises 5.18–5.19.)

7.32 If T is an operator on $\mathscr{H}$ such that $\sigma_e(T) = \mathbb{T}$ and $T^*T - TT^*$ is compact, then the C^*-algebra generated by T is an extension of $\mathfrak{LC}$ by $C(\mathbb{T})$ with $\Psi(T) = \chi_1$. Which one is it?

7.33 If B^2 is the closure of the polynomials in $L^2(\mathbb{D})$ with planar Lebesgue measure and P_B is the projection of $L^2(\mathbb{D})$ onto B^2, then the C^*-algebra generated by the operators $\{R_\varphi : \varphi \in C(\overline{\mathbb{D}})\}$, where $R_\varphi = P_\mathscr{B} M_\varphi | B^2$, is an extension of $C(\mathbb{T})$ by $\mathfrak{LC}$ with $\Psi(R\hat{\chi}_1) = \chi_1$. Which one is it?

7.34 If $\mathfrak{A}$ is an extension of $\mathfrak{LC}$ by $C(\mathbb{T})$, determine the range of the mapping from $\mathrm{Aut}(\mathfrak{A})$ to $\mathrm{Aut}(C(\mathbb{T}))$.*

7.35 If λ is in T and φ is in $L^\infty(\mathbb{T})$, then $\hat{\varphi}(F_\lambda) \subset \sigma(\Phi_\lambda(T_\varphi)) \subset h(\hat{\varphi}(F_\lambda))$, where $\hat{\varphi}$ is the harmonic extension of φ to the Šilov boundary of H^∞.

7.36 (*Widom*) If φ is in PC, $\varphi^\#$ is the curve obtained from the range of φ by filling in the line segments joining $\varphi(e^{it^-})$ to $\varphi(e^{it^+})$ for each discontinuity, then T_φ is a Fredholm operator if and only if $\varphi^\#$ does not contain the origin. Moreover, in this case the index of T_φ is minus the winding number of $\varphi^\#$.*
(Hint: Use Section 7.51 to show that T_φ is a Fredholm operator in this case and that the index is minus the winding number of $\varphi^\#$. If $\varphi^\#$ passes through the origin, then small perturbations of φ produce Fredholm operators of different indexes.)

7.37 (*Gohberg–Krupnik*) The quotient algebra $\mathfrak{T}(PC)/\mathfrak{LC}(H^2)$ is a commutative C^*-algebra. Show that its maximal ideal space can be identified as a cylinder with an exotic topology.*

7.38 If X is an operator on H^2 such that $T_\varphi^* X T_\varphi - X$ is compact for each inner function φ, then is $X = T_\psi + K$ for some ψ in H^∞ and compact operator K?**

7.39 If for each z in $\mathbb{C}$, φ_z is a function in $L^\infty(\mathbb{T})$ such that $\varphi_z(e^{it})$ is an entire function in z having at most N zeros for each e^{it} in $\mathbb{T}$, then the set of z for which T_{φ_z} fails to be invertible is a closed subset of $\mathbb{C}$ having at most N components.* (Repeat the whole proof of Theorem 7.45.)

References

1. M. B. Abrahamse, Toeplitz operators in multiply connected domains, *Bull. Amer. Math. Soc.* **77**, 449–454 (1971).
2. N. I. Akheizer and I. M. Glazman, *Theory of linear operators in Hilbert space.* Ungar, New York, 1961.
3. M. Atiyah, Algebraic topology and operators in Hilbert space, *Lectures in Analysis,* vol. 103, pp. 101–121. Springer-Verlag, New York, 1969.
4. F. V. Atkinson, The normal solvability of linear equations in normed spaces, *Mat. Sb.* **28** (70), 3–14 (1951).
5. S. Banach, *Théorie des opérations linéares.* Monografie Matematyczne, Warsaw, 1932.
6. A. Beurling, On two problems concerning linear transformations in Hilbert space, *Acta Math.* **81**, 239–255 (1949).
7. N. Bourbaki, *Espaces Vectoriels Topologiques, Éléments de mathématique, livre V.* Hermann, Paris, 1953, 1955.
8. M. Breuer, Fredholm Theories in von Neumann Algebras I, *Math. Ann.* **178**, 243–354 (1968).
9. M. Breuer, Fredholm Theories in von Neumann Algebras II, *Math. Ann.* **180**, 313–325 (1969).
10. A. Browder, *Introduction to Function Algebras.* Benjamin, New York, 1968.
11. A. Brown and P. R. Halmos, Algebraic properties of Toeplitz operators, *J. Reine Angew. Math.* **213**, 89–102 (1964).
12. J. Bunce, The joint spectrum of commuting non-normal operators, *Proc. Amer. Math. Soc.* **29**, 499–504 (1971).
13. A. Calderon, F. Spitzer, and H. Widom, Inversion of Toeplitz matrices, *Illinois J. Math.* **3**, 490–498 (1959).
14. L. Carleson, Interpolation by bounded analytic functions and the corona problem, *Ann. Math.* **76**, 542–559 (1962).
15. L. Carleson, The corona theorem, *Proceedings of the 15th Scandinavian Congress Oslo 1968,* vol. 118, pp. 121–132. Springer-Verlag, New York, 1970.

16. L. A. Coburn, Weyl's Theorem for non-normal operators, *Michigan Math. J.* **13**, 285–286 (1966).

17. L. A. Coburn, The C^*-algebra generated by an isometry I, *Bull. Amer. Math. Soc.* **73**, 722–726 (1967).

18. L. A. Coburn, The C^*-algebra generated by an isometry II, *Trans. Amer. Math. Soc.* **137**, 211–217 (1969).

19. L. A. Coburn and R. G. Douglas, Translation operators on the half-line, *Proc. Nat. Acad. Sci. U.S.A.* **62**, 1010–1013 (1969).

20. L. A. Coburn and R. G. Douglas, On C^*-algebras of operators on a half-space I, *Inst. Hautes Etudes Sci. Publ. Math.* **40**, 59–67 (1972).

21. L. A. Coburn, R. G. Douglas, D. G. Schaeffer, and I. M. Singer, On C^*-algebras of operators on a half-space II: Index Theory, *Inst. Hautes Etudes Sci. Pub. Math.* **40**, 69–79 (1972).

22. L. A. Coburn and A. Lebow, Algebraic theory of Fredholm operators, *J. Math. Mech.* **15**, 577–584 (1966).

23. H. O. Cordes and J. P. Labrouse, The invariance of the index in the metric space of closed operators, *J. Math. Mech.* **12**, 693–720 (1963).

24. A. Devinatz, Toeplitz operators on H^2 spaces, *Trans. Amer. Math. Soc.* **112**, 304–317 (1964).

25. A. Devinatz, On Wiener-Hopf Operators in *Functional analysis* (B. Gelbaum, ed.). Thompson, Washington, 1967.

26. A. Devinatz and M. Shinbrot, General Wiener-Hopf operators, *Trans. Amer. Math. Soc.* **145**, 467–494 (1969).

27. J. Dixmier, *Les algèbres d'opérateurs dans l'espace hilbertien* (*Algèbres de von Neumann*). Gauthier-Villar, Paris, 1957.

28. J. Dixmier, *Les C^*-algèbres et leurs representations.* Gauthier-Villar, Paris, 1964.

29. R. G. Douglas, On the spectrum of a class of Toeplitz operators, *J. Math. Mech.* **18**, 433–436 (1968).

30. R. G. Douglas, Toeplitz and Wiener-Hopf operators in $H^\infty + C$, *Bull. Amer. Math. Soc.* **74**, 895–899 (1968).

31. R. G. Douglas, On the spectrum of Toeplitz and Wiener-Hopf operators, in *Abstract Spaces and Approximation Theory* (P. L. Butzer and B. Sz.-Nagy, ed.). Birkhauser Verlag, Basel and Stuttgart, 1969.

32. R. G. Douglas and Roger Howe, On the C^* algebra of Toeplitz operators on the quarter-plane, *Trans. Amer. Math. Soc.* **158**, 203–217 (1971).

33. R. G. Douglas and C. Pearcy, Spectral theory of generalized Toeplitz operators, *Trans. Amer. Math. Soc.* **115**, 433–444 (1965).

34. R. G. Douglas and W. Rudin, Approximation by inner functions, *Pacific J. Math.* **31**, 313–320 (1969).

35. R. G. Douglas and D. E. Sarason, Fredholm Toeplitz Operators, *Proc. Amer. Math. Soc.* **26**, 117–120 (1970).

36. R. G. Douglas and D. E. Sarason, A class of Toeplitz operators, *Indiana U. Math. J.* **20**, 891–895 (1971).

37. R. G. Douglas and J. L. Taylor, Wiener-Hopf operators with measure kernel, *Proceedings of Conference on Operator Theory, Hungary, 1970.*

38. R. G. Douglas and H. Widom, Toeplitz operators with locally sectorial symbol, *Indiana U. Math. J.* **20**, 385–388 (1970).

39. P. Duren, H^p spaces. Academic Press, New York, 1970.

40. T. W. Gamelin, *Uniform algebras.* Prentice-Hall, Englewood Cliffs, New Jersey, 1969.

41. I. M. Gelfand, Normierte rings, *Mat. Sb.* (*N.S.*) **9** (51), 3–24 (1941).

42. I. M. Gelfand, D. A. Raikov and G. E. Šilov, Commutative normed rings, *Usp. Mat. Nauk* 1, 48–146 (1946); *Amer. Math. Soc. Transl.* 5 (2), 115–220 (1951).

43. A. Gleason and H. Whitney, The extension of linear functionals defined on H^∞, *Pacific J. Math.* 12, 163–182 (1962).

44. C. Goffman and G. Pedrick, *First course in functional analysis*. Prentice-Hall, Englewood Cliffs, New Jersey, 1965.

45. I. C. Gohberg and I. A. Fel'dman, Projection methods for solving Wiener-Hopf equations, Moldavian Academy of Sciences, Kishinev, 1967 (Russian).

46. I. C. Gohberg and I. A. Fel'dman, On Wiener-Hopf integral difference equations, *Dokl. Akad. Nauk SSSR* 183, 25–28 (1968) (Russian); *Soviet Math. Dokl.* 9, 1312–1316 (1968).

47. I. C. Gohberg and I. A. Fel'dman, Wiener-Hopf integral-difference equations, *Acta. Sci. Math.* 30, 199–224 (1969) (Russian).

48. I. C. Gohberg and M. G. Kreĭn, Fundamental theorems on deficiency numbers, root number, and indices of linear operators, *Usp. Mat. Nauk* 12, 43–118 (1957) (Russian); *Amer. Math. Soc. Transl.* 13 (2), 185–265 (1960).

49. I. C. Gohberg and M. G. Kreĭn, Systems of integral equations on a half line with kernels depending on the difference of arguments, *Usp. Mat. Nauk* 13, 3–72 (1958) (Russian); *Amer. Math. Soc. Transl.* 14 (2), 217–287 (1960).

50. I. C. Gohberg and N. Ya. Krupnik, On an algebra generated by Toeplitz matrices, *Functional Anal Prilozen.* 3, 46–56 (1969) (Russian); *Functional Anal. Appl.* 3, 119–127 (1969).

51. Seymour Goldberg, *Unbounded linear operators*. McGraw-Hill, New York, 1966.

52. L. S. Goldenstein, Multi-dimensional integral equations of Wiener-Hopf type, *Bull. Akad. Stiince RSS Mold.* no. 6, 27–38 (1964) (Russian).

53. L. S. Goldenstein and I. C. Gohberg, On a multi-dimensional integral equation on a half-space whose kernel is a function of the difference of the arguments and on a discrete analogue of this equation, *Dokl. Akad. Nauk SSSR* 131, 9–12 (1960) (Russian); *Soviet Math. Dokl.* 1, 173–176 (1960).

54. U. Grenander and G. Szegö, *Toeplitz forms and their applications*. University of California Press, 1958.

55. P. R. Halmos, *Introduction to Hilbert space and the theory of spectral multiplicity*. Chelsea, New York, 1951.

56. P. R. Halmos, Shifts on Hilbert spaces, *J. Reine Angew. Math.* 208, 102–112 (1961).

57. P. R. Halmos, A glimpse into Hilbert space. *Lectures on modern mathematics*. Vol. 1, 1–22. Wiley, New York, 1963.

58. P. R. Halmos, *A Hilbert space problem book*. van Nostrand-Reinhold, Princeton, New Jersey, 1967.

59. P. Hartman and A. Wintner, On the spectra of Toeplitz's matrices, *Amer. J. Math.* 72, 359–366 (1950).

60. P. Hartman and A. Wintner, The spectra of Toeplitz's matrices, *Amer. J. Math.* 76, 867–882 (1954).

61. H. Helson, *Lectures on invariant subspaces*. Academic Press, New York, 1964.

62. H. Helson and D. Lowdenslager, Prediction theory and Fourier series in several variables, *Acta Math.* 99, 165–202 (1958).

63. H. Helson and D. E. Sarason, Past and future, *Math. Scand.* 21, 5–16 (1967).

64. H. Helson and G. Szëgo, A problem in prediction theory, *Am. Mat. Pura Appl.* 51, 107–138 (1960).

65. E. Hewitt and K. Stromberg, *Real and abstract analysis*. Springer-Verlag, New York, 1965.

66. K. Hoffman, *Banach spaces of analytic functions*. Prentice–Hall, Englewood Cliffs, New Jersey, 1962.
67. Sze-Tsen Hu, *Homotopy theory*. Academic Press, New York, 1959.
68. R. S. Ismagilov, The spectrum of Toeplitz matrices, *Dokl. Akad. Nauk SSSR* **149**, 769–772 (1963); *Soviet Math. Dokl.* **4**, 462–465 (1963).
69. M. Kac, Theory and applications of Toeplitz forms, in *Summer institute on spectral theory and statistical mechanics*. Brookhaven National Laboratory, 1965.
70. T. Kato, *Perturbation theory for linear operators*. Springer-Verlag, New York, 1966.
71. J. L. Kelley, *General topology*. van Nostrand–Reinhold, Princeton, New Jersey, 1955.
72. M. G. Kreĭn, Integral equations on half line with kernel depending upon the difference of the arguments, *Usp. Mat. Nauk* **13**, 3–120 (1958) (Russian); *Amer. Math. Soc. Transl.* **22** (2), 163–288 (1962).
73. N. H. Kuiper, The homotopy type of the unitary group of Hilbert space, *Topology* **3**, 19–30 (1965).
74. Serge Lang, *Analysis II*. Addison-Wesley, Reading, Massachusetts, 1969.
75. P. Lax, Translation invariant subspaces, *Acta. Math.* **101**, 163–178 (1959).
76. P. Lax, Translation invariant spaces, *Proc. Internat. Symp. Linear Spaces, Jerusalem, 1960*, pp. 251–262. Macmillan, New York (1961).
77. M. Lee and D. E. Sarason, The spectra of some Toeplitz operators, *J. Math. Anal. Appl.* **33**, 529–543 (1971).
78. V. A. Malyšev, On the solution of discrete Wiener-Hopf equations in a quarter-plane, *Dokl. Akad. Nauk SSSR* **187**, 1243–1246 (1969) (Russian); *Soviet Math. Dokl.* **10**, 1032–1036 (1969).
79. K. Maurin, *Methods of Hilbert spaces*. Polish Scientific Publishers, Warsaw, 1967.
80. M. A. Naimark, *Normed rings*. Noordhoff, Groningen, 1959.
81. J. von Neumann, Eigenwerttheorie Hermitescher Funktionaloperatoren, *Math. Ann.* **102**, 49–131 (1929).
82. J. von Neumann, Zur Algebra der Funktionaloperationen und Theorie der Normalen Operatoren, *Math. Ann.* **102**, 370–427 (1929).
83. S. J. Osher, Systems of difference equations with general homogeneous boundary conditions, *Trans. Amer. Math. Soc.* **137**, 177–201 (1969).
84. S. J. Osher, On certain Toeplitz operators in two variables, *Pacific J. Math.* **37**, 123–129 (1970).
85. R. Palais, *Seminar on the Atiyah-Singer index theorem*. Annals of Math. Studies, Princeton, 1965.
86. J. D. Pincus, The spectral theory of self-adjoint Wiener-Hopf operators, *Bull. Amer. Math. Soc.* **72**, 882–887 (1966).
87. H. R. Pousson, Systems of Toeplitz operators on H^2 II, *Trans. Amer. Math. Soc.* **133**, 527–536 (1968).
88. M. Rabindranathan, On the inversion of Toeplitz operators, *J. Math. Mech.* **19**, 195–206 (1969).
89. C. E. Rickart, *Banach algebras*. van Nostrand-Reinhold, Princeton, New Jersey, 1960.
90. F. Riesz, Über die Randwert einer analytischen Funktion, *Math. Z.* **18** (1922).
91. F. Riesz and M. Riesz, Über die Randwert einer analytischen Funtion *4e Congr. des Math. Scand.* 27–44 (1916).
92. F. Riesz and B. Sz.-Nagy, *Functional analysis*. Ungar, New York, 1955.
93. M. Rosenblum, On a theorem of Fuglede and Putnam, *J. London Math. Soc.* **33**, 376–377 (1958).
94. M. Rosenblum, A concrete spectral theory for self-adjoint Toeplitz operators, *Amer. J. Math.* **87**, 709–718 (1965).

95. W. Rudin, *Real and complex analysis.* McGraw-Hill, New York, 1966.
96. David Ruelle, Integral representations of states on a C^*-algebra, *J. Functional Analysis* **6**, 116–151 (1970).
97. D. E. Sarason, Generalized interpolation on H^∞, *Trans. Amer. Math. Soc.* **127**, 179–203 (1967).
98. D. E. Sarason, On products of Toeplitz operators (to be published).
99. D. E. Sarason, Invariant subspaces, *Studies in operator theory*, Math. Assoc. Amer., Prentice-Hall, Englewood Cliffs, New Jersey (in press).
100. I. J. Schark, The maximal ideals in an algebra of bounded analytic functions, *J. Math. Mech.* **10**, 735–746 (1961).
101. I. B. Simonenko, Operators of convolution type in cones, *Mat. Sb.* **74** (116) (1967) (Russian); *Math. USSR Sb.* **3**, 279–293 (1967).
102. I. B. Simonenko, Some general questions in the theory of the Riemann boundary problem, *Izv. Akad. Nauk SSSR* **32** (1968) (Russian); *Math USSR Izv.* **2**, 1091–1099 (1968).
103. J. G. Stampfli, On hyponormal and Toeplitz operators, *Math. Ann.* **183**, 328–336 (1969).
104. M. H. Stone, *Linear transformations in Hilbert space and their applications to analysis.* Amer. Math. Soc., New York, 1932.
105. M. H. Stone, Applications to the theory of Boolean rings to general topology, *Trans. Amer. Math. Soc.* **41**, 375–481 (1937).
106. G. Strang, Toeplitz operators in the quarter-plane, *Bull. Amer. Math. Soc.* **76**, 1303–1307 (1970).
107. B. Sz.-Nagy and C. Foiaş, *Harmonic analysis of operators on Hilbert space.* Akademiai Kiado, Budapest, 1970.
108. J. L. Taylor, The cohomology of the spectrum of a measure algebra, *Acta Math.* **126**, 195–225 (1971).
109. O. Toeplitz, Zur theorie der quadratischen Formen von unendlichvielen Veränderlichen, *Math. Ann.* **70**, 351–376 (1911).
110. H. Widom, Inversion of Toeplitz matrices II, *Illinois J. Math.* **4**, 88–99 (1960).
111. H. Widom, Inversion of Toeplitz matrices III, *Notices Amer. Math. Soc.* **7**, 63 (1960).
112. H. Widom, Toeplitz matrices, *Studies in real and complex analysis*, Math. Assoc. Amer., Prentice-Hall, Englewood Cliffs, New Jersey, 1965.
113. H. Widom, On the spectrum of Toeplitz operators, *Pacific J. Math.* **14**, 365–375 (1964).
114. H. Widom, Toeplitz operators on H_p, *Pacific J. Math.* **19**, 573–582 (1966).
115. N. Wiener and E. Hopf, Über eine Klasse singulären Integral-gleichungen, *S.-B. Preuss Akad. Wiss. Berlin, Phys.-Math. Kl.* **30/32**, 696–706 (1931).
116. A. Wintner, Zur theorie der beschränkten Bilinear formen, *Math. Z.* **30**, 228–282 (1929).
117. K. Yoshida, *Functional analysis.* Springer-Verlag, Berlin, 1966.

Index

Pure and Applied Mathematics

A Series of Monographs and Textbooks

Editors **Paul A. Smith and Samuel Eilenberg**

Columbia University, New York